FLORIDA STATE
UNIVERSITY LIBRARIES

MAY 0 9 2001

TALLAHASSEE, FLORIDA

Fluvial Processes and Environmental Change

British Geomorphological Research Group Symposia Series

Geomorphology in Environmental Planning
Edited by **J. M. Hooke**

Floods
Hydrological, Sedimentological and
Geomorphological Implications
Edited by **Keith Beven** and **Paul Carling**

Soil Erosion on Agricultural Land
Edited by **J. Boardman, J. A. Dearing** and **I. D. L. Foster**

Vegetation and Erosion
Processes and Environments
Edited by **J. B. Thornes**

Lowland Floodplain Rivers
Geomorphological Perspectives
Edited by **P. A. Carling** and **G. E. Petts**

Geomorphology and Sedimentology of Lakes and Reservoirs
Edited by **J. McManus** and **R. W. Duck**

Landscape Sensitivity
Edited by **D. S. G. Thomas** and **R. J. Allison**

Process Models and Theoretical Geomorphology
Edited by **M. J. Kirkby**

Environmental Change in Drylands
Biogeographical and Geomorphological Perspectives
Edited by **A. C. Millington** and **K. Pye**

Rock Weathering and Landform Evolution
Edited by **D. A. Robinson** and **R. B. G. Williams**

Geomorphology and Land Management in a Changing Environment
Edited by **D. F. M. McGregor** and **D. A. Thompson**

Advances in Hillslope Processes
Volumes 1 and 2
Edited by **M. G. Anderson** and **Sue M. Brooks**

Landform Monitoring, Modelling and Analysis
Edited by **S. N. Lane, K. S. Richards** and **J. H. Chandler**

Fluvial Processes and Environmental Change
Edited by **A. G. Brown** and **T. A. Quine**

Fluvial Processes and Environmental Change

Edited by

A. G. Brown and T. A. Quine

*Department of Geography,
University of Exeter, UK*

JOHN WILEY & SONS
Chichester • New York • Weinheim • Brisbane • Singapore • Toronto

Copyright © 1999 by John Wiley & Sons Ltd,
          Baffins Lane, Chichester,
          West Sussex PO19 1UD, England

National   01243 779777
International   (+44) 1243 779777
e-mail (for orders and customer service enquiries): cs-books@wiley.co.uk
Visit our Home Page on http://www.wiley.co.uk
          or http://www.wiley.com

All rights reserved. No part of this publication may be reproduced, stored in a retrieval system, or transmitted, in any form or by any means, electronic, mechanical, photocopying, recording, scanning or otherwise, except under the terms of the Copyright, Designs and Patents Act 1988 or under the terms of a licence issued by the Copyright Licensing Agency, 90 Tottenham Court Road, London, UK W1P 9HE, without the permission in writing of John Wiley and Sons Ltd., Baffins Lane, Chichester, West Sussex, UK PO19 1UD.

Other Wiley Editorial Offices

John Wiley & Sons, Inc., 605 Third Avenue, New York, NY 10158-0012, USA

WILEY-VCH Verlag GmbH, Pappelallee 3,
D-69469 Weinheim, Germany

Jacaranda Wiley Ltd, 33 Park Road, Milton,
Queensland 4064, Australia

John Wiley & Sons (Canada) Ltd, 22 Worcester Road,
Rexdale, Ontario M9W 1L1, Canada

John Wiley & Sons (Asia) Pte Ltd, 2 Clementi Loop #02-01,
Jin Xing Distripark, Singapore 129809

Library of Congress Cataloging-in-Publication Data

Fluvial processes and environmental change/edited by A.G. Brown and T. A. Quine
    p.   cm. — (British Geomorphological Research Group symposia series)
   Includes bibliographical references and index.
   ISBN 0-471-98548-1 (alk. paper)
    1. Rivers—Congresses.   2. Geomorphology—Congresses.   3. Climatic
changes—Environmental aspects—Congresses.   I. Brown, A.G.
II. Quine, T.A.   III. Series.
GB1201.2.F59   1999
551.48′3—dc21   98-37167
                                                                                                        CIP

British Library Cataloguing in Publication Data

A catalogue record for this book is available from the British Library

ISBN 0 471 98548 1

Typeset in 10/12pt Times by Vision Typesetting, Manchester
Printed and bound in Great Britain by Bookcraft Ltd, Midsomer Norton, Somerset
This book is printed on acid-free paper responsibly manufactured from sustainable forestry, in which at least two trees are planted for each one used for paper production.

Contents

List of Contributors		viii
Preface		xi
Chapter 1	Fluvial Processes and Environmental Change: An Overview A. G. Brown and T. A. Quine	1
Section 1	**THE SLOPE–CATCHMENT SCALE**	29
Chapter 2	Modelling the Impacts of Holocene Environmental Change in an Upland River Catchment, Using a Cellular Automaton Approach T. J. Coulthard, M. J. Kirkby and M. G. Macklin	31
Chapter 3	Stream Bank and Forest Ditch Erosion: Preliminary Responses to Timber Harvesting in Mid-Wales T. Stott	47
Chapter 4	Slope and Gully Response to Agricultural Activity in the Rolling Loess Plateau, China T. A. Quine, D. E. Walling and X. Zhang	71
Chapter 5	River Activity in Small Catchments over the Last 140 ka, North-east Mallorca, Spain J. Rose and X. Meng	91
Section 2	**CHANNEL RESPONSE**	103
Chapter 6	Impact of Major Climate Change on Coarse-grained River Sedimentation: A Speculative Assessment Based on Measured Flux I. Reid, J. B. Laronne and D. M. Powell	105
Chapter 7	Modelling and Monitoring River Response to Environmental Change: The Impact of Dam Construction and Alluvial Gravel Extraction on Bank Erosion Rates in the Lower Alfios Basin, Greece A. P. Nicholas, J. C. Woodward, G. Christopoulos and M. G. Macklin	117

vi   Contents

Chapter 8   Significance of River Bank Erosion as a Sediment Source in
             the Alternating Flood Regimes of South-eastern Australia   139
                W. D. Erskine and R. F. Warner

Chapter 9   Middle to Late Holocene Environments in the Middle to
             Lower Trent Valley   165
                A. J. Howard, D. N. Smith, D. Garton, J. Hillam and M. Pearce

Section 3   **FLOODPLAIN PROCESSES**   179

Chapter 10   Alluvial Microfabrics, Anisotropy of Magnetic Susceptibility
             and Overbank Processes   181
                C. Ellis and A. G. Brown

Chapter 11   Changing Rates of Overbank Sedimentation on the
             Floodplains of British Rivers During the Past 100 Years   207
                D. E. Walling and Q. He

Chapter 12   Floodplain Evolution and Sediment Provenance
             Reconstructed from Channel Fill Sequences: The Upper
             Clyde Basin, Scotland   223
                J. S. Rowan, S. Black and C. Schell

Chapter 13   Siberian-type Quaternary Floodplain Sedimentation: The
             Example of the Yenisei River   241
                A. F. Yamskikh, A. A. Yamskikh and A. G. Brown

Section 4   **FLOODPLAIN RESPONSE**   253

Chapter 14   Long-Term Episodic Changes in Magnitudes and Frequencies
             of Floods in the Upper Mississippi River Valley   255
                J. C. Knox

Chapter 15   High Resolution Palaeochannel Records of Holocene Valley
             Floor Environments in the North Tyne Basin, Northern
             England   283
                A. J. Moores, D. G. Passmore and A. C. Stevenson

Chapter 16   Fluvial Processes, Land Use and Climate Change 2000 Years
             Ago in Upper Annandale, Southern Scotland   311
                R. Tipping, P. Milburn and S. Halliday

Chapter 17	A 1000 Year Alluvial Sequence as an Indicator of Catchment/Floodplain Interaction: The Ruda Valley, Sub-Carpathians, Poland *K. Klimek*	329
Chapter 18	Historic River Response to Extreme Flooding in the Yorkshire Dales, Northern England *S. P. Merrett and M. G. Macklin*	345
Section 5	**GLACIERISED BASINS**	361
Chapter 19	Environmental Change and Sediment Yield from Glacierised Basins: The Role of Fluvial Processes and Sediment Storage *J. Warburton*	363
Chapter 20	The Impact of Recent Climate Change on River Flow and Glaciofluvial Suspended Sediment Loads in South Iceland *D. M. Lawler and L. J. Wright*	385
Index		409

List of Contributors

S. Black PRIS, University of Reading, Whiteknights, PO Box 227, Reading RG6 6AD, UK

A. G. Brown Department of Geography, Amory Building, Rennes Drive, University of Exeter, Exeter EX4 4RJ, UK. <a.g.brown@exeter.ac.uk>

G. Christopoulos School of Geography, University of Leeds, Leeds LS2 9JT, UK

T. J. Coulthard School of Geography, University of Leeds, Leeds LS2 9JT, UK

C. Ellis AOC (Scotland) Ltd., The Old School House, Pipe Street, Leith, Edinburgh, EH6 8BR, UK

W. D. Erskine School of Geography, University of New South Wales, Sydney, NSW 2052, Australia

D. Garton Trent & Peak Archaeological Trust, University Park, Nottingham NG7 2RD, UK

S. Halliday RCAHMS, John Sinclair House, 16 Bernard Terrace, Edinburgh EH8 9NX, UK

Q. He Department of Geography, Amory Building, Rennes Drive, University of Exeter, Exeter EX4 4RJ, UK

J. Hillam Dendrochronology Laboratory, Research School of Archaeology and Archaeological Sciences, Department of Archaeology and Prehistory, University of Sheffield, Sheffield S1 4DT, UK

A. J. Howard School of Geography, University of Leeds, Leeds LS2 9JT, UK

M. J. Kirkby School of Geography, University of Leeds, Leeds LS2 9JT, UK

K. Klimek Earth Sciences Faculty, University of Silesia, ul.Bedzinska 60, 41-200 Sosnowiec, Poland

J. C. Knox Department of Geography, University of Wisconsin-Madison, 384 Science Hall, 550 North Park Street, Madison, WI 53706-1491, USA

J. B. Laronne Department of Geography and Environmental Development, Ben Gurion University of the Negev, Beer Sheva 84105, Israel

D. M. Lawler School of Geography, The University of Birmingham, Edgbaston, Birmingham B15 2TT, UK. <D.M.Lawler@bham.ac.uk>

M. G. Macklin School of Geography, University of Leeds, Leeds LS2 9JT, UK

X. Meng Department of Geography, Royal Holloway, University of London, Egham, Surrey TW20 0EX, UK. <x.meng@rhbnc.ac.uk>

S. P. Merrett School of Geography, University of Leeds, Leeds LS2 9JT, UK

P. Milburn Department of Environmental Science, University of Stirling, Stirling FK9 4LA, UK

A. J. Moores Department of Geography, University of Newcastle upon Tyne, Newcastle upon Tyne NE1 7RU, UK

A. P. Nicholas Department of Geography, Amory Building, Rennes Drive, University of Exeter, Exeter EX4 4RJ, UK. <a.p.nicholas@exeter.ac.uk>

D. G. Passmore Department of Geography, University of Newcastle upon Tyne, Newcastle upon Tyne NE1 7RU, UK

M. Pearce Department of Archaeology, University of Nottingham, Nottingham NG7 2RD, UK

D. M. Powell Department of Geography, University of Leicester, Leicester LE1 7RH, UK

T. A. Quine Department of Geography, Amory Building, Rennes Drive, University of Exeter, Exeter EX4 4RJ, UK. <t.a.quine@exeter.ac.uk>

I. Reid Department of Geography, Loughborough University, Loughborough, Leicestershire LE11 3TU, UK. <ian.reid@lboro.ac.uk>

J. Rose Department of Geography, Royal Holloway, University of London, Egham, Surrey TW20 0EX, UK. <j.rose@rhbnc.ac.uk>

J. S. Rowan Department of Geography, University of Dundee, Dundee, DD1 4HN, UK

C. Schell School of Geography, University of Leeds, Leeds LS2 9JT, UK

D. N. Smith Department of Ancient History and Archaeology, University of Birmingham, Edgbaston, Birmingham B15 2TT, UK

A. C. Stevenson Department of Geography, University of Newcastle upon Tyne, Newcastle upon Tyne NE1 7RU, UK

T. Stott School of Education and Community Studies, Liverpool John Moores University, I. M. Marsh Campus, Barkhill Road, Liverpool L17 6BD, UK. <T.A.STOTT@lLIVJM.ac.uk>

R. Tipping Department of Environmental Science, University of Stirling, Stirling FK9 4LA, UK

D. E. Walling Department of Geography, Amory Building, Rennes Drive, University of Exeter, Exeter EX4 4RJ, UK

J. Warburton Department of Geography, University of Durham, Science Laboratories, South Road, Durham DH1 3LE, UK. <jeff.warburton@durham.ac.uk>

R. F. Warner Department of Geography, University of Sydney, Sydney, NSW 2006, Australia

J. C. Woodward School of Geography, University of Leeds, Leeds LS2 9JT, UK

L. J. Wright School of Geography, The University of Birmingham, Edgbaston, Birmingham B15 2TT, UK.

A. A. Yamskikh Department of Ecology, Krasnoyarsk State Pedagogical University, Krasnoyarsk, Russia

A. F. Yamskikh Laboratory of Paleogeography, Krasnoyarsk State Pedagogical University, Krasnoyarsk, Russia

X. Zhang Institute of Mountain Disasters and Environment, Chinese Academy of Sciences, Chengdu, Sichuan Province, China

Preface

This volume originated in a British Geomorphological Research Group Session at the annual conference of the Royal Geographical Society and Institute of British Geographers at the University of Exeter in January of 1997. The aim was to try and assess how far fluvial geomorphologists could go in predicting and understanding the likely changes to fluvial systems that could result from environmental change, particularly coupled climatic and anthropogenic change. This was in line with one of the themes of the conference, namely Global Environmental Change. The session attracted over 30 papers from 50 contributors and was well attended. We would like to thank all those who spoke or gave posters whether or not they have contributed to this volume. We must also thank the organisers of the annual conference and BGRG for its support.

The editors are indebted to the referees who numbered over 40 and whom in the vast majority of cases responded quickly and decisively to the refereeing process. For their own contributions, and advice and help with submitted papers, the editors must thank Helen, Terry and Andrew of the Geography Department at Exeter. Finally our thanks also go to John Wiley & Sons Ltd and in particular Sally Wilkinson, Mandy Collison and Emma Bottomley for their forbearance and assistance.

1 Fluvial Processes and Environmental Change: An Overview

A. G. BROWN and T. A. QUINE
Department of Geography, University of Exeter, UK

GEOMORPHOLOGY AND ENVIRONMENTAL CHANGE

The potential contribution of geomorphologists to debates concerning global environmental change has been a regular theme in the British Geomorphological Research Group (BGRG) symposia series. In the introduction to *Vegetation and Erosion: Processes and Environments*, Thornes (1990) identified the two most important environmental issues as the enhanced greenhouse effect and desertification, and he identified two fundamental research needs: (1) to identify the processes by which change is brought about; (2) to establish the extent to which such changes are part of the natural state of affairs as revealed by the history of past changes. Thornes (1990) suggested that both environmental issues, and both research needs, lay within the remit of geomorphology. In January 1993, the theme of the BGRG symposium, *Geomorphology and Land Management in a Changing Environment* (McGregor and Thompson, 1995a), made explicit reference to the potential contribution of geomorphology to the environmental debate. However, in the symposium volume, Jones (1995) suggested that while research into potential effects of global environmental change was well established in the atmospheric, hydrospheric and biospheric sciences, geomorphological research had been fitful and fragmentary (with the exception of coastal studies), and that "whilst explanations of past changes in environmental conditions and their effect on landform development are important, the utility of this information to decision-makers whose main concern is with future changes is not always clear, especially as the circumstances of the past appear to them wholly different to those of the near future" (Jones, 1995, p.13). The editors of the same volume expressed a more positive view of existing research, suggesting that "successful forecasting of the effects of land use change requires a secure geomorphological database and a reliable working model – however, defined – of the process-response system within which land use change is taking place" (McGregor and Thompson, 1995b, p.4). This statement may provide a justification of much geomorphological research; however, if it is accepted that it is important for geomorphologists to make a contribution to the debates concerning global environmental change, it is clearly important to assess whether current research methodologies are providing or can provide the following:

Fluvial Processes and Environmental Change. Edited by A. G. Brown and T. A. Quine.
© 1999 John Wiley & Sons Ltd.

(1) the necessary "secure geomorphological database";
(2) the understanding required to develop "working models" of geomorphological process-response systems;
(3) the data necessary to demonstrate the reliability of such models over a range of time and spatial scales.

This volume provides geomorphologists with an opportunity to evaluate progress in these areas in relation to fluvial environments and this introductory chapter reviews some of the issues of relevance to this process of evaluation. Understanding the linkages between environmental change and fluvial processes is becoming increasingly important in the last decades of the 20th century due to a variety of factors:

(1) climate changes resulting from the enhanced greenhouse effect;
(2) pressure on land and water resources resulting from an increased global population (Greenland et al., 1997);
(3) pressure on land and water resources resulting from the following social changes:
    - changes to household number/size in the Developed World,
    - rural to urban attraction (Developing World) and urban to rural attraction (Developed World),
    - intra- and inter-state migrations,
    - large-scale environmental projects (e.g. the Three Gorges Dam on the Yangtse River).

The recognition of these pressures has led to a heightened concern regarding future sustainability of agricultural production (e.g. Lal, 1997; Syers, 1997) and the potential impact of population pressure on water resources (e.g. Tinker, 1997), both issues being highlighted at a recent Royal Society symposium, *Land Resources at the Malthusian Precipice* (Greenland et al., 1997). In the light of these pressures and concerns, there is a clear need for a sound understanding of catchment–river linkages and responses to environmental change based on process knowledge. The degree to which this understanding has been and can be gained is addressed in this introductory chapter. The treatment of the issues broadly follows the structure of the book, focusing firstly on erosion at the slope scale, before examining floodplain processes and sedimentation, and finally considering questions of scale and causality.

CATCHMENT EROSION AND ENVIRONMENTAL CHANGE

In relation to erosion processes, the last three decades have witnessed both a cumulative increase in detailed short-term, relatively small-scale, process knowledge and the development of process-based modelling approaches. Advances in these areas have been reflected in recent BGRG symposia volumes (Anderson, 1988; Boardman et al., 1990; Kirkby, 1994; Anderson and Brookes, 1996) and contributed papers. Thus, there has been significant progress towards the establishment of "a secure geomorphological

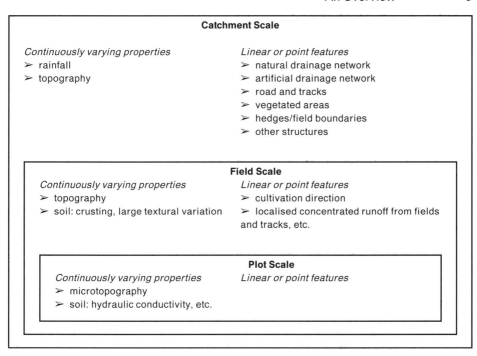

Figure 1.1 Controls on flow paths at plot (c. 10 m^2), field and catchment scales

database" and the development of the "working models" referred to by McGregor and Thompson (1995b). Nevertheless, when the focus of attention moves from the individual slope unit to the slope as a whole or to the wider catchment then it appears that the level of understanding declines significantly, even when examining short timescales. This is exemplified by the poor agreement between both field-scale and catchment-scale process-based soil erosion models and observational data for individual storm events (cf. De Roo and Walling, 1994; De Roo, 1996; Boardman and Favis-Mortlock, 1998). This poor agreement may not be a result of deficient understanding of the physical processes, but instead may reflect a shift in the dominant controls on erosion when moving from the plot scale to the field and catchment scale. This may be illustrated by considering controls on flow paths and consequent erosion and deposition within agricultural catchments (Figure 1.1). At the plot scale, the dominant controls on the generation and, more importantly, the flow direction of runoff are found in the intrinsic properties of the plot, the soil and topography. The key properties may be measured effectively, if rather laboriously, and may be described by continuous distributions that are readily converted to the raster formats used for data input in many spatially distributed models (De Roo, 1993). At the field scale, even where an isolated geomorphic unit is represented, overland flow paths may be controlled as much by management controls as by intrinsic topography. For example, furrows up and down slopes will preferentially channel water along straight flow paths rather than following topographic gradients. In contrast, cross-slope furrows tend to channel overland flow directly into concavities, leading to significant flow convergence at

(a)

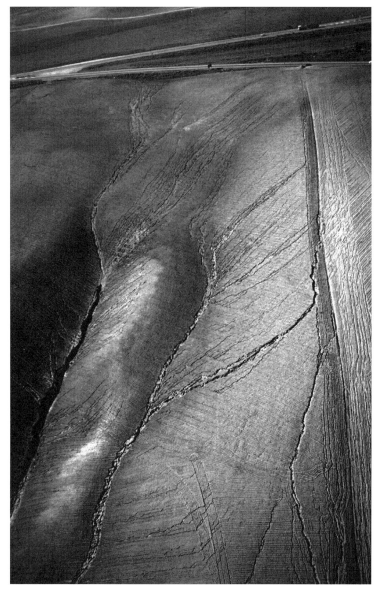

Figure 1.2 Patterns of erosion on cultivated land near Cordoba, southern Spain, in November 1997, following 150 mm rain in 24 hours. (a) Locations of rill and gully initiation controlled by topography, tillage direction and transient boundary in the tilled area. (b) On the left of the photograph, tillage up and down the slope has led to parallel zones of incision controlled by tillage direction. On the right, tillage across the slope has led to convergence of flow in the thalweg with consequent more localised, but deeper incision. This ephemeral gully has also been fed by runoff from the adjacent cultivated area

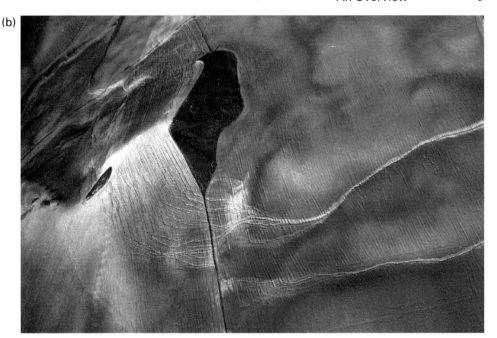

points upslope of those that would be identified on the grounds of topography alone. Furthermore, less predictable discontinuities in the land surface, such as transient edges of cultivation or isolated wheelings, may act to concentrate overland flow and thereby induce erosion (Figure 1.2). These are well-known field phenomena and part of the configurational state of the system (see later discussion), but are difficult to incorporate into predictive models. However, it may be instructive to incorporate such factors into data sets used in model validation exercises.

At the catchment scale, problems for spatially distributed modelling increase because controls on overland flow pathways and sediment transport shift towards artificial drainage networks and small, often linear, zones of the catchment take on disproportionate significance. For example, the condition of field boundaries – fence, hedge with no "understorey", raised bank and hedge – will control flow paths and, more significantly for erosion models, sediment transport (Shanahan, 1998). Other linear features include compacted earth paths and trackways, which, in both temperate and tropical environments, may be the major sources of runoff and sediment (Froehlich *et al.*, 1993; Brown *et al.*, 1996; Brown and Schneider, 1998). Therefore, spatially variable, linear features of very low width may become dominant controls on sediment delivery and flow paths. Such features are very difficult to represent effectively within the raster data structure used for most spatially distributed erosion modelling. Therefore, if the development of models with "real world" applicability is considered a valuable goal, two needs may be identified. First, a greater understanding of overland flow movement and sediment transport in field boundary zones is needed. Secondly, it may be necessary to develop new data structures which allow effective simulation of processes operating within these boundary zones.

When longer timescales are considered, deficiencies in understanding become more apparent. To a certain extent the focus on short-term, small-scale studies reflects the need to establish the fundamental process-response relationships before progressing to examination of the longer term and larger scale. Nevertheless, there remains a need for studies of a wider scope, in particular when facing the challenge of validating models over longer timescales, in order to be able to use them to predict erosional response to environmental change. The scarcity of relevant studies reflects the problems associated with assembling long-term and large-scale data or suitable surrogates and this may be illustrated by considering the approaches used to investigate erosional responses to environment change. These can be broadly categorised as follows:

(a) space–time substitution,
(b) inference from deposits,
(c) repeat observation,
(d) radioisotope measurement.

Space–Time Substitution

Space–time substitution (sometimes referred to as the ergodic approach, but see Paine, 1985) is probably the most widely used approach to the study of erosional responses to contemporary environmental change. In this approach geomorphic processes are studied on comparable slopes under differing land use and it is assumed that differences in slope behaviour are driven by the differences in land use. Therefore, if land use is changed at an individual site it may be expected to result in a shift in process-response which reflects the between-site contrasts. Recent explicit examples of the application of this approach to the investigation of environmental change impacts include a study of the effects of deforestation on slope and channel evolution in the Himalayas (Froehlich and Starkel, 1987) and an examination of changing fluvial and slope processes in Singapore in response to deforestation and urbanisation (Chatterjea, 1994). Many other studies have an implicit reliance on space–time substitution, for example those which examine increased rates of erosion as a result of agricultural intensification (cf. Boardman et al., 1990; Wicherek, 1993). Rarely are data available to indicate rates of erosion for a study site prior to the environmental change represented by the change in land management.

While space–time substitution is a widely used approach, it is not without its problems (Paine, 1985). Two problems, in particular, are apposite here. First, the approach rests on the assumptions that the studied sites had equivalent initial conditions and that differences in site behaviour may be attributed to differences in treatment. Such assumptions may not always be defensible, as was illustrated by De Rose et al. (1993) in a study of post-deforestation soil loss as a result of water erosion and landslides from steep hillslopes near Taranaki, New Zealand. They found that the previous erosion history of hillslopes had an important influence on the spatial distribution of landslides, i.e. the state of the system was an important control on process-response. Secondly, even where the assumptions of the approach are accepted, space–time substitution has limited spatial scope and will usually only provide an indication of the range and rate of processes that are likely to occur within individual landscape elements after a particular change in management. The approach is more limited when applied to process-response at a larger-scale because this response will be dependent on the configuration of landscape elements (cf. Quine *et al.*, this volume) and, therefore, may not be simply derived from the summed response of individual elements.

The Inference of Erosion Rates from Deposits

Holocene and historical erosion rates have been estimated from sediment deposition in order to give *some approximation* of natural or re-impact erosion rates (Trimble, 1981; Brown and Barber, 1985; Bork, 1989; Bell and Boardman, 1992). These estimates may sometimes be used to approximate the natural-climatic baseline (e.g. in parts of the New World), or baselines contingent on both the land-use history and the prevailing climate (e.g. in most of the Old World). The evidence of deposits has also been used to derive information concerning erosional response to specific environmental conditions and changes in a wide range of environments. For example, Haberle *et al.* (1991) have used charcoal in slopewash to identify the initiation of burning and slope degradation in Indonesia, while Blum *et al.* (1994) have postulated episodic changes in slope stability on the basis of processes of floodplain construction in Texas. In Britain, sediments examined in dry valley excavations have been used to reconstruct land-use histories for adjacent hillslopes (Allen, 1992; Bell, 1992) on the assumption that the close physical proximity of the sediment source area to the depositional environment will be reflected in a close temporal linkage between the stratigraphic sequence and slope processes. Under these conditions, the assumption may be valid; however, sediment transport from source slope to current depositional environment, more often, occurs via intermediate storage, and significant quantities of sediment may be transported beyond those depositional environments which are readily subject to investigation. The quantitative significance of such sediment delivery processes can be established by measuring (rather than estimating) delivery ratios. In studies of the rivers Culm (Lambert and Walling, 1987) and Severn (Walling and Quine, 1993), as much as 75% of the suspended sediment was found to bypass floodplains and was transported to the outlet. Clearly, when only a relatively small percentage of sediment is deposited in the depositional environment, deconvolution of catchment erosional history from the depositional record is problematic. This is illustrated by studies by Walling and He (1994; Chapter 11, this volume) who have shown that 40-year and 100-year estimates of overbank sedimentation rates, derived from caesium-137 and fallout lead-210 for several lowland floodplains, are not significantly different despite a widely held recognition that the last four decades have seen significant increases in erosion on agricultural land. This lack of sensitivity of floodplain sedimentation to environmental change in the catchment occurs because the rate of deposition on the floodplain is a product of overbank events and overbank processes, as discussed later in this chapter, and is not necessarily related to the rate of sediment production on slopes (Prosser *et al.*, 1994). Nevertheless, whilst studies of recent overbank sedimentation may cast doubts over the sensitivity of the overbank accretion rate to subtle changes in catchment erosion rate, major variations in overbank accretion rates have been shown to exist and have been related to both land use and climate change (Robinson and Lambrick, 1984; Brown, 1987a, 1987b; Macklin *et al.*, 1992a; Moores *et*

Table 1.1 Environmental radionuclides used in geomorphological research on contemporary processes

Radionuclide	Source	Half-life	Fallout	Applications
Caesium-137	Nuclear weapons testing	30 years	1950s–1970s	SE, LS, FS, SF
Lead-210	Uranium series	22 years	continuous	SE, LS, FS, SF
Beryllium-7	Cosmogenic	53 days	continuous	SF

Applications: SE = soil erosion; LS = lake sedimentation; FS = floodplain sedimentation; SF = sediment fingerprinting.

al., Chapter 15, this volume). In the light of the potential value of floodplain sediments for investigating sediment dynamics, there is a clear need for improved understanding of floodplain processes. For example, a factor that has hitherto received little attention is the role of floodplain vegetation in determining sedimentation rates. Changing vegetation will alter floodplain roughness and the surface area exposed to the flow and sedimentation (Brown and Brookes, 1997; see later discussion). Improved understanding of sediment dynamics will also require the development of full sediment budgets for floodplains, including within-channel sedimentation and the sedimentation of secondary channels and cut-offs.

Radioisotope Measurements

Radioisotopes offer considerable potential for the retrospective assembly of information concerning net soil redistribution within the landscape. Caesium-137 (^{137}Cs) has been widely used to investigate erosion rates over the period since the initiation of fallout in the mid-1950s (e.g. Menzel *et al.*, 1987; Ritchie and McHenry, 1990; Walling and Quine, 1991; Quine *et al.*, 1992, 1993, 1997; Zhang *et al.*, 1994, 1998). Furthermore, the combination of ^{137}Cs-derived soil redistribution data with topographic-based erosion modelling has permitted an increased understanding of erosion processes on agricultural land (Quine *et al.*, 1994, 1996, 1997; Govers *et al.*, 1996). Applications of this combined approach have highlighted the role of tillage erosion in soil redistribution and provided hitherto unavailable medium-term (> 40 year) estimates of water erosion rates (Quine *et al.*, 1994, 1996, 1997; Govers *et al.*, 1996). However, in isolation, ^{137}Cs can provide only a long-term average erosion rate and therefore, in the examination of erosional response to environmental change, suffers the same limitations as other space–time substitution approaches. Nevertheless, when combined with other independent data representing erosion over different time periods, it may be possible to examine temporal variation in erosion rates and, thereby, begin to examine response to environmental change (Quine *et al.*, 1996). This potential for deriving information concerning temporal variation in erosion rates using radioisotope data is being investigated further by use of multiple isotopes. In particular there is growing interest in the use of beryllium-7 (^{7}Be) and fallout or excess lead-210 (^{210}Pb$_{ex}$) in combination with ^{137}Cs to investigate erosion at different timescales (Walling and Quine, 1995; Walling *et al.*, 1995; Wallbrink and Murray, 1996a, 1996b). As Table 1.1

An Overview 9

illustrates, these isotopes have different sources, fallout patterns and half-lives. This adds complexity to the task of interpreting erosional history from the radioisotope signatures, but potential richness in the data so derived.

Repeat Observations

Continuous or regularly repeated monitoring of erosion under field conditions can provide invaluable data concerning erosional responses to specific land management strategies (e.g. Evans, 1988; Boardman, 1990; Auzet et al., 1993). However, such monitoring programmes may be expensive to continue over extended periods and this has led to the discontinuation of one of the most valuable examples (Evans, 1988). Such strategies, while providing detailed contemporary data, are unlikely to be applicable to the development of medium- to long-term databases. A novel approach to establishing net landscape change over longer periods of time, using detailed archive aerial photography, has been demonstrated by Vandaele et al. (1996). Stereoscopic aerial photographs taken in 1947 and 1991 were used to create detailed digital terrain models (DTMs) of an area of agricultural land in the Huldenberg region of central Belgium. These DTMs were then compared and the spatial distribution of net soil loss and gain was thereby established. Clearly such an approach is only possible where suitable imagery is available and can only provide information concerning net landscape change. However, it does provide long-term data inaccessible using many other approaches. As the spatial resolution of satellite imagery approaches levels which are appropriate for the construction of sufficiently detailed DTMs, then sequential remote observation may provide a means of establishing patterns of landscape change. However, in view of the rate of improvement of resolution and the rates of landscape change in many areas, it will be sometime before this is realistic for investigation of all but the largest magnitude geomorphic processes.

In the slopes–small catchments section of this volume, Coulthard et al. (Chapter 2) and Quine et al. (Chapter 4) illustrate in rather different ways the non-linearity of sediment discharge event magnitude/frequency relationships. The causes lie in lagged feedback processes and the changing balance between different sediment generation mechanisms (e.g. pipe/overland flow). The role of distinct and semi-independent mechanisms is echoed by Stott (Chapter 3) who shows how sediment response to logging is partially controlled by frost activity. Over the much longer timescale considered by Rose and Meng (Chapter 5) the response of small catchments to climate change is still not simple, being mediated by vegetational processes and previous sedimentary and erosional history.

FLOODPLAINS AND SEDIMENT CONVEYANCE

The observation that floodplain reaches differ in their sedimentary history and/or sensitivity to external factors highlights the problem of the diachronic nature of fluvial responses due to variations in boundary variables such as bedrock characteristics and tectonics (Yamskikh et al., Chapter 13, this volume), and due to the variable efficiency of sediment conveyance by river channels. This had led to attempts to model sediment conveyance and alluvial response as sediment waves (or slugs) (Church and Jones, 1982). However, without precise dating, diachrony of sediment deposition can be regarded as noise or just a poorly defined complex response. Floodplain form and channel conditions can be regarded as part of a temporal sequence related through the transfer and storage of sediment from reach to reach (Ferguson, 1981). Sediment waves can occur at temporal and spatial scales that are one to three orders of magnitude larger than those employed in process studies, operating over hundreds to thousands of years and at the mega-reach to stream-order or regional scale. It is assumed that sediment waves originate in generation areas, such as the hillslopes and headwater channels, and proceed through a propagation area, the floodplain and coastal zone. In fact the situation is rather more complex, since the generation zone is governed by both hydrometeorology, inherent characteristics of the catchment area, and the distribution of available sediment.

The most readily available coarse sediments in most catchments in the temperate zones are Pleistocene fluvial and glacial sediments. For British rivers this means that the potential sources are predominantly in the piedmont and lowland zones, but sediment supply from these sources is governed by stream power through competence and bank erosion rates. The sediment generation in these reaches is therefore governed by reach-scale variables which affect stream power, including channel slope which is itself often influenced by the distribution and characteristics of terrace remnants. The piedmont zone is therefore typically the zone with maximum rates of both erosion and deposition and the zone which is highly sensitive to external factors such as climate change. This is exemplified by studies of several UK rivers in this zone, including the Rheidol (Macklin and Lewin, 1986), Tyne (Macklin and Lewin, 1989; Macklin *et al.*, 1992b), Dane (Hooke *et al.*, 1990) and the Bollin (Knighton, 1973) as well as many European rivers such as the Vistula (Starkel, 1990; Kalicki, 1991). For fine sediment the situation is different only in that there is also a non-channel storage component on the floodplain at least over the short to medium timescale.

The sediment wave concept can be applied at the reach scale and may be particularly applicable to large relatively unstable gravel-bedded rivers such as the Trent in Midland England. Gravel quarrying in the Trent has revealed adjacent reaches where the major gravel bodies are identical in lithological composition, texture and altitude but are of very different ages (Brown, 1996a). In contrast to other larger British lowland rivers, where a horizontally bedded or parcel-like stratigraphy is common (Burrin and Scaife, 1984; Needham, 1992; Brown *et al.*, 1994), the Trent displays laterally accreted sand and gravel beds predominantly formed during one, albeit prolonged, period of channel activity. This has been elucidated at Hemington, where various dating methods have been used to piece together the history of channel change over the last 1300 years (Ellis and Brown, 1998; Chapter 10 this volume). From a single-channel highly sinuous meandering state in the 6th century AD, the Trent straightened then bifurcated, underwent a period of anastomosing and braiding with flow switching from the north-west to the south-east side of the floodplain, before returning to a single low-sinuosity channel through a transitional anastomosing phase (Figure 1.3). This puts into perspective questions of the stability or transitional nature of channel forms, as each form lasted for, on average, *c.* 250 years, although these phases were in reality not of equal length. Change was clearly episodic with avulsion,

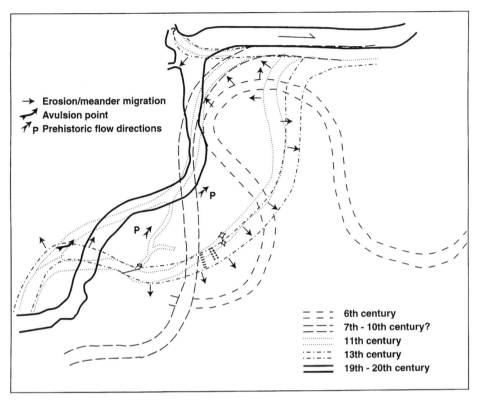

Figure 1.3 A generalised model of the late Holocene changes in channel pattern of the Trent at Hemington, Leicestershire

depositional units and artifacts associated with historically recorded floods. However, the complete cycle of channel change spans periods of altered flood frequency and is fundamentally conditioned by the reach and floodplain conditions. In only 40 km four tributaries enter the Middle Trent, with two draining uplands. The result is that discharge ratios can vary from an average of 35% contribution from the largest tributary, the Derwent, to 60% during some of the largest floods (e.g. the 10 December 1965 flood; Brown, 1996a). These tributaries, particularly those from the uplands (the Dove and the Derwent), contribute bedload directly to the main channel in addition to the input from the Pleistocene gravels that constrain the Holocene floodplain. Since the floodplain shows no change in slope in this zone, adjustment to the additional sediment input has been achieved through channel change and associated lateral gravel deposition. A very similar cycle of channel change occurred 20 km downstream at Colwick, but between 100 and 200 years later than at Hemington (Salisbury *et al.*, 1984). This example illustrates the importance of dating control for determining the rates of response to external variables, and the possible problems associated with small systems where rapid perturbations may obscure longer-term directional responses.

Lateral instability is more common in central European rivers such as the Main, Danube, Vistula and Maas where the large scale of the system allows aggregate responses to external forcing to be seen in the sedimentary record (Becker and

Schirmer, 1977; Starkel, 1991; Tornquist, 1993; Kalicki, 1996). These floodplains also reveal many landforms and processes that are not observable on smaller or more maritime floodplains (e.g. floodplain dunes, ice-dams and deflation hollows).

The chapters on channels focus on the non-linearity of morphological response to hydrological change. In an example from the arid zone, Reid *et al.* (Chapter 6) show how bedload response to increased discharges will be conditioned by whether the channels are ephemeral and un-armoured, or perennial and armoured. Nicholas *et al.* (Chapter 7) also highlight the importance of armouring in determining the rate and magnitude or morphological response to external factors, and Erskine and Warner (Chapter 8) show how channels respond quite differently to periods of flood-dominated regime and drought-dominated regime. The latter study also illustrates why short-term studies may not be able to predict the range of responses a channel may exhibit to environmental change.

FLOODPLAINS, SEDIMENT CHARACTERISTICS AND PROCESS INFERENCE

Fluvial processes on floodplains can be direct or indirect indicators of environmental change. There are also autogenic adjustments which may be part of the dynamic equilibrium of the system and which can occur without any environmental change. In most cases it is changes in process rates and dominant processes which can be used to infer changes in the hydrological regime. Our knowledge of floodplain processes comes largely from studies of contemporary floodplain change. Whilst this is appropriate for floodplains highly modified by human activities, it has serious limitations where this is not the case. An extreme example is the British Isles which has no floodplains with primary (natural) vegetation, very few without channel management (weirs, etc.) and none without some regime management through reservoirs or land drainage. Most British rivers have also seen a rise in overbank sedimentation during the last millennium due to accelerated soil erosion and this has profoundly altered both channel and floodplain morphology (Brown, 1997a). The result of all these factors is that lowland British floodplains generally exhibit a greatly reduced set of floodplain processes and relatively low rates of channel erosion and floodplain change, having in effect entombed themselves. Therefore, in order to interpret the observed sedimentary record, it is not only necessary to utilise the range of contemporary processes observed in different climatic zones, but also some processes which are either rare or non-existent today. Examples include megafloods, large-scale floodplain deflation, floodplain dune formation, alluvial tufa deposition and a number of biophysical processes associated with forested floodplains.

An important sedimentary process for many floodplains is tufa deposition. Tufa is formed by both the carbonate degassing process, and organically assisted deposition through photosynthesis changing the CO_2 balance of waters and by Cyanobacteria and Charaophyta (Baker and Sims, 1998). Tufa precipitation is favoured by increasing calcium levels in groundwater, increased temperatures, abundant organic matter such as plant stems which form a substrate for the algae, and clear water with low levels of suspended sediment. These conditions are most frequently met when calcium and bicarbonate rich groundwater enters a floodplain from a spring, but tufa can also form in abandoned channels (Pedley, 1990). It has been recorded in many valley fills in south-eastern England including the Gipping (Rose et al., 1980), the Kennet (Churchill, 1962; Evans et al., 1988), the Brue (G. Aalbersberg, pers. comm. 1997) and the Dour (Bates and Barham, 1993) where it was found both *in situ* and reworked. Several authors have proposed a late Holocene decline in tufa deposition (Goudie et al., 1993; Griffiths and Pedley, 1995; Pentecost, 1995) but this may have been exaggerated as Baker and Sims (1998) have shown there to be considerable under-reporting of contemporary sites with active tufa deposition. Studies of tufa geochemistry and inclusions (including pollen) may reveal important information on the surrounding environmental conditions such as vegetation type and precipitation variations, especially if combined with radiocarbon dating of inclusions in order to provide a chronology. If, however, a maximum of deposition (larger active sites) in the mid-Holocene is accepted then there are several possible causes: a slightly higher mean annual temperature, but possibly just as important the prevalence of relatively wide and shallow anastomosing channels containing abundant growth of aquatic plants and mineral-rich waters of low turbidity.

The most researched processes on floodplains are the geomorphological processes of erosion and sedimentation. The balance between the two provides the basis for sediment budget models and models of floodplain formation. The distribution of excess energy (the energy required to transport the sediment load) and distribution of resistance provide the template for channel/floodplain type and the mode of channel change (e.g. meandering, braiding, anastomosing, etc.). Whilst the distribution of fluid stress is determined by the energy gradient and the channel/floodplain morphology, the distribution of shear resistance is determined by the nature and distribution of fluvial deposits (in an alluvial floodplain) which itself is the product of past sedimentary processes and post-depositional processes. This is why rivers are evolutionary systems with a history influenced by inherited characteristics. One result is variations in fluvial sensitivity and response to environmental change (Brown, 1990).

There are also many biogeochemical processes which can both reflect environmental change and influence fluvial response. Sediments are altered after deposition by a number of processes. The first set of processes are associated with consolidation, de-watering if above the floodplain watertable, and bioturbation. Consolidation and de-watering can be approximated by standard methods using loading, grain size and porosity, but bioturbation is dependent upon climate, watertable height and the sedimentation rate. A second set of processes involve the alteration of sediments by biogeochemical processes. Floodplain geochemistry is influenced by groundwater chemistry, sediment mineralogy and rainwater chemistry. Groundwater chemistry varies with catchment lithology and land use. Since floodplains are generally zones with low groundwater transmission rates there is often ample residence time for complex geochemical reactions to occur.

In the context of this volume we can categorise these biochemical and biophysical processes into those which do or do not increase sediment shear resistance. An attempt to do this is given in Table 1.2 along with the conditions under which the process operates. Some of these post-depositional changes are reversible; others are not.

All the chapters in the floodplain processes section of this volume are concerned with sediment storage on floodplains. Ellis and Brown (Chapter 10) illustrate how a wide range of biogeochemical processes are involved in both sediment deposition and transformation, and this can affect both stability, being one of the causes of non-linear dynamics, and environmental quality. Rowan *et al.* (Chapter 12) show how the geochemistry of sediments can also be used to partition sources and along with Walling and He (Chapter 11) illustrate the use of short-lived radionuclides for estimating floodplain sedimentation rates. As Walling and He point out, although seemingly constant, the sedimentation rates may hide considerable complexity in the sediment delivery system. This complexity is almost certainly even greater in more continental climates and Yamskikh *et al.* (Chapter 13) propose a significantly different conceptual model of Siberian-type floodplain sedimentation in order to encompass processes such as ice- and landslide-dams.

CHANNEL FORM AND SEDIMENTARY MODELLING

Over the last 20 years it has become clear that we have underestimated the full range of variation of channel morphologies and types that can occur in nature. A wider classification of channels has been suggested by Croke and Nanson (1991) which, whilst based on stream power, includes channel types additional to straight meandering and braided. Work on anastomosing and anabranching channels has shown them to be stable forms fulfilling efficiency criteria (Nanson and Huang, 1998) as well as transitional forms. This realisation has come from studies of the world's few remaining natural systems such as Arctic Canada (Smith and Smith, 1980) and Central Australia (Nanson and Knighton, 1996) and, in Europe, the spur has been a desire to restore channels and floodplains to a natural-look state for both aesthetic and ecological reasons (Brown *et al.*, 1997; Brown, 1997b). The multiple channel state has been modelled by Mackay and Bridge (1995), the exercise revealing the critical nature of the cross-valley floodplain surface slope parameter. This slope is dependent on spatial variation in overbank accretion rates. Marriott (1992) has shown how these rates may conform to the diffusion model proposed by Pizzutto (1987) but so far this model has only been tested on relatively flat floodplains with a low and uniform roughness. Direct measurements of overbank sedimentation using ^{137}Cs suggest more locally and topographically sensitive variation but still with the highest rates adjacent to the channel (Walling *et al.*, 1996; Nicholas and Walling, 1997). Over long time periods, and in the absence of a high migration rate, this should lead to avulsion which is the major cause of secondary channel formation in anastomosing and anabranching systems (Mackay and Bridge, 1995). There are as yet very few data on the effects of changes in floodplain vegetation/land use on overbank accretion rates; however, some preliminary observations suggest that different plant species may be associated with different sedimentation rates and the importance of convective currents on floodplains makes the diffusion model unsuited to floodplains with old channel scars and palaeochannels (Brown and Brookes, 1997).

There is now a narrowing gap between the geological scale modelling of alluvial architecture and the physically based modelling of channel processes (e.g. Howard, 1992, 1996; Nicholas *et al.*, Chapter 7, this volume). But it is necessary to add the

Table 1.2 Some less researched processes effecting channel change. Those in italics are essentially non-reversible at a millennial timescale

Process	Effect on bank resistance	Resultant landform/feature
Fe/Mn cementation	Greatly increased	Armoured beds, bank benches, steep stable banks
Bar colonisation by vegetation	Increased	Stable islands, anastomosis?
Tree-throw	Decreased	Bank embayments, debris dams, channel change
Peat formation (groundwater fed)	Increased	Bank benches
Peat formation (ombrotrophic)	Increased	Channel diversion (palaeochannels) and anastomosing channels
Faunal activities (e.g. beaver dams, bank burrowing, wallows, etc.)	Decreased	Flooding deposits and avulsion
Groundwater rise	Decreased	Channel morphology change (increased bank failure) and flood deposits, avulsion, mud-springs/cones
Groundwater fall	Increased	Incised channels, dry channels
Freezing/ice	Decreased	Bank benches, alteration of channel morphology, channel change

complexity of anisotropic conditions to process-based models. This includes floodplain topography that is a function of past channel change, spatial variations in bank and bed sediments and variations in surface roughness/trap efficiency. This will go some way to simulating the complexity of response which systems may exhibit following a uniform change in extrinsic controls, including variations in the number of functioning channels (Brown, 1995). It does, however, raise an important question, namely what should the starting conditions be? Very few rivers have been initiated unconstrained, on a pseudo-isotropic plain, which is why most Quaternary scientists, as opposed to engineers and geomorphologists, have generally held an evolutionary view of river change (Rose, 1995). This does not mean that there are no meaningful starting points: an obvious starting point for modern rivers is the condition of floodplains at the end of the Lateglacial, about which we have some knowledge from many European rivers (Frenzel, 1995).

CHANGING ENVIRONMENT AND FLUVIAL RESPONSE

There is a need to critically evaluate concepts of landscape sensitivity, landscape equilibrium and thresholds (Allen, 1974). It is not always clear that they have analytical meaning and the criteria for their recognition are often only clear at the smallest spatial and temporal scales. As Lewin *et al.* (1988) have pointed out, there is a contradiction between regime theory and environmental change. To some extent this is a product of differing timescales of enquiry, but as the rate of climate change accelerates, the

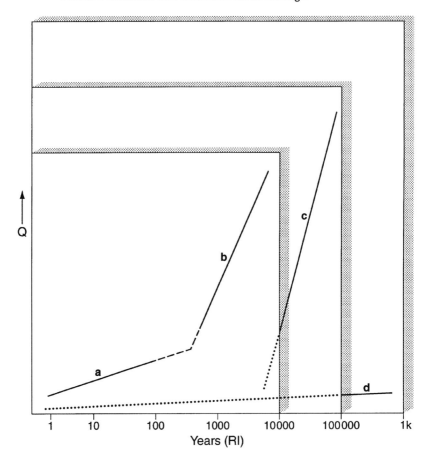

Figure 1.4 The nested frequency concept of magnitude–frequency relationships. Each slope is caused by a different mechanism, e.g. slope a represents westerly air masses crossing the UK and rain-generated floods (or predominant Lamb circulation type); slope b results from a rarer circulation type leading to rain or snow and larger flood; slope c could result from a change in thermal and precipitation regime; and slope d represents a totally different cold climate regime. The result of a, b and even c occurring during a specific time period will produce anomalies or outliers from the flood frequency curve which may become more (or less) probable as the mix of types changes

problem becomes more acute. If we conceptualise environmental change as nested frequency distributions with associated fluvial regimes (Brown, 1991; Bevan, 1993; see Figure 1.4) then regime must be both unstable, due to overlapping frequency domains, and subject to more systematic changes as the dominating synoptic conditions change (Rumsby and Macklin, 1994). This is of immediate importance in relation to the potential fluvial effects of greenhouse gas induced or naturally induced (e.g. through solar activity) climate change which is now believed to be occurring (DoE, 1996; IPCC,

1996). There are two components to the climatic-forcing scenario: first is the hydrometeorological/synoptic effects of global warming and the second is catchment response to hydrometeorological change. Both are up-line of fluvial response and neither are understood any better than fluvial dynamics. In particular, although a rise in temperatures is agreed for the UK (0.2 °C per decade; DoE, 1996), the effects on precipitation are far less certain (Arnell, 1996), and despite the difficulty of accurately predicting seasonal or annual runoff totals (Arnell, 1996), palaeohydrology and process-based interpretation of fluvial sediments do have a role in improving our understanding of climate dynamics over different timescales and the variability of fluvial response to climatic events such as El Niño/Southern Oscillation (Wells, 1990; Brown, 1996b), or changes in the North Atlantic Oscillation Index (e.g. Lawler and Wright, Chapter 20, this volume).

A separate section on glacierised basins is included in this volume because these basins are particularly sensitive to climate change through changes to the mass balance of glaciers. Warburton (Chapter 19) provides a valuable framework for assessing fluvial sediment storage in glacierised basins and therefore how sediment output may respond to environmental change. In their study of three undisturbed basins, Lawler and Wright (Chapter 20) show how, through seasonal effects, changes in the North Atlantic Oscillation Index have affected suspended sediment output. They also highlight the role of "autogressive memory behaviour" which is part of the configurational state of the system.

THE IMPLICATIONS OF SPATIAL–TEMPORAL SCALE DEPENDENCY

We intuitively recognise that small systems such as rill networks change appreciably with every event and between events. Whilst their formation is to some extent related to the magnitude of the event, there is considerable scatter, not all of which is attributable to measurement error. The identification of morphological response due to a change in event magnitude and frequency is rarely straightforward and system response is rarely linear. At the opposite end of the fluvial spectrum, large rivers do not necessarily show a morphological response to every flood; a threshold may be exceeded and since this can be an inherent property of the system (e.g. shear resistance), the response may be dependent on the state of the system. This can be illustrated with reference to the previously discussed channel change of the River Trent during the late Holocene. During the 11th to 14th centuries AD a series of large floods caused an episodic cycle of channel change involving a shift of the channel from one side of the floodplain to the opposite side and back again. However, in 1795 a flood that was probably of greater magnitude than those of the 11th to 14th centuries caused no observed morphological response, the most likely reason being that since the medieval period the channel had become stabilised by incision and the deposition of cohesive silt–clay banks and levees on top of the channel gravels. Whilst remarkable stability has been recorded for some lowland rivers in the UK, such as the Great Ouse, which was unaffected by a 1000 year flood (Dury, 1981), the present stability is not the typical state of the Trent over the last few thousand years. These observations conform with Lane and Richards' (1997) argument that response depends upon the configurational state of the system. They go on to argue that this necessitates the study of processes at ever finer spatial and temporal scales. However, given the relative rarity of a combination of suitable (unstable) internal conditions and an event of critical magnitude, it is difficult to see how studies over small timescales can identify the controls on the evolution of larger systems. It is the combination of configurational dependency of response and spatial–temporal dependency of rates that necessitates observations (or reconstructions) over long time periods in order to identify the critical processes involved in channel change and floodplain formation. Just as ever more detailed knowledge of human genetic make-up will never fully explain or predict human behaviour, so the ever more precise description of the configurational component of one meander cannot explain or predict fluvial response at the reach scale or above. This is not to question the importance of detailed process work, but it is a justification for complementary studies at broader temporal and spatial scales. Another reason for observations over long time periods is the need to incorporate biological processes or phenomena which are often critical in determining the configurational state. Examples include bar and bank stabilisation by vegetation (Hughes *et al.*, 1997), the formation of debris dams (Gurnell, 1997), changing floodplain roughness due to forest growth, and biological stabilisation of stream beds by aquatic plants, algae and even freshwater sponges (Brown *et al.*, 1995).

One of the most obvious implications of the configurational dependency of fluvial systems is the lack of a synchronous response of even adjacent catchments to any external factor, such as climate. It could be argued that as catchment size increases so configurational controls become more varied, in some cases even cancelling each other out. In addition, the likelihood of preservation of evidence of past fluvial events also increases with catchment size. This may explain why it appears to be the piedmont zone of many rivers that is most sensitive to changes in external factors, particularly climate, whilst the lower reaches seem to display a frequently human-induced, relative insensitivity through floodplain alluviation and channel stabilisation (Brown, 1997a). Several papers in this volume echo this landscape-scale variation in fluvial sensitivity. Knox (Chapter 14) shows how the Mississippi has responded to climate change and most recently human impact, and how the change in sediment discharge is frequently not proportional to flood discharge. He also illustrates the non-stationarity in flood behaviour that is implied in several other chapters and must now be taken to be the normal fluvial condition. Moores *et al.* (Chapter 15) reveal non-synchronous fluvial histories of tributaries in the northern Tyne and between the north and south Tyne; the cause being differing catchment conditions and land-use history. A similar conclusion is also reached by Tipping *et al.* (Chapter 16) for upper Annandale. Klimek (Chapter 17) also uses heavy metal pollution to show changes in sedimentation in the absence of climate change. However, this anthropogenically related diachrony is not universal as Merrett and Macklin (Chapter 18) show, using lichenometry, that 26 small streams in the Yorkshire Dales, UK, were all subject to large floods towards the end of the Little Ice Age. The evidence for this is coarse boulder sheets on the floodplains, which implies that the fine sediment was transported downstream and must have constituted a climatically driven sediment slug.

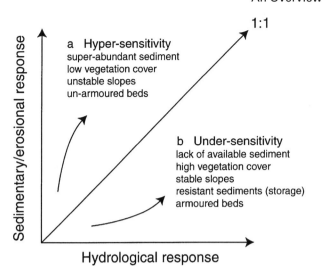

Figure 1.5 A simple diagrammatic representation of disproportionate response (cf. hysteresis) and some of the factors involved. A switch from state b to state b may be controlled by thresholds of resistance and negative feedback, both tending towards cyclicity of system behaviour

All chapters in this volume are concerned with the non-linearity of fluvial response to environmental change. In some cases there may be a disproportionately large sediment flux from a small hydrological change (hyper-sensitivity); in others there seems to be a disproportionately small response (under-sensitivity; see Figure 1.5). It is also clear that this sensitivity is crucial to discussions as to the "fundamental" cause of fluvial change and whether this response can be characterised as climatically driven but culturally blurred (*sensu* Macklin and Needham, 1992; Macklin and Lewin, 1993) or the opposite; *culturally driven but climatically blurred*; both situations are implied in studies contained in this volume. Other than in extreme situations (i.e. totally natural basins) it would seem likely that the key to which cause dominates is controlled by their relative magnitudes (as generally assumed) *and/or* the state of the system. For example, hyper-sensitivity in the discharge of suspended sediment is likely to result from significant land-use change (producing a result that could be characterised as culturally driven; Figure 1.5, state a) whereas, once cleared and under heathland, upland catchments may produce a relatively constant morphological and sedimentary response to changes in the flood series. Future research aimed at improving our predictive capabilities must consider all the processes that can influence response dynamics; some of which are currently included such as sediment supply, but also biological processes, sediment transformations, and most important of all the initial state of the system and how it is modified during change (i.e. feedback in fully dynamic modelling). As modelling is the most appropriate method for attempting to predict the impact of environmental change on fluvial systems, it follows that process understanding is crucial to understanding the response of fluvial systems to future change in both climate and land use. Whilst it may be possible to explain fluvial behaviour at these scales by incorporating both extrinsic and intrinsic factors, there remains a fundamental challenge to extend this understanding to the floodplain, slope–river, or catchment scale. It is also not axiomatic that the reductionist approach is necessarily appropriate, or the most efficient for modelling complex non-linear systems, irrespective of whether it does or does not provide a deeper understanding of nature. Whilst it has been accepted since Classical times that every fluvial event must be unique ("you cannot put your toe in the same river twice"), this is not necessarily a positive foundation for the understanding, modelling or prediction of fluvial systems. Modelling and perhaps even understanding requires generalisations, and the discrimination of the critical controls on system behaviour. This is particularly relevant to understanding the possible effects of a major change in extrinsic controls such as climate and/or land-use change.

CONCLUDING POINTS

The purpose of this introduction has been to set a broader context for the chapters in this volume. Traditional engineering approaches such as regime theory, and traditional geomorphological approaches, such as hydraulic geometry, may not be suitable for the prediction of the effects of environmental change on fluvial systems. In this introduction we have emphasised two key points. First, there is increasing evidence that the response of fluvial systems is not only or necessarily dependent on the direction, magnitude and frequency of the extrinsic factor, but upon the configurational state of the system, and this in turn is a product of the geomorphic history of the system. One result of this is the frequent (and problematic) non-linearity of response of fluvial systems to environmental change. The magnitude of the problem is highlighted when one remembers that the aim is to predict the response of a non-linear system to the non-stationary behaviour of flood series. This calls into question both the theoretical validity and practical utility of concepts of fluvial equilibrium. It also implies that models of fluvial behaviour must be fully dynamic with interaction between forms and fluxes, such that, for example, bed morphology is fully coupled with sediment transport rates.

Secondly, there are a wider range of biophysical processes, which are part of the configurational state and can affect the response of a system to environmental change, than are generally modelled, and channels may be affected indirectly as well as through a change in regime. Examples include the increased growth of aquatic weed due to increased stream temperatures and changing influent/effluent conditions due to floodplain watertable changes. A greater understanding of these processes may be particularly important if we desire to return rivers and floodplains to more "natural" conditions (Brown et al., 1997). Indeed it is somewhat ironic that there is considerable interest in "re-naturalising" rivers, which in geomorphic terms means returning them to greater environmental control, just as environmental factors are changing at an increasing rate. Laudable though its aims are, the Scottish Wild Rivers Project obviously cannot return any Scottish river to a "wild" state and states less constrained by human intervention may be more sensitive to the trajectory and rates of environmental control. The natural variability of rivers in both space and time has been reduced, both physically through accelerated alluviation, regulation and channelisation, and in our quantitative descriptions of the fluvial system. Sustainable richer rivers in both geomorphic and ecological terms can only be achieved by re-coupling rivers with functioning floodplains and functioning floodplains to catchments. The greater sensitivity of such systems to environmental change provides both opportunities and hazards. Hopefully the chapters in this volume point the way to the deeper understanding of fluvial processes and the role of environmental change in fashioning the fluvial environment.

REFERENCES

Allen, J. R. L. 1974. Reaction, relaxation and lag in natural sedimentary systems: general principles, examples and lessons. *Earth Science Reviews*, **10**, 263–342.

Allen, M. J. 1992. Products of erosion and the prehistoric land-use of the Wessex chalk. In Bell, M. and Boardman, J. (Eds) *Past and Present Soil Erosion*. Oxbow Monograph 22, Oxbow Books, Oxford, 37–52.

Anderson, M. G. (Ed.) 1988. *Modelling Geomorphological Systems*. John Wiley, Chichester.

Anderson, M. G. and Brooks, S. M. (Eds) 1996. *Advances in Hillslope Processes*. John Wiley, Chichester.

Arnell, N. W. 1996. Palaeohydrology and future climate change. In Branson, J., Brown, A. G. and Gregory, K. J. (Eds) *Global Continental Changes: The Context of Palaeohydrology*. Geological Society Special Publication No. 115, Geological Society, London, 19–26.

Auzet, A. V., Boiffin, J., Papy, F., Ludwig, B. and Mancorps, J. 1993. Rill erosion as a function of the characteristics of cultivated catchments in the North of France. *Catena*, **20**, 41–62.

Baker, A. and Sims, M. J. 1998. Active deposition of calcareous tufa in Wessex, UK, and its implications for the "late-Holocene tufa decline". *The Holocene*, **8**, in press.

Bates, M. R. and Barham, A. J. 1993. Recent observations of tufa in the Dour valley, Kent. *Quaternary Newsletter*, **71**, 11–25.

Becker, B. and Schirmer, W. 1977. Palaeoecological study on the Holocene valley development of the river Main, Southern Germany. *Boreas*, **6**, 303–321.

Bell, M. 1992. The prehistory of soil erosion. In Bell, M. and Boardman, J. (Eds) *Past and Present Soil Erosion*. Oxbow Monograph 22, Oxbow Books, Oxford, 21–36.

Bell, M. and Boardman, J. (Eds) 1992. *Past and Present Soil Erosion*. Oxbow Monograph 22, Oxbow Books, Oxford.

Bevan, K. 1993. Riverine flooding in a warmer Britain. *The Geographical Journal*, **159**, 157–161.

Blum, M. D., Tooimey, R. S. and Valastro, S. 1994. Fluvial response to late Quaternary climatic and environmental change, Edwards Plateau, Texas. *Palaeogeography, Palaeoclimatology, Palaeoecology*, **108**, 1–21.

Boardman, J. 1990. Soil erosion on the South Downs: a review. In Boardman, J., Foster, I. D. L. and Dearing, J. A. (Eds) *Soil Erosion on Agricultural Land*. John Wiley, Chichester, 87–105.

Boardman, J. and Favis-Mortlock, D. 1998. *Modelling Soil Erosion by Water*. NATO-ASI Global Environmental Change Series, I-55, Springer Verlag, Berlin.

Boardman, J., Foster, I. D. L. and Dearing, J. A. (Eds) 1990. *Soil Erosion on Agricultural Land*. John Wiley, Chichester.

Bork, H.-F. 1989. Soil erosion during the past millennium in Central Europe and its significance within the geomorphodynamics of the Holocene. In Ahnert, F. (Ed.) *Landforms and Landform Evolution in West Germany*. Catena, Supplement, **15**, 121–132.

Brown, A. G. 1987a. Holocene floodplain sedimentation and channel response of the lower River Severn, United Kingdom. *Zeitschrift für Geomorphologie*, **31**, 293–310.

Brown, A. G. 1987b. Long-term sediment storage in the Severn and Wye catchments. In Gregory, K. J., Lewin, J. and Thornes, J. B. (Eds) *Palaeohydrology in Practice*. John Wiley, Chichester, 307–332.

Brown, A. G. 1990. Holocene floodplain diachrony and inherited downstream variations in fluvial processes: a study of the river Perry, Shropshire, England. *Journal of Quaternary Science*, **5**, 39–51.

Brown, A. G. 1991. Hydrogeomorphology and palaeoecology of the Severn basin during the last 15,000 years: orders of change in a maritime catchment. In Gregory, K. J., Starkel, L. and Thornes, J. B. (Eds) *Fluvial Processes in the Temperate Zone During the Last 15,000 Years*. John Wiley, 147–169.

Brown, A. G. 1995. Lateglacial–Holocene sedimentation in lowland temperate environments: floodplain metamorphosis and multiple channel systems. In Frenzel, B. (Ed.) European River Activity and Climate Change During the Lateglacial and Early Holocene. *Palaeoclimate Research*, Special Issue, **14**, 21–36.

Brown, A. G. 1996a. Human dimensions of palaeohydrological change. In Branson, J., Brown, A. G. and Gregory, K. J. (Eds) *Palaeohydrology and Global Changes: The Context of Palaeohydrology*. Geological Society Special Publication No. 115, Geological Society, London, 57–72.

Brown, A. G. 1996b. Palaeohydrology: prospects and future advances. In Branson, J., Brown, A. G. and Gregory, K. J. (Eds) *Palaeohydrology and Global Changes: The Context of Palaeohydrology*. Geological Society Special Publication No. 115, Geological Society, London, 257–265.

Brown, A. G. 1997a. *Alluvial Environments: Geoarchaeology and Environmental Change*. Cambridge University Press, Cambridge.

Brown, A. G. 1997b. Biogeomorphology and diversity in multiple-channel river systems. *Global Ecology and Biogeography Letters*, **6**, 179–186.

Brown, A. G. and Barber, K. E. 1985. Late Holocene palaeoecology and sedimentary history of a small lowland catchment in Central England. *Quaternary Research*, **24**, 87–102.

Brown, A. G. and Brookes, A. 1997. Floodplain vegetation and overbank erosion and sedimentation. In Large, A. (Ed.) *Floodplain Rivers: Hydrological Processes and Ecological Significance*. British Hydrological Society Occasional Paper No. 8, 30–39.

Brown, A. G. and Schneider, H. 1998. From plots to basins: the scale problem in studies of soil erosion and sediment yield. In Harper, D. M. and Brown, A. G. (Eds) *The Sustainable Management of Tropical Catchments*. John Wiley, Chichester, in press.

Brown, A. G., Keough, M. K. and Rice, R. J. 1994. Floodplain evolution in the East Midlands, United Kingdom: the Lateglacial and Flandrian alluvial record from the Soar and Nene valleys. *Philosophical Transactions of the Royal Society Series A*, London, 261–293.

Brown, A. G., Stone, P. and Harwood, K. 1995. *The Biogeomorphology of a Wooded Anastomosing River: The Gearagh on the River Lee in County Cork, Ireland*. Occasional Papers in Geography, No. 32, University of Leicester.

Brown, A. G., Schneider, H. and Harper, D. M. 1996. Multi-scale estimates of erosion and sediment yields in the Upper Tana basin, Kenya. In Walling, D. E. and Webb, B. (Eds) *Erosion and Sediment Yield: Global and Regional Perspective*. IAHS Publication No. 236, 49–54.

Brown, A. G., Harper, D. and Peterken, G. F. 1997. European floodplain forests: structure, functioning and management. *Global Ecology and Biogeography Letters*, **6**, 169–178.

Burrin, P. J. and Scaife, R. G. 1984. Aspects of Holocene valley sedimentation and floodplain development in southern England. *Proceedings of the Geologists Association*, **95**, 81–96.

Chatterjea, K. 1994. Dynamics of fluvial and slope processes in the changing geomorphic environment of Singapore. *Earth Surface Processes and Landforms*, **19**, 585–607.

Church, M. and Jones, D. 1982. Channel bars in gravel-bed rivers. In Hey, R. D., Bathurst, J. C. and Thorne, C. R. (Eds) *Gravel-Bed Rivers*. John Wiley, Chichester, 291–338.

Churchill, D. M. 1962. The stratigraphy of the Mesolithic sites III and V at Thatcham, Berkshire, England. *Proceedings Prehistoric Society*, **14**, 362–370.

Croke, J. and Nanson, G. C. 1991. Floodplains: their character and classification on the basis of stream power, sediment type, boundary resistance and antecedency. *Geomorphology*, **3**, 13–26.

De Roo, A. P. J. 1993. Modelling surface runoff and soil erosion in catchments using Geographical Information Systems. *Netherlands Geographical Studies*, **157**, Utrecht.

De Roo, A. P. J. 1996. Validation problems of hydrologic and soil-erosion catchment models: examples from a Dutch erosion project. In Anderson, M. G. and Brooks, S. M. (Eds) *Advances*

in Hillslope Processes. John Wiley, Chichester, 669–683.
De Roo, A. P. J. and Walling, D. E. 1994. Validating the "ANSWERS" soil erosion model using ^{137}Cs. In Rickson, R. J. (Ed.) *Conserving Soil Resources: European Perspectives*. CAB International, Cambridge, 246–263.
De Rose, R. C., Trustrum, N. A. and Blaschke, P. M. 1993. Post-deforestation soil loss from steepland hillslopes in Taranaki, New Zealand. *Earth Surface Processes and Landforms*, **18**, 131–144.
DoE (Department of the Environment) 1996. *Review of the Potential Effects of Climate Change in the UK, 1996*. HMSO, London.
Dury, G. H. 1981. Magnitude and frequency analysis and channel morphology. In Morrisawa, M. (Ed.) *Fluvial Geomorphology*. George Allen and Unwin, London, 91–121.
Ellis, C. E. and Brown, A. G. 1998. Archaeomagnetic dating and palaeochannel sediments: data from the Medieval channel fills at Hemington, Leicestershire. *Journal of Archaeological Science*, **25**, 149–163.
Evans, J. G., Limbrey, S., Mate, I. and Mount, R. J. 1988. Environmental change and land-use history in a Wiltshire river valley in the last 14000 years. In Barrett, J. C. and Kinnes, I. A. (Eds) *The Archaeology of Context in the Neolithic and Bronze Age: Recent Trends*. Department of Archaeology and Prehistory, University of Sheffield, 97–104.
Evans, R. 1988. *Water Erosion in England and Wales 1982–1984*. Report for Soil Survey and Land Research Centre, Silsoe.
Ferguson, R. I. 1981. Channel form and channel changes. In Lewin, J. (Ed.) *British Rivers*. Allen and Unwin, London, 90–125.
Frenzel, B. (Ed.) 1995. European River Activity and Climate Change During the Lateglacial and Early Holocene. *Palaeoclimate Research*, Special Issue, **14**.
Froehlich, W. and Starkel, L. 1987. Normal and extreme monsoon rains – their role in the shaping of the Darjeeling Himalaya. *Studia Geomorphologica Carpatho-Balcanica*, **21**, 129–160.
Froehlich, W., Higgitt, D. L. and Walling, D. E. 1993. The use of caesium-137 to investigate soil erosion and sediment delivery from cultivated slopes in the Polish Carpathians. In Wicherek, S. (Ed.) *Farm Land Erosion in Temperate Plains and Hills*. Elsevier, Amsterdam, 271–283.
Goudie, A., Viles, H. A. and Pentacost, A. 1993. The late-Holocene tufa decline in Europe. *The Holocene*, **3**, 181–186.
Govers, G., Quine, T. A., Desmet, P. J. J. and Walling, D. E. 1996. The relative contribution of soil tillage and overland flow erosion to soil redistribution on agricultural land. *Earth Surface Processes and Landforms*, **21**, 929–946.
Greenland, D. J., Gregory, P. J. and Nye, P. H. (Eds) 1997. Land resources: on the edge of the Malthusian Precipice? *Philosophical Transactions of the Royal Society, Series B*, **352**, 859–1033.
Griffiths, H. I. and Pedley, H. M. 1995. Did changes in the Last Glacial and early Holocene atmospheric CO_2 concentrations control rates of tufa precipitation? *The Holocene*, **5**, 238–242.
Gurnell, A. 1997. The hydrological and geomorphological significance of forested floodplains. *Global Ecology and Biogeography Newsletter*, **6**, 219–230.
Haberle, S. G., Hope, G. S. and Defretes, Y. 1991. Environmental change in the Baliem Valley, montane Irian Jaya, Republic of Indonesia. *Journal of Biogeography*, **18**, 25–40.
Hooke, J. M., Harvey, A. M., Miller, S. Y. and Redmond, C. E. 1990. The chronology and stratigraphy of the alluvial terraces of the river Dane valley, Cheshire. *Earth Surface Processes and Landforms*, **15**, 717–737.
Howard, A. D. 1992. Modelling channel migration and floodplain development in meandering streams. In Carling, P. A. and Petts, G. E. (Eds) *Lowland Floodplain Rivers*. John Wiley, Chichester, 1–42.
Howard, A. D. 1996. Modelling channel evolution and channel morphology. In Anderson, M. G., Walling, D. E. and Bates, P. D. (Eds) *Floodplain Processes*. John Wiley, Chichester, 15–62.
Hughes, F. M. R., Harris, T., Richards, G. P., El Hames, A., Barsoum, N., Girel, J., Peiry, J.-L. and Foussadier, R. 1997. Woody riparian species' response to different soil moisture condi-

tions: laboratory experiments on *Alnus incana* (L.) Moecn. *Global Ecology and Biogeography Newsletter*, **6**, 247–256.

IPCC (Intergovernmental Panel on Climate Change) 1996. *Second Assessment Report*, 3 vols. Cambridge University Press.

Jones, D. K. C. 1995. Environmental change, geomorphological change and sustainability. In McGregor D. F. M. and Thompson, D. A. (Eds) *Geomorphology and Land Management in a Changing Environment*. John Wiley, Chichester, 11–34.

Kalicki, T. 1991. The evolution of the Vistula river valley between Cracow and Niepolamice in late Vistulan and Holocene times. In Starkel, L. (Ed.) *Evolution of the Vistula River Valley During the Last 15,000 Years*. Part IV, *Geographical Studies*, Special Issue, **6**, 11–37.

Kalicki, T. 1996. Climatic or anthropogenic alluviation in Central European valleys during the Holocene? In Branson, J., Brown, A. G. and Gregory, K. J. (Eds) *Global Continental Changes: The Context of Palaeohydrology*. Geological Society Special Publication 115, Geological Society, London, 205–216.

Kirkby, M. J. (Ed.) 1994. *Process Models and Theoretical Geomorphology*. John Wiley, Chichester.

Knighton, A. D. 1973. Riverbank erosion in relation to streamflow conditions, River Bollin-Dean, Cheshire. *East Midlands Geographer*, **6**, 416–426.

Lal, R. 1997. Degradation and resilience of soils. *Philosophical Transactions of the Royal Society, Series B*, **352**, 997–1010.

Lambert, C. P. and Walling, D. E. 1987. Floodplain sedimentation: a preliminary investigation of contemporary deposition within the lower reaches of the river Culm, Devon, UK. *Geografiska Annaler*, **69A**, 47–59.

Lane, S. N. and Richards, K. S. 1997. Linking river channel form and process: time, space and causality revisited. *Earth Surface Processes and Landforms*, **22**, 249–260.

Lewin, J., Macklin, M. G. and Newson, M. D. 1988. Regime theory and environmental change – irreconcilable concepts? In White, W. R. (Ed.) *International Conference on River Regime*. John Wiley, Chichester, 431–445.

Mackay, S. D. and Bridge, J. S. 1995. Three-dimensional model of alluvial stratigraphy: theory and application. *Journal of Sedimentary Research*, **B65**, 7–31.

Macklin, M. G. and Lewin, J. 1986. Terraced fills of Pleistocene and Holocene age in the Rheidol valley, Wales. *Journal of Quaternary Science*, **1**, 21–34.

Macklin, M. G. and Lewin, J. 1989. Sediment transfer and transformation of an alluvial valley floor: the river Tyne, Northumbria, UK. *Earth Surface Processes and Landforms*, **14**, 233–246.

Macklin, M. G. and Lewin, J. 1993. Holocene river alluviation in Britain. *Zeitschrift für Geomorphologie*, **88**, 109–122.

Macklin, M. G. and Needham, S. 1992. Studies of British alluvial archaeology: potential and prospect. In Needham, S. and Macklin, M. G. (Eds) *Alluvial Archaeology in Britain*. Oxbow Monograph 27, Oxbow Books, Oxford, 9–26.

Macklin, M. G., Passmore, D. G. and Rumsby, B. T. 1992a. Climatic and cultural signals in Holocene alluvial sequences: the Tyne basin, northern England. In Needham, S. and Macklin, M. G. (Eds) *Alluvial Archaeology in Britain*. Oxbow Monograph 27, Oxbow Books, Oxford, 123–140.

Macklin, M. G., Rumsby, B. T. and Heap, T. 1992b. Flood alluviation and entrenchment: Holocene valley floor development and transformation in the British Uplands. *Geological Society of America Bulletin*, **104**, 631–643.

Marriott, S. 1992. Textural analysis and modelling of flood deposits: river Severn, UK. *Earth Surface Processes and Landforms*, **17**, 687–698.

McGregor, D. F. M. and Thompson, D. A. (Eds) 1995a. *Geomorphology and Land Management in a Changing Environment*. John Wiley, Chichester.

McGregor, D. F. M. and Thompson, D. A. 1995b. Geomorphology and land management in a changing environment. In McGregor, D. F. M. and Thompson, D. A. (Eds) *Geomorphology and Land Management in a Changing Environment*. John Wiley, Chichester, 3–10.

Menzel, R. G., Jung, P., Ryu, K. and Um, K. 1987. Estimating soil erosion losses in Korea with fallout cesium-137. *Journal of Applied Radiation and Isotopes*, **38**, 451–454.

Nanson, G. C. and Huang, H. Q. 1998. Anabranching rivers: divided efficiency leading to fluvial diversity. In Miller, A. and Gupta, A. (Eds) *Varieties of Fluvial Form*. John Wiley, New York, in press.

Nanson, G. C. and Knighton, A. D. 1996. Anabranching rivers: their cause, character and classification. *Earth Surface Processes and Landforms*, **21**, 217–239.

Needham, S. 1992. Holocene alluviation and interstratified settlement evidence in the Thames valley at Runneymede Bridge. In Needham, S. and Macklin, M. G. (Eds) *Alluvial Archaeology in Britian*. Oxbow Books, Oxford, 249–260.

Nicholas, A.P. and Walling, D. E. 1997. Modelling flood hydraulics and overbank deposition on river floodplains. *Earth Surface Processes and Landforms*, **22**, 59–77.

Paine, A. D. M. 1985. Ergodic reasoning in geomorphology – time for a review of the term? *Progress in Physical Geography*, **9**, 1–15.

Pedley, H. M. 1990. Classification and environmental models of freshwater tufas. *Sedimentary Geology*, **68**, 143–154.

Pentecost, A. 1995. The Quaternary travertine deposits of Europe and Asia Minor. *Quaternary Science Reviews*, **14**, 1005–1028.

Pizzuto, J. E. 1987. Sediment diffusion during overbank flows. *Sedimentology*, **34**, 304–317.

Prosser, I. P., Chappell, J. and Gillespie, R. 1994. Holocene valley aggradation and gully erosion in headwater catchments, south-eastern Australia. *Earth Surface Processes and Landforms*, **19**, 465–480.

Quine, T. A., Walling, D. E., Zhang, X. and Wang, Y. 1992. Investigation of soil erosion on terraced fields near Yanting, Sichuan Province, China, using caesium-137. In *Erosion, Debris Flows and Environment in Mountain Regions*, Proceedings of the Chengdu Symposium, July 1992. IAHS Publication No. 209, 155–168.

Quine, T. A., Walling, D. E. and Mandiringana, O. T. 1993. An investigation of the influence of edaphic, topographic and land-use controls on soil erosion on agricultural land in the Borrowdale and Chinamora areas, Zimbabwe, based on caesium-137 measurements. In Hadley, R. F. and Mizuyama, T. (Eds) *Sediment Problems: Strategies for Monitoring, Prediction and Control*, Proceedings of the Yokohama Symposium, July 1993. IAHS Publication No. 217, 185–196.

Quine, T. A., Desmet, P. J. J., Govers, G., Vandaele, K. and Walling, D. E. 1994. A comparison of the roles of tillage and water erosion in landform development and sediment export on agricultural land near Leuven, Belgium. In Olive, L., Loughran, R. J. and Kesby, J. A. (Eds) *Variability in Stream Erosion and Sediment Transport*. IAHS Publication No. 224, 77–86.

Quine, T. A., Walling, D. E. and Govers, G. 1996. Simulation of radiocaesium redistribution on cultivated hillslopes using a mass-balance model: an aid to process interpretation and erosion rate estimation. In Anderson, M. G. and Brooks, S. M. (Eds) *Advances in Hillslope Processes*. John Wiley, Chichester, 561–588.

Quine, T. A., Govers, G., Walling, D. E., Zhang, X., Desmet, P. J. J. and Zhang, Y. 1997. Erosion processes and landform evolution on agricultural land – new perspectives from caesium-137 data and topographic-based erosion modelling. *Earth Surface Processes and Landforms*, **22**, 799–816.

Ritchie, J. C. and McHenry, J. R. 1990. Application of radioactive fallout cesium-137 for measuring soil erosion and sediment accumulation rates and patterns: a review. *Journal of Environmental Quality*, **19**, 215–233.

Robinson, M. and Lambrick, G. H. 1984. Holocene alluviation and hydrology in the Upper Thames basin. *Nature*, **308**, 809–814.

Rose, J. 1995. Lateglacial and early Holocene river activity in lowland Britain. In Frenzel, B. (Ed.) European River Activity and Climate Change During the Lateglacial and Early Holocene. Palaeoclimate Research, Special Issue, **14**, 51–74.

Rose, J., Turner, C., Coope, G. R. and Bryan, M. D. 1980. Channel changes in a lowland river catchment over the last 13,000 years. In Cullingford, R. A., Davidson, D. A. and Lewin, J. (Eds) *Timescales in Geomorphology*. John Wiley, Chichester, 159–176.

Rumsby, B. T. and Macklin, M. G. 1994. Channel and floodplain response to recent abrupt

climate change: the Tyne basin, Northern England. *Earth Surface Processes and Landforms*, **19**, 499–515.
Salisbury, C. R., Whitley, P. J., Litton, C. D. and Fox, J. L. 1984. Flandrian courses of the river Trent at Colwick, Nottingham. *Mercian Geologist*, **9**, 189–207.
Shanahan, J. 1998. *Sediment delivery in small agricultural catchments*. Unpublished PhD thesis, University of Exeter.
Smith, D. G. and Smith, N. D. 1980. Sedimentation in anastomosing river systems: examples from alluvial valleys near Banff, Alberta. *Journal of Sedimentary Petrology*, **50**, 157–164.
Starkel, L. 1990. *Evolution of the Vistula River Valley During the Last 15,000 Years*, Vol. III and IV. Geographical Studies Issues 5 and 6, Polish Institute of Geography.
Starkel, L. 1991. The Vistula valley: a case study for Central Europe. In Starkel, L., Gregory, K. J. and Thornes, J. B. (Eds) *Temperate Palaeohydrology*. John Wiley, Chichester, 171–188.
Syers, J. K. 1997. Managing soils for long-term productivity. *Philosophical Transactions of the Royal Society, Series B*, **352**, 1011–1022.
Thornes, J. B. (Ed.) 1990. *Vegetation and Erosion: Processes and Environments*. John Wiley, Chichester.
Tinker, P. B. 1997. The environmental implications of intensified land use in developing countries. *Philosophical Transactions of the Royal Society, Series B*, **352**, 1023–1033.
Tornquist, T. 1993. *Fluvial Sedimentary Geology and Chronology of the Rhine–Meuse Delta, the Netherlands*. University of Utrecht, Utrecht.
Trimble, S. W. 1981. Changes in sediment storage in the Coon Creek basin, Driftless Area, Wisconsin, 1853 to 1975. *Science*, **214**, 181–183.
Vandaele, K., Vanommeslaeghe, J., Muylaert, R. and Govers, G. 1996. Monitoring soil redistribution patterns using sequential aerial photographs. *Earth Surface Processes and Landforms*, **21**, 353–362.
Wallbrink, P. J. and Murray, A. S. 1996a. Determining soil loss using the inventory ratio of excess lead-210 to cesium-137. *Soil Science Society of America Journal*, **60**, 1201–1208.
Wallbrink, P. J. and Murray, A. S. 1996b. Distribution and variability of ^{7}Be in soils under different surface cover conditions and its potential for describing soil redistribution processes. *Water Resources Research*, **32**, 467–476.
Walling, D. E. and He, Q. 1994. Rates of overbank sedimentation on the floodplains of several British rivers during the past 100 years. In Olive, L., Loughran, R. J. and Kesby, J. A. (Eds) *Variability in Stream Erosion and Sediment Transport*. IAHS Publication No. 224, 203–210.
Walling, D. E. and Quine, T. A. 1991. The use of caesium-137 measurements to investigate soil erosion on arable fields in the UK: potential applications and limitations. *Journal of Soil Science*, **42**, 147–165.
Walling, D. E. and Quine, T. A. 1993. Using Chernobyl-derived fallout radionuclides to investigate the role of downstream conveyance losses in the suspended sediment budget of the River Severn, UK. *Physical Geography*, **14**, 239–253.
Walling, D. E. and Quine, T. A. 1995. The use of fallout radionuclide measurements in soil erosion investigations. In *Nuclear Techniques in Soil–Plant Studies for Sustainable Agriculture and Environmental Preservation*. International Atomic Energy Agency, IAEA-SM-334/35, 597–619.
Walling, D. E., He, Q. and Quine, T. A. 1995. Use of caesium-137 and lead-210 as tracers in soil erosion investigations. In Leibundgut, Ch. (Ed.) *Tracer Technology for Hydrological Systems*. IAHS Publication No. 229, 163–172.
Walling, D. E., He, Q. and Nicholas, A. P. 1996. Floodplains as suspended sediment sinks. In Anderson, M. G., Walling, D. E. and Bates, P. D. (Eds) *Floodplain Processes*. John Wiley, Chichester, 399–440.
Wells, L. E. 1990. Holocene history of the El Niño phenomenon as recorded in flood sediments in northern Peru. *Geology*, **18**, 1134–1137.
Wicherek, S. (Ed.) 1993. *Farm Land Erosion in Temperate Plains and Hills*. Elsevier, Amsterdam.
Zhang, X., Quine, T. A., Walling, D. E. and Li, Z. 1994. Application of the caesium-137 technique in a study of soil erosion on gully slopes in a yuan area of the Loess Plateau near Xifeng, Gansu Province, China. *Geografiska Annaler*, **76A**, 103–120.

Zhang, X., Quine, T. A. and Walling, D. E. 1998. Soil erosion rates on sloping cultivate land on the Loess Plateau near Ansai, Shaanxi Province, China: an investigation using ^{137}Cs and rill measurements. *Hydrological Processes*, in press.

Section 1

THE SLOPE–CATCHMENT SCALE

2 Modelling the Impacts of Holocene Environmental Change in an Upland River Catchment, Using a Cellular Automaton Approach

T. J. COULTHARD, M. J. KIRKBY and M. G. MACKLIN
School of Geography, University of Leeds, UK

INTRODUCTION

Two main factors, climate change and anthropogenic activity, can be identified as shaping the morphology of Britain's uplands during the Holocene (Macklin and Lewin, 1993). The relative impact of these factors on valley floor evolution, however, is still uncertain. Questions exist regarding the influences of changing flood frequency and magnitude, impact of catastrophic flood events, and the hydrological effects of deforestation (Macklin *et al.*, 1992). Are alluvial fans and terrace sequences the results of long-term Holocene alluviation or the product of one large flood event? Modelling could help us understand the past, present and future behaviour of our river systems in response to environmental changes, as well as the importance of the processes and feedback mechanisms operating within a river catchment.

There are a wide range of fluvial and catchment based models. These include detailed two-dimensional finite-element grids to determine water surface profiles (Bates *et al.*, 1997; Nicholas and Walling, 1997) and the more "classic" one-dimensional approaches, such as HEC II (Feldman, 1981). These appear successful, but their application can be limited. The complexity of solving the complex Navier-Stokes equation and the need to define finite-element lattices, mean they are computationally restricted to operating in a confined area. They largely fail to account for processes outside of this study reach, such as mass movement, hydrology and changes in upstream sediment supply, and have difficulty in dealing with highly dynamic reaches.

Howard (1994) and Polarski (1997) take an alternative approach, placing the emphasis on the slope processes. Howard simplifies channel processes to operating within a grid cell, and the values for width and depth are calculated using empirical relationships. This approach allows the aggradation and degradation of the channel, in the context of the whole catchment, but does not allow the formation of terraces, a floodplain stratigraphy, or differing channel patterns which geomorphologists use to interpret past environmental change.

Whilst both of these approaches are fruitful, the former, hydraulic approach trades

Fluvial Processes and Environmental Change. Edited by A. G. Brown and T. A. Quine.
© 1999 John Wiley & Sons Ltd.

catchment-scale realism for local floodplain accuracy, whereas the latter sacrifices channel accuracy for realism at the catchment/basin scale. In the evolution of a catchment, the interactions between the channel, floodplain and hillslopes are of central importance. The channel is the principal transport mechanism, the floodplain the store, and channel heads and hillslopes the provider of fresh sediment. However, these processes act over a wide range of spatial and temporal scales, ranging from the rapid movement of bedload during a flood, to the slow creep of a hillslope. A methodology is required to integrate all these processes and their scales.

In this chapter a Cellular Automaton (CA) model, developing the work of Coulthard et al. (1996), is applied to the catchment of Cam Gill Beck, a tributary of the River Wharfe, North Yorkshire. This model reconciles these scale issues by dividing the catchment into uniform $1 m^2$ grid cells. This resolution is small enough to allow representation of fluvial processes, yet large enough to encompass a whole catchment. Furthermore, to resolve temporal scale problems a variable time step is used which is dependent upon the erosion rates. This allows the representation of small-scale processes such as fluvial erosion, yet incorporates the long-term effects of soil creep and mass movement.

Cellular automaton models have a substantial background in ecological and population modelling (Coulthard et al., 1996; Green, 1998), yet have rarely been applied in geomorphology. Examples of these include Murray and Paola (1994), Favis-Mortlock (1996) and Coulthard et al. (1996, 1997). Wolfram (1984) identifies five key factors that characterise CA models:

(1) they consist of a discrete lattice of cells,
(2) they evolve in discrete time steps,
(3) each cell takes on a finite set of possible variables,
(4) the value of each cell evolves according to the same deterministic laws,
(5) the laws for the cell's evolution depend only on the local neighbourhood of cells around it.

Though the concept and laws of a cellular automata may be simple, the interaction between the cells has been shown to give rise to complex, emergent behaviour, often consistent with the systems modelled. Murray and Paola's (1994) model of braided rivers reproduces bars and channels that remain statistically stable yet never reach a steady state. Favis-Mortlock (1996) models rill growth, showing emergent growth of a "realistic" rill pattern, and an apparently "chaotic" sediment discharge per iteration. Coulthard et al. (1996, 1997) show how previous applications of this model have produced emergent "pool–riffle" like structures along with braids, bars and alluvial fans. Furthermore, the lattice structure of a CA lends itself ideally to raster GIS and DEM data.

STUDY AREA

Cam Gill Beck has a catchment area of $4.2 km^2$, with its headwaters lying in the uplands near Buckden Pike, underlain by Millstone Grit. The middle and lower part of the catchment runs over Carboniferous Limestone and coarse glacial deposits. The channel is partly bedrock based, with extensive gravel/cobble bed reaches where the valley widens. This area was chosen for two reasons: first it is a small, discrete and uncomplicated catchment; and secondly, in 1686, the village of Starbotton at the bottom of Cam Gill Beck, was severely affected by a catastrophic flood.

> On the 18th feb, the whole of England was visited by a tempest, accompanied with thunder, which committed general devastation. The inhabitants of Kettlewell and Starbotton, in Craven, were almost all drowned in a violent flood. These villages are situate under a great hill, whence the rain descended with such violence for an hour and a half, that the hill on one side opened and casting up water into the air to the height of an ordinary church steeple, demolished several houses, and carried away the stones entirely!
> (Mayhall, 1860)

The extreme magnitude of the event is emphasised by a mention in parliamentary records. This catastrophic event appears to be one of several in the area around this time (Merrett and Macklin, Chapter 18, this volume), linked to a period of extreme weather during the "Little Ice Age" (Lamb, 1977).

Field evidence supports these qualitative accounts. Two boulder berms in the catchment have been dated to pre-1750 by lichenometry (Merrett, pers. comm., 1998). Examination of a dissected berm approximately 1 km up from Starbotton village reveals a 3 m thick, poorly sorted flood unit overlying boulder clay. Lichen dating, along with low lead concentrations, show it predates late 18th century mining activity, suggesting that it was probably formed during the 1686 flood event. From this and other deposits, palaeohydraulic reconstructions (Merrett, pers. comm.) can calculate the magnitude of sediment and water discharges during this flood, against which future model outputs can be validated.

DESCRIPTION OF CA MODEL

For this model, the catchment of Cam Gill Beck is represented by a grid of 1 m by 1 m cells. Each grid cell is then assigned initial values for elevation, water discharge, water depth, drainage area and grain-size fractions. For each time step or iteration, these values are updated in relation to the immediate neighbours according to laws applied to every cell. These laws fall into four groups covering hydrological, hydraulic, erosion and slope process modelling (Figure 2.1).

Hydrological and hydraulic operations are lumped together and are described in Coulthard *et al.* (1996). The input of water at each grid cell is calculated by taking the hydrograph produced by a storm, dividing the total discharge at the required time, by the size of the catchment and inputting this evenly to every cell. The flow is then routed according to a multiple flow algorithm as discussed by Desmet and Govers (1996) (equation 1).

$$Q_i = Q_0 \frac{S_i}{\sum S_i} \quad (1)$$

Here Q_i is the fraction of discharge delivered to the neighbouring cell i from the total cell discharge (Q_0) in m^3 s^{-1}. According to the slope S between the cell and its relative

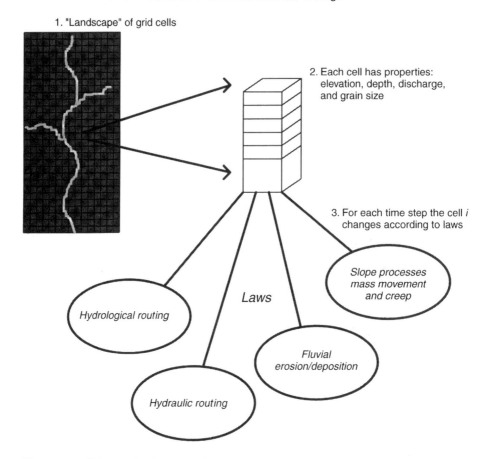

Figure 2.1 Schematic diagram of the key processes operating in the CA model

neighbours i numbering from 1–x (x ranging from 3 to 8 depending on the number of neighbours). When the discharge exceeds a threshold (set at $0.01 \text{ m}^3 \text{ s}^{-1}$) it is treated as fluid flow, and a depth is calculated. This is done using an adaptation of Manning's equation (2).

$$d = \left(\frac{Qn}{S^{0.5}}\right)^{3/5} \qquad (2)$$

Here d is depth (m), Q is discharge, n is Manning's roughness coefficient, and S is the slope. Water is then routed according to equation (3) where the depth of water as well as cell elevation is considered.

$$Q_i = Q_o \frac{[(e+d) - e_i]^x}{\sum [(e+d) - e_i]^x} \qquad (3)$$

Here e is the elevation and d is the depth of water (in metres) for each neighbouring cell i. This allows the flow to be routed over and around obstacles. In both these expressions, differences in slope between diagonal neighbours are accounted for by dividing by root 2. The calculation of depth is an important approximation to allow discharges to be routed over as well as around obstacles. The ability of this procedure to effectively mimic channel flow routing and depth characteristics at this meso scale was demonstrated by Coulthard et al. (1996).

The model negates the need to sort the elevations of cells as required by other grid-based hydrological models in order to work from highest to lowest cell, by scanning across the catchment. This novel procedure is described further in Coulthard et al. (1996). To summarise, the model scans across the catchment from left to right, right to left, up then down, effectively "pushing" the water only to the three cells immediately in front, keeping a tally of the maximum discharge. This gives near identical results to a method based on sorting the altitudes, with a substantial saving in computer time, even in complex situations such as tributaries.

The model accounts for fluvial erosion of five different grain sizes, using the Einstein (1950) sediment transport relationship. To implement this the model contains an "active" layer (Parker, 1990) to allow transference between the bedload and substrate. This is carried out in a similar manner to that used by Parker (1990), Hoey and Ferguson (1994) and Cui et al. (1996), with three layers being used: bedload, active layer and substrate. The active layer is divided into five grain-size fractions according to whole phi class sizes. Material is eroded from and to this layer in proportion to the relative grain size, the power of flow and the sediment transport laws employed. The thickness of this active layer is defined as equal to D_{90} (Cui et al., 1996). The transfer of material to and from this active layer is currently poorly understood. During degradation computation of the active layer is fairly straightforward, with material incorporated directly from the subsurface to the active layer. If material is lifted from the subsurface to the active layer in the same proportions, then the winnowing of fines from the active layer leads to armouring, as observed in nature. During aggradation, however, the situation is less clear. Observations indicate that a proportion of the material deposited on to the active layer becomes directly incorporated into the substrate instead of all being added to the active layer. Here, the active layer is acting as a "filter" (Toro-Escobar et al., 1996). The exact nature of this process is unclear. Hoey and Ferguson (1994) overcome this by introducing an exchange parameter, so only a fraction (30%) is left in the active layer; the rest is moved to the substrate. More recent physical and numerical modelling by Toro-Escobar et al. (1996) investigating this process supports this proportion. However, the figure of Toro-Escobar et al. (1996) was calculated from a flume experiment and is the result of several hours of degradation and aggradation over a 20 m reach. This term would appear to be dependent on time and volume of material transported. If such a term is applied to this model then 70% of the material would be lost every metre it moved, every iteration. At present this is ignored but may need to be scaled for future application.

This complete procedure allows the development of an armour bed, as well as coarse and fine deposits. It cannot, however, give a true stratigraphic record of the flood deposits. This would require many "layers" of storage which is currently computationally unfeasible.

The amount eroded by fluvial action from cell to cell is determined using the Einstein (1950) formulation. This takes the following form:

$$\psi = \frac{(\rho_s - \rho)D}{\rho dS} \tag{4}$$

where ψ is the balance between the forces moving and restraining the particle, $\rho_s - \rho$ is the relative density of the submerged sediment, D is the grain size (m), d is the flow depth, and S is the energy slope. A dimensionless bedload transport rate ϕ is then calculated:

$$\phi = q_s \sqrt{\frac{\rho}{(\rho_s - \rho)gD^3}} \tag{5}$$

ϕ is then related to ψ by the relationship plotted by Brown (1950):

$$\phi = 40(1/\psi)^3 \tag{6}$$

A rearrangement of equations (5) and (6) then allows q_s, the volumetric sediment load in m^3 s^{-1}, to be calculated. For each grid cell, the amount in each grain-size class which can be eroded is calculated, removed from the active layer of the cell in question, and deposited to the active layer of the downstream cell.

The model generates enough data per cell to drive any of the contemporary sediment transport equations (Bathurst et al., 1987), but the Einstein equation is selected for several reasons. First, Bathurst et al. (1987) compare several major sediment transport relationships (Meyer-Peter and Muller, 1948; Einstein, 1950; Bagnold, 1980; Parker et al., 1982) against the Schoklitsch 1943 formula (Schoklitsch, 1950) for a steep mountain stream. He finds that the Schoklitsch results give the best fit to most data, followed closely by Parker's equation. However, Bathurst et al. (1987) recommend that in areas where much information is available on the channel geometry and flow regime, as in this model, a more detailed approach such as Einstein is used. Secondly, Einstein accounts directly for different size fractions, whilst most of the other approaches are "total load" formulations. With total load formulae, bedload transport is calculated for all fractions lumped together, and volumes of the separate fractions transported (as required by this model) have to be back calculated from a distribution. Thirdly, the Einstein formula has no threshold for entrainment. There is no critical limit for stream power from which sediment transport occurs. If the depth/slope product and sediment discharge is plotted, a log–log relationship is shown. This results in a sudden increase in bedload transport, not unlike a threshold, but a small amount of activity always remains. This possibly represents the situation in a stream more accurately than threshold formulae, as whilst at low flow situations there is very little sediment transport, there is always a small amount. This implies that fractional amounts of boulders are moved between events, which does happen, but in minute proportions. However, if this were not the case, then incision would not be possible, as the channel would always remain armoured to the extent of the last major flood.

Landslides and episodes of mass movement are represented as follows. There is a slope threshold, above which mass movement occurs, moving material from an "uphill" cell to a lower cell. This process is itself iterative as a small slide in a cell at the base of a slope may trigger more movement uphill. The model accounts for this by continuing to check the rows and columns after a slide until there is no more movement. One uncertainty which this section addresses is that of the interaction of the material from a landslide with the river's active bed layer. This is incorporated here by adding the material from the sliding cells' active layer to the active layer of the receiving cell. If the amount moved is greater than the active layer thickness, then material from the subsurface is also added.

MODEL IMPLEMENTATION

A 1 m resolution DEM of the Starbotton catchment was created by digitising the contours of a 1:10 000 scale OS map. Extra detail of the channel position and dimensions, along with supplementary valley floor morphology such as terrace and berm positions, was added to the DEM from an EDM field survey. The DEM was then created by combining this line and point data using ARC-INFO's TOPOGRID function. This function is designed specifically for the interpolation of a hydrological DEM from contour data, and also removes topographic sinks and hollows. This is necessary as many artificial sinks are produced by errors interpolating the DEM from contour data (Goodchild and Mark 1987; Hutchinson, 1989).

This operation created a DEM of 4.2 million points (Figure 2.2). However, for most of the model's operation time, many areas are dormant but remain important. The model therefore only needs to concentrate where there is activity, i.e. running water. This is achieved by scanning the whole catchment every 100 iterations, selecting the cells which have water running in them and those within a 5 m proximity of running water. These cells are then used for the next 100 iterations. This operation is fraught with complexity, since it is easy for the model to ignore a small section, e.g. in the middle of the main channel, halting the flow. This procedure also has to account for the expanding dimensions of the drainage network during the passage of a flood. To combat this, the model checks to see if the channel is trying to push out of the previously selected area; if so, the area is re-selected. The net result of this operation is that for 98 to 99% of the model's operating time, only up to 100 000 cells are checked, yet periodically all 4.2 million are. This provides a huge computational time saving of several orders of magnitude, making the operation feasible at a catchment scale. The landslide section is optimised by checking slide conditions at each iteration only for cells in the immediate proximity of the channel. The whole catchment is then checked every 100 iterations. For all other functions, a variable time step is used. This is adjusted so the maximum amount that can be removed or deposited from one cell to another is a small proportion of the average slope. A maximum time step of 120 seconds is introduced, allowing significantly rapid progress in low flow situations, yet not too long as to miss any part of a storm. The boundary conditions were fixed so that none of the edge cells can be altered yet material can pass over them, similar to the bottom of a flume.

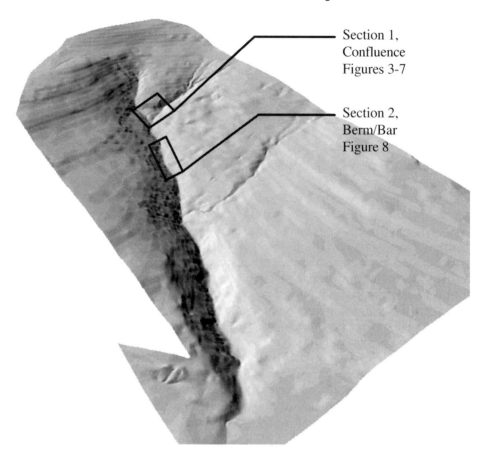

Figure 2.2 Draped image of Starbotton DEM. Scale: 1600 m by 2800 m

RESULTS

Figure 2.3 shows the effects of simulating 16 floods of approximately bankfull discharge on the upper part of the catchment. The initial conditions were set with every cell having the same grain-size distribution. Because of the catchment's size, initial conditions cannot feasibly be defined for every square metre; therefore, a "spin up" time is required to allow the initial conditions to evolve. This meant that for the first few runs there was a large sediment discharge, until an armoured layer had developed on the channel bed. Subsequently, the catchment displays a deterministic non-linear pattern of behaviour (Coulthard *et al.*, 1997), with irregular peaks in the sediment discharge. This may be attributed to the movement of "slugs" (Nicholas *et al.*, 1995) of sediment downstream and the consequent re-mobilisation of these in later floods. Other observations show that these peaks in activity can also be linked to the input of landslides – mass movement producing an input of hillslope fines to the armoured

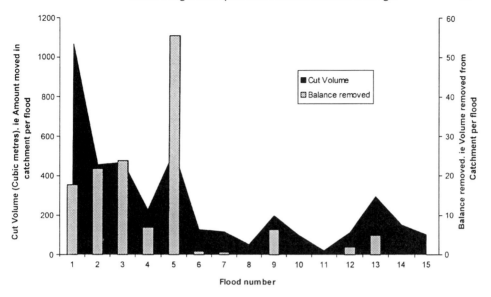

Figure 2.3 Volume of sediment moved and removed from the catchment for each flood. Flood length 1 hour; hydrograph peak after 15 minutes

channel. Episodes of fluvial erosion and deposition correspond largely with the rising and falling limbs, respectively, of the hydrograph. There are, however, sporadic episodes of activity during periods of low flow. This results from the input of mass movement from the slopes, demonstrating a partial de-coupling of fluvial and slope processes (Coulthard et al., 1997).

Figures 2.4–2.7 show valley floor evolution at the confluence of the two main upland channels. Figure 2.4 shows the initial conditions, where a small discharge has been applied to the catchment, resulting in the formation of channels. Figure 2.5 shows the same region after the 16 floods, and Figure 2.6 shows the same area but after one large flood of approximately five-year return interval. These three views show the development of several features. The series of flood events has led to the development of a fan at the mouth of the right-hand tributary, formed of fines eroded from upland areas. This has caused the widening of the channel opposite and downstream. During the rising and falling limb of the hydrograph, a multiple channel forms, as the large sediment influx causes the channel to diverge and converge. Figure 2.7 corroborates these observations, showing the grain-size distribution for the section after the 16 floods. This shows an "armouring" down the centre of the multiple channels and fine material deposited at the base of the fan.

Figure 2.8 details the section outlined in Figure 2.2. Figure 2.8(a) shows a shaded plan view, Figure 2.8(b) the grain-size, and Figure 2.8(c) four cross-sections. Here flow (from top to bottom) emerges from a narrow section into a wider section of the valley floor, resulting in deposition and the formation of a coarse deposit on the right bank of the channel. Some 30 m downstream the tail of this deposit is being eroded as the valley floor narrows and steepens, forming a deposit of fines on the left bank. These features

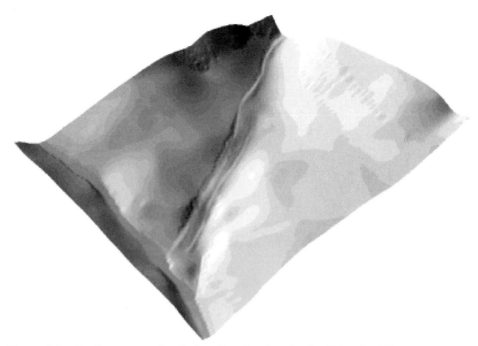

Figure 2.4 Confluence section before flood series. Scale: 100 m by 100 m

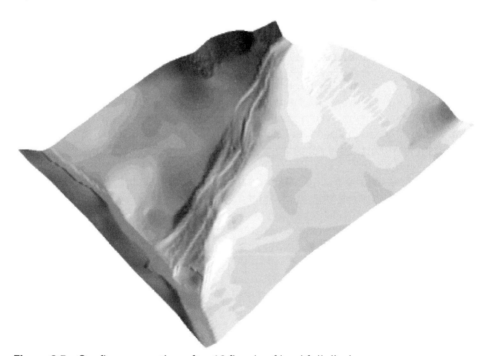

Figure 2.5 Confluence section after 16 floods of bankfull discharge

Figure 2.6 Confluence section after a 'five-year' flood event

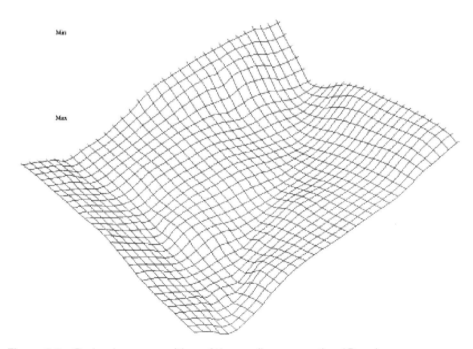

Figure 2.7 Grain-size composition of the confluence section (D_{50} m)

are similar to a boulder berm and side bar/terrace in plan form, elevation and grain-size. Although this is only a preliminary simulation, cross-sections from the model (Figure 2.8(c)) compare favourably with those from a field survey (Figure 2.8(d)).

DISCUSSION

Initial runs of this model show the simulation of landslides, berms, bars, braids, terraces and alluvial fans, all of similar magnitude and form to that observed in the study area. These landforms evolved over the 15 floods, from the featureless valley floor with even distributions of sediment.

The preliminary results are already beginning to show a difference between the effects of floods of different frequency and magnitude. Figures 2.4–2.8 demonstrate the different impacts of 16 small floods as well as one large five-year event, suggesting that large floods have the greatest influence on valley floor development. The model, however, needs to be applied to the entire catchment, along with longer runs to assess the impacts of sustained periods of aggradation and incision. Future model runs, incorporating different flood frequencies, magnitudes and vegetation cover (forested/ deforested), may enable the impact of climate change and human activity on landscape evolution to be evaluated.

The unpredictable sediment discharge (Figure 2.3) is an unexpected result, but such behaviour is typical of many upland catchments (Evans, 1996; Lane and Richards, 1997). The results strongly suggest that the non-linear behaviour of the basin is not just the result of "random" flood events, but reflects an inherent instability within the system. For example, the deposition of one clast may deflect enough flow to initiate re-mobilisation of stored sediment, or a small amount of incision may trigger the release of a considerable volume of sediment through a landslide propagating upslope. The spatial and temporal instability of such sediment inputs is further represented in the non-linear structure and position of the landforms produced (Coulthard *et al.*, 1997). This reinforces the findings of Coulthard *et al.* (1997), that non-linear inputs from the whole catchment must be incorporated when modelling fluvial systems.

To the authors' knowledge, this is the first time a two-dimensional model of grain-size distribution at this scale has been presented, and preliminary results give insights into the dynamics of sediment dispersion, storage and re-mobilisation. In Figure 2.7, areas of storage are apparent, as sections of fines in the braided section above the confluence. Figures 2.4–2.6 show the re-working of these fines. In Figure 2.8, the grain-size pattern is again as observed, with a coarse "berm" type deposit and the finer "bar/terrace" downstream. Cut–fill analysis shows that the berm has formed in a depositional section and the bar where there is erosion. This is the sequence of formation observed in the field.

One factor which the model has highlighted is the importance of mass movement. Removal of the landslide module from the model slows sediment discharge and also increases the rate of incision. Analysis of cut–fill sequences show two areas to be

43

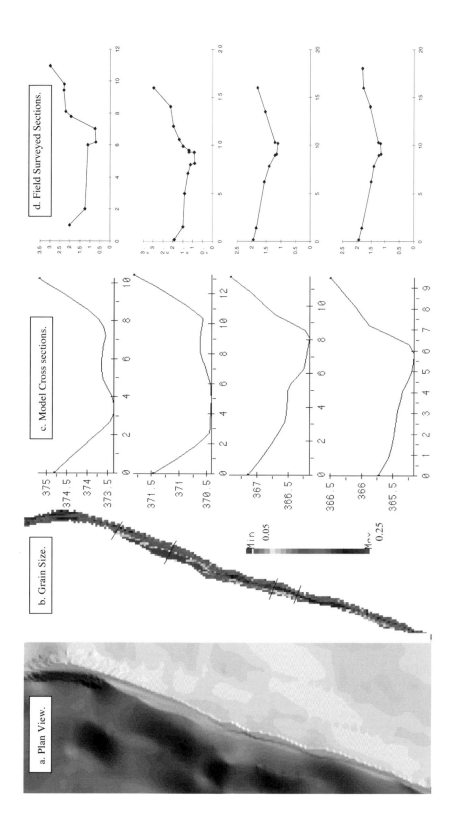

Figure 2.8 Section 2. Views (a) and (b) are 120 m by 30 m. All units are in metres

important. The dominant supply area is the head of the stream and its tributaries. Here a feedback is apparent where the channel head incises, increasing the area drained by the head, so increasing erosion. Secondly, where there has been deposition, this has caused the channel to widen, resulting in erosion of the valley walls, adding fresh fines to the system. The importance of this is the addition of fine material from the bank, not the re-working of coarser sediment. Because of the importance of mass movement, a better representation is required. Some integration of hydrology and soil depth on landslide thresholds may well be necessary, possibly in a similar manner to Brookes *et al.* (1993). A wetter climate regime may have a large effect on slope stability, with a one or two degree change in failure angle significantly increasing sediment delivery (Brookes *et al.*, 1993).

To effectively quantify human impact on the catchment, the effect of vegetation on erosion and runoff needs to be included. The lack of vegetation cover in the present model means that catchment behaviour is more typical of an arid than a cool humid environment, with no restriction on the movement of sediment by vegetation. This results in a high sediment yield. Some authors (e.g. Harvey, 1987, 1996) comment that sediment availability may well be the most important feature in the development of British upland landscapes. Therefore, vegetation needs to be represented, at least as a more resistant initial layer to erode. One of the largest impacts of vegetation change on the catchment is on runoff. Deforestation results in reduced interception and evapotranspiration, providing greater runoff and a "flashier" hydrograph. Integrating a more effective hydrological approximation, such as TOPMODEL (Beven and Kirkby, 1979), would enable this, as well as allowing the model to be driven by an hourly rainfall record.

This model is not designed to provide a precise sediment budget, or channel flow structure, but to simulate the dynamic response of an upland catchment to a series of flood events. These initial results have shown how the application of relatively simple laws on a *whole* catchment can yield subjectively realistic simulations of catchment morphology, sediment discharge and non-linear responses. We anticipate that this holistic approach will provide many insights into how changes in the environment have affected upland river catchments, and the relative roles and importance of climatic and natural factors in landscape development.

ACKNOWLEDGEMENTS

One of the authors (T.J.C.) would like to thank Stephen Merrett for all his help, as well as the Natural Environment Research Council for the provision of the research studentship no. GT4/95/147/F.

REFERENCES

Bagnold, R. A. 1980. An empirical correlation of bedload transport rates in flumes and natural rivers. *Proceedings of the Royal Society of London*, **A372**, 453–473.

Bates, P. D., Anderson, M. G., Hervouet, J. M. and Hawkes, J. C. 1997. Investigating the behaviour of two-dimensional finite element models of compound channel flow. *Earth Surface Processes and Landforms*, **22**, 3–17.

Bathurst, J. C., Graf, W. H. and Cao, H. H. 1987. Bed load discharge equations for steep mountain rivers. In Thorne, C. R., Bathurst, J. C. and Hey, R. D. (Eds) *Sediment Transport in Gravel-bed Rivers*. John Wiley, Chichester, 453–491.

Beven, K. J. and Kirkby, M. J. 1979. A physically based variable contributing-area model of catchment hydrology. *Hydrological Science Bulletin*, **24**(1), 43–69.

Brookes, S. M., Richards, K. S. and Anderson, M. G. 1993. Shallow failure mechanisms during the Holocene: utilisation of a coupled slope hydrology–slope stability model. In Thomas, D. S. G. and Allison, R. J. (Eds) *Landscape Sensitivity*. John Wiley, Chichester, 149–175.

Brown, C. B. 1950. Sediment transportation. In Rouse, H. (Ed.) *Engineering Hydraulics*. John Wiley, London, 769–857.

Coulthard, T. J., Kirkby, M. J. and Macklin, M. G. 1996. A cellular automaton landscape evolution model. In Abrahart, R. J. (Ed.) *Proceedings of the First International Conference on GeoComputation*, Volume 1. School of Geography, University of Leeds, 248–281.

Coulthard, T. J., Kirkby, M. J. and Macklin, M. G. 1997. Modelling hydraulic, sediment transport and slope processes, at a catchment scale, using a cellular automaton approach. In Pascoe, R. T. (Ed.) *Proceedings of the Second Annual Conference: GeoComputation 97*. University of Otago, Dunedin, New Zealand, 309–318.

Cui, Y., Parker, G. and Paola, C. 1996. Numerical simulation of aggradation and downstream fining. *Journal of Hydraulic Research*, **34**(2), 185–204.

Desmet, P. J. J. and Govers, G. 1996. Comparison of routing algorithms for digital elevation models and their implications for predicting ephemeral gullies. *International Journal of Geographical Information Systems*, **10**(3), 311–331.

Einstein, H. A. 1950. The bed-load function for sediment transport on open channel flows. *Technical Bulletin No. 1026*. USDA, Soil Conservation Service.

Evans, R. 1996. Hydrological impact of a high-magnitude rainfall event. In Anderson, M. G. and Brooks, S. M. (Eds) *Advances in Hillslope Processes*, Volume 1. John Wiley, Chichester, 98–127.

Favis-Mortlock, D. T. 1996. An evolutionary approach to the simulation of rill initiation and development. In Abrahart, R. J. (Ed.) *Proceedings of the First International Conference on GeoComputation*, Volume 1. School of Geography, University of Leeds, 248–281.

Feldman, A. D. 1981. HEC models for water resources system simulation: theory and experience. *Advances in Hydroscience*, **12**, 298–396.

Goodchild, M. F. and Mark, D. M. 1987. The fractal nature of geographic phenomena. *Annals of Association of American Geographers*, **77**(2), 265–278.

Green, D. G. 1998. Connectivity and complexity in landscapes and ecosystems. *Pacific Conservation Biology*, in press.

Harvey, A. M. 1987. Sediment supply to upland streams: influence on channel adjustment. In Thorne, C. R., Bathurst, J. C. and Hey, R. D. (Eds) *Sediment Transport in Gravel-bed Rivers*. John Wiley, Chichester, 121–149.

Harvey, A. M. 1996. Holocene hillslope gully systems in the Howgill Fells, Cumbria. In Anderson, M. G. and Brooks, S. M. (Eds) *Advances in Hillslope Processes*, Volume 2. John Wiley, Chichester, 731–752.

Hoey, T. and Ferguson, R. 1994. Numerical simulation of downstream fining by selective transport in gravel bed rivers: model development and illustration. *Water Resources Research*, **30**(7), 2251–2260.

Howard, A. 1994. A detachment limited model of drainage basin evolution. *Water Resources Research*, **30**(7), 2261–2285.

Hutchinson, M. F. 1989. A new procedure for gridding elevation and stream line data with automatic removal of spurious pits. *Journal of Hydrology*, **106**, 211–232.

Lamb, H. H. 1977. *Climate: Present, Past and Future 2: Climatic History and the Future*. Methuen, London.

Lane, S. N. and Richards, K. S. 1997. Linking river channel form and process: time, space and causality revisited. *Earth Surface Processes and Landforms*, **22**, 249–260.

Macklin, M. G. and Lewin, J. 1993. Holocene river alluviation in Britain. *Zeitschrift für Geomorphologie* (Supplement), **88**, 109–122.

Macklin, M. G., Rumsby, B. T. and Heap, T. 1992. Flood alluviation and entrenchment: Holocene valley-floor development and transformation in the British uplands. *Geological Society of America Bulletin*, **104**, 631–643.

Mayhall, J. 1860. *The Annals of Yorkshire: From the Earliest Period to the Present Time. Volume 1, 1856 BC to 1859 AD*. H. C. Johnson, Leeds.

Meyer-Peter, E. and Muller, R. 1948. Formulas for bed-load transport. *Proceedings 2nd Meeting International Assoc. Hydrautic Structures Research*, Stockholm, 39–64.

Murray, A. B. and Paola, C. 1994. A cellular model of braided rivers. *Nature*, **371**, 54–57.

Nicholas, A. P. and Walling, D. E. 1997. Modelling flood hydraulics and overbank deposition on river floodplains. *Earth Surface Processes and Landforms*, **22**, 59–77.

Nicholas, A. P., Ashworth, P. J., Kirkby, M. J., Macklin, M. G. and Murray, T. 1995. Sediment slugs: large scale fluctuations in fluvial sediment transport rates and storage volumes. *Progress in Physical Geography*, **19**(4), 500–519.

Parker, G. 1990. Surface based bedload transport relation for gravel rivers. *Journal of Hydraulic Research*, **28**(4), 417–436.

Parker, G., Klingeman, P. C. and McLean, D. G. 1982. Bedload and size distribution in paved gravel-bed streams. *Proceedings of the American Society of Civil Engineers, Journal of the Hydraulic Division*, **108**, 544–571.

Polarski, M. 1997. Distributed rainfall–runoff model incorporating channel extension and gridded digital maps. *Hydrological Processes*, **11**, 1–11.

Schoklitsch, A. 1950. *Handbuch des Wasserbaues*. Springer Verlag, New York.

Toro-Escobar, C. M., Parker, G. and Paola, C. 1996. Transfer function for the deposition of poorly sorted gravel in response to stream bed aggradation. *Journal of Hydraulic Research*, **34**(1), 35–51.

Wolfram, S. 1984. Preface. *Physica*, **10D**.

3 Stream Bank and Forest Ditch Erosion: Preliminary Responses to Timber Harvesting in Mid-Wales

TIM STOTT
School of Education and Community Studies, Liverpool John Moores University, UK

INTRODUCTION

In terms of fluvial erosion of soils and re-mobilisation of Quaternary sediment deposits, land-use changes in the British uplands over the past two decades have made a significant contribution to environmental change (Newson, 1984). For example, moorland improvement and in particular the afforestation of rough grazing land has produced a growing body of evidence that has raised concerns over the detrimental effects of land-use conversion for forestry and subsequent timber harvesting on stream runoff (Law, 1956); water quality (Ormerod and Edwards, 1985; Leeks and Roberts, 1987); channel erosion (Murgatroyd and Ternan, 1983; Stott, 1997a); drainage ditch erosion (Newson, 1980, 1984); salmon fisheries (Graesser, 1979); soil erosion and suspended sediment yields (Austin and Brown, 1982; Robinson and Blyth, 1982; Burt *et al.*, 1984; Stretton, 1984; Stott *et al.*, 1986; Ferguson and Stott, 1987; Greene, 1987; Stott, T.A., 1987, 1989; Miller *et al.*, 1988; Francis and Taylor, 1989; Soutar, 1989; Ferguson *et al.*, 1991; Johnson, 1993) and bedload yields (Newson, 1980; Leeks, 1992; Stott, 1997b). The downstream effects of such changes on channel stability (Newson and Leeks, 1985; Newson, 1986), fisheries (Graesser, 1979) and reservoir sedimentation rates (Stott, A. P., 1987) have, and continue to be, the subject of other enquiries.

The environmental factors controlling stream bank erosion processes have attracted attention from geomorphologists, hydrologists and river engineers for several decades (e.g. Wolman, 1959; Slaymaker, 1972; Hill, 1973; McGreal and Gardiner, 1977; Lawler, 1978, 1982, 1986, 1991, 1992, 1993a, 1994, 1995, Dickinson and Scott, 1979; Hooke, 1979, 1980; Thorne, 1981, 1982, 1990; Gardiner, 1983; Hagerty *et al.*, 1983; Murgatroyd and Ternan, 1983; Stott, 1984, 1997a; Nanson and Hickin, 1986; Lawler and Leeks, 1992; Davis and Gregory, 1994; Madej *et al.*, 1994; Ashbridge, 1995). In some landscapes, bank erosion may be an important, if not the dominant process in terms of its contribution to river sediment loads, and can supply well in excess of 50% of the catchment sediment output (e.g. Coldwell, 1957; Carson *et al.*, 1973; Imeson and

---

Fluvial Processes and Environmental Change. Edited by A. G. Brown and T. A. Quine.
© 1999 John Wiley & Sons Ltd.

Jungerius, 1974; Lewin et al., 1974; Bello et al., 1978; Grimshaw and Lewin, 1980; Curr, 1984; Duijsings, 1986, 1987; Stott et al., 1986; Church and Slaymaker, 1989; Walling and Woodward, 1992). Indeed, investigations in south-west Scotland, for example, revealed that 93% of the total sediment removed from the Water of Deugh, a mountain grassland drainage basin, resulted from erosion of river bluffs (Kirkby, 1967). Despite these numerous studies, direct measurements of stream bank erosion in upland streams have been relatively uncommon (Blacknell, 1981; Stott, 1997a).

This study is a contribution to our understanding of stream bank erosion processes in afforested upland streams undergoing the radical and extremely rapid environmental change of timber harvesting. This phase of the plantation forest cycle in Britain is becoming widespread as the timber in forests planted after the First World War becomes mature. There is clearly a need to identify and monitor the geomorphological effects that result from such rapid deforestation and, to date, studies of the geomorphological effects of timber harvesting in the British uplands have been relatively scarce (Ferguson et al., 1991; Leeks, 1992; Johnson, 1993). This study contributes to the relatively small data set from which we can assess the initial impacts of plantation clearfelling in the British uplands on erosion and sediment yields.

AIMS

The aims of this study are as follows:

(i) to monitor channel, tributary and forest ditch erosion rates before, during and after timber harvesting and to assess temporal and spatial variations;
(ii) to monitor hydrometeorological variables and use these to produce a simple model to predict mean bank erosion rates;
(iii) to examine bank erosion processes at an increased temporal resolution by means of a photo-electronic erosion pin; and
(iv) to investigate the relationship between air and bank surface temperatures before and after timber harvesting.

HYPOTHESES

(a) Erosion rates will be higher on channel banks than fluvially "inactive" forest ditches.
(b) Timber harvesting will alter the hydrometeorological conditions which control bank erosion in the following ways:

- remove the "insulating effect" of trees on channel and ditch banks thereby increasing susceptibility to subaerial preparation processes (in particular frost action);
- reduce canopy interception to zero thereby increasing runoff, flood magnitude, fluvial entrainment of bank material and altering bank moisture regime (banks would become wet more readily but also may dry out faster due to lack of shade from the trees);

- cause an increase in organic debris entering the channel which could result in the formation of organic debris jams (e.g. Mosley, 1981; Megahan, 1982; Gregory et al., 1985; Stott et al., 1986; Hedin et al., 1988) and lead to changes in channel geometry, bank erosion and bank stability in the longer term.

STUDY AREA

The Plynlimon Catchments Experiment (Figure 3.1) was established by the UK Institute of Hydrology (IH hereafter) in 1968 with the initial aim of investigating the effects of upland afforestation on water yields (Newson, 1979; Hudson and Gilman, 1993) but subsequently for sediment transport and water quality research (Kirby et al., 1991). The physiography, deposits and vegetation of the catchments are described by Newson (1976). The Upper Severn, which is dominantly forested, is 8.7 km² in area and was 68% mature coniferous forest prior to the start of timber harvesting in the late 1980s. The Upper Wye catchment covers 10.55 km² and is dominantly grassland (98%). The monitoring of flow, suspended sediments and bedload from the sub-catchments of the Afon Tanllwyth (forested, 0.89 km²) and Afon Cyff (grassland, 3.13 km²) as indicated in Figure 3.1(a) began in 1973 and various studies have reported results (e.g. Painter et al., 1974; Newson, 1980; Moore and Newson, 1986; Leeks and Roberts, 1987). Newson (1980) compared the physiographic characteristics of the Cyff and Tanllwyth catchments and on the basis of this the two channels were selected for comparative study.

Stream banks of the Afon Tanllwyth, its tributaries and forest ditches are primarily composed of fine, cohesive sediments. The main channel and tributary streams surveyed in this study have gravel beds, but the two forest ditches (A and B on Figure 3.1(b)) have not incised into the glacial drift below the soils. Ditch A has a continuous, though small, discharge (a mean of less than $0.002 \, \text{m}^3 \, \text{s}^{-1}$) which can rise to around $0.25 \, \text{m}^3 \, \text{s}^{-1}$ for short periods during storm events. Ditch B has virtually no discharge and has been partially infilled with pine needles and fine sediment and is essentially "fluvially inactive". Tributaries C and D (Figure 3.1(b)) both drain ditched areas of the catchment (tributary D is referred to as LT2 in Newson's earlier study of Tanllwyth forest ditches; Newson, 1980) and have discharges in the range 0.01–$0.5 \, \text{m}^3 \, \text{s}^{-1}$. The main channel mean discharge over this study period was $0.06 \, \text{m}^3 \, \text{s}^{-1}$, with a maximum discharge of $2.32 \, \text{m}^3 \, \text{s}^{-1}$ in winter 1995 followed by peak discharges of 2.31 and $2.25 \, \text{m}^3 \, \text{s}^{-1}$ in autumn 1994 and autumn 1995, respectively.

Part of the Afon Tanllwyth catchment (16% in area terms) was clearfelled about half way through this study in March–April 1996. Sites T1–T12 were in the clearfelled area, whereas ditches A and B, tributaries C and D and sites T13–T17 remained unaffected by the timber harvesting operations (see Figure 3.1(b)).

METHODS

Monitoring of the main channel bank, tributary banks and forest ditch erosion rates was carried out in two ways. First, on the main Tanllwyth channel, erosion pins were

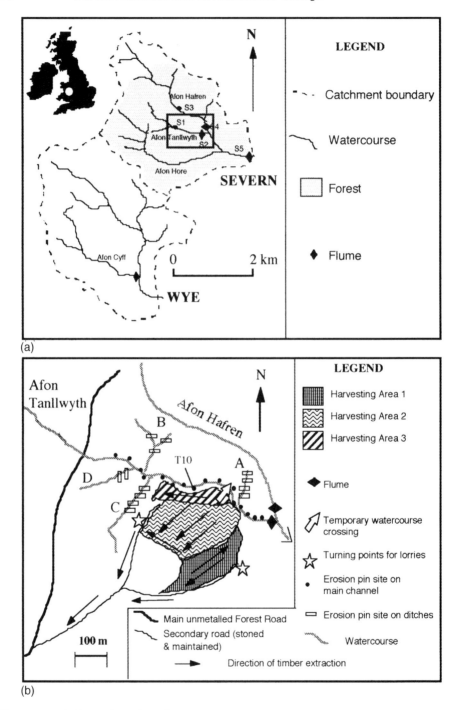

Figure 3.1 (a) Location of the Plynlimon catchments showing study reaches on Afon Tanllwyth and Afon Cyff. (b) Plan of the study area on Afon Tanllwyth, Hafren Forest, Plynlimon

located at 16 sites (see Figure 3.1(b)) where 105 erosion pins (300 mm × 3 mm welding rod) were installed in vertical lines (100 mm spacing) at 30 m intervals along the main channel, on either the right or left bank, at a point on the channel where the cohesive bank sediment was exposed and not vegetated. On tributary streams C and D, pins were installed in the same way at four and two sites respectively (a total of 42 pins). On forest ditches A and B, pins were installed at seven sites (a total of 90 pins), but this time on both left and right banks of the ditch so as to allow any changes in ditch cross-section to be observed. All pins were resurveyed, measurements being taken with callipers which read to 0.1 mm accuracy, on 12 occasions between 22 October 1994 and 15 February 1997 which represents approximately 2400 individual pin readings. A further 30 pins were installed in the same way on five sites on the Afon Cyff (Figure 3.1(a)) and these were resurveyed on four occasions between 24 April 1995 and 15 February 1997 (approximately 120 individual pin readings). In order to assess the possibility of pins being heaved out of the banks by frost (R. Evans, pers. comm., 1997), a further 32 pins were installed at 11 of the Tanllwyth sites on 13 January 1997 where the pins were pushed into the bank until they hit a wooden stake installed vertically some 250 mm from the bank face and these were resurveyed on 15 February 1997 after 36 days.

Secondly, a photo-electronic erosion pin (PEEP) (Lawler, 1991) was employed to gain a more detailed insight into the temporal variability of channel bank and ditch erosion (Lawler, 1994). This was first installed in a forest ditch (A in Figure 3.1(b)) for a 12-month period and then moved to site T10 on the main channel in July 1996, following clearfelling. The PEEP scanned at 1-minute intervals and the 15-minute mean was stored in a datalogger. A thermistor was also connected which logged temperature on the bank surface at 15-minute intervals. Hydrometeorological records were provided by IH; hourly air temperature and precipitation records were available from the nearby Moel Cynedd automatic weather station (AWS) 200 m away (see Figure 3.1(b)) and 15-minute streamflow records were also provided by IH from the Afon Tanllwyth flume (see Figure 3.1(b)).

RESULTS

A test was conducted where seven erosion pins at one site were surveyed using random numbers to dictate which of the seven pins was measured. When three of the pins had accumulated 10 measurements, standard errors were computed and these ranged from 0.17 to 0.33 mm with a mean of 0.26 mm. This suggests that the measurement technique would easily be able to detect changes of $\pm$ 0.5 mm. Nevertheless, despite this reasonable level of measurement accuracy, a proportion of erosion pin measurements after each survey showed negative changes (Table 3.1). Deposition of bank material in cohesive banks, particularly in the lower bank zone, is well recognised (e.g. Thorne and Lewin, 1979; Lawler, 1994) but in this study on headwater streams, with the exception of the survey at the end of period 11 on 13 January 1997, field observations did not confirm such deposition to be important. The possibility of pin "movement" by frost heaving, for example, was investigated over a 36-day period (13 January 1997–15 February 1997) as described earlier. During this period there were

Table 3.1 Proportions of erosion pin readings in each survey period showing negative changes

Period	1	2	3	4	5	6	7	8	9	10	11	12	Mean
Mainstream													
Count	35	76	102	104	95	92	92	90	95	104	101	101	90.6
Count < 0	0	10	37	29	18	29	3	18	16	61	28	11	21.7
% -ve	0	13.2	36.3	27.9	18.9	31.5	3.3	20	16.8	58.7	27.2	13.8	22.3
Ditches													
Count	11	11	89	90	139	139	136	134	135	136	134	140	107.8
Count < 0	1	6	33	19	22	44	0	21	30	32	41	45	24.5
% -ve	9.1	54.5	37.1	21.1	15.8	31.7	0	15.7	22.2	23.5	30.6	32.1	24.4

three frost cycles on the banks (as measured by the thermistor located on the bank at site T10) but none of the 32 pins installed showed any movement at all. This result does not completely rule out the possibility of pins moving by such mechanisms as frost heave, but it suggests that such a possibility is relatively unimportant in this study and adds some confidence to the results reported.

The proportion of "negative" erosion readings was 22% and 24% for pins on the main channel and forest ditches respectively (Table 3.1). No clear seasonal pattern exists and there is a poor relationship between the number of negative readings in ditches and the number of main channel erosion pins (correlation coefficient = 0.30). One important methodological problem which arose was how to deal with these "negative" erosion readings. Three strategies were considered:

(1) include all negative readings in mean calculations,
(2) ignore all negatives and treat as "missing data",
(3) replace all negatives with "0".

The third of these strategies was adopted in this study although the nature and timing of negative readings is the subject of continuing investigations.

Main Channel, Tributary and Ditch Erosion: Temporal and Spatial Patterns

Figure 3.2(a) shows the temporal changes in mean erosion rates through the study period with sites grouped according to the effects of clearfelling on them. Erosion rates between periods vary over almost two orders of magnitude and a clear seasonal pattern is apparent, with peaks in erosion rates occurring in the winter periods. Some 89% of the total erosion in the study period occurred in the winter periods. Table 3.2 shows that the mean erosion rate of the Tanllwyth main channel is 70.5 ± 1.1 mm year^{-1}, which is about twice the rate on the tributaries (30.3 ± 0.5 mm year^{-1}), which in turn is almost twice that on the ditches (17.0 ± 0.2 mm year^{-1}). These differences are all highly significant as shown by the t-test. However, in the pre-clearfelling phase of the study, mean erosion rates on the Tanllwyth (34.6 ± 0.5 mm year^{-1}) and Cyff (31.2 ± 0.8 mm year^{-1}) main channels showed no significant difference. In the post-clearfelling phase, however, mean erosion rates have generally doubled on both the Tanllwyth and Cyff main channels. The mean erosion rate on the Cyff increased from 31.2 ± 0.8 to 65.5 ± 1.1 mm year^{-1} in the post-clearfelling phase, the mean erosion rate on the unclearfelled section of the Tanllwyth channel (sites T13–T17) showed an increase to 70.5 ± 1.1 mm year^{-1}, while the mean erosion rate at those Tanllwyth sites in the clearfelled area (T1–T12) increased to 95.8 ± 0.8 mm year^{-1}. The higher erosion rates in the winter of 1997, even at the unclearfelled sites (T13–T17) and particularly in the period 13 January 1997 to 15 February 1997, is thought to be due to local climatic factors (i.e. a greater amount of frost activity). Nevertheless, following clearfelling the difference between the erosion rate on those Tanllwyth sites in the clearfelled area (T1–T12, mean is 95.8 ± 0.8 mm year^{-1}) and the unaffected "control" sites in the Cyff (65.5 ± 1.1 mm year^{-1}) was significant at the 0.05 level (see Table 3.2), which suggests that the increased erosion rates measured at sites on those parts of the Tanllwyth

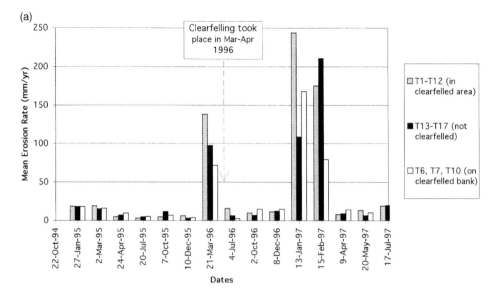

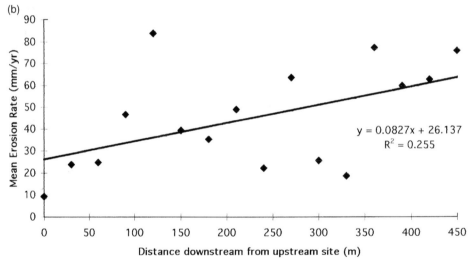

Figure 3.2 (a) Temporal variation of mean erosion rate for sites clearfelled and unaffected on the Afon Tanllwyth (22 October 1994–17 July 1997). (b) Scattergraph of mean erosion rate at each site on the Tanllwyth main channel over the 31-month study period versus distance downstream from T17 (the most upstream site)

channel which are in the clearfelled area are real. Mean erosion rates for individual sites of up to 600 mm year^{-1} occur at the downstream sites in the winter following timber harvesting. The maximum erosion rate measured at one individual pin was 961 mm year^{-1}, recorded at site T3 during the last survey (13 January 1997–15 February 1997). The higher erosion rates seen in the winter periods in particular are associated with the greatest variability. Figure 3.2(b) indicates that the mean erosion

Table 3.2 Comparisons (*t*-test results) of mean erosion rates on Afon Tanllwyth forest ditches, tributary streams, mainstream channel banks and Afon Cyff mainstream before and after timber harvesting

Stream	Mean erosion rate (mm year^{-1})	n	Stream	Mean erosion rate (mm year^{-1})	n	p-value
Pre-clearfelling (27/4/95 to 2/4/96)						
Tanllwyth mainstream	34.6 ± 0.5	376	Cyff mainstream	31.2 ± 0.8	70	ns
Post-clearfelling (2/4/96 to 15/2/97)						
Tanllwyth sites T1–T12 (in clearfelled area)	95.8 ± 0.8	348	Cyff mainstream	65.5 ± 1.1	72	$p < 0.05$
Tanllwyth sites T13–17 (not clearfelled)	70.5 ± 1.1	139	Cyff mainstream	65.5 ± 1.1	72	ns
Tanllwyth (forested) catchment (all periods)						
Tanllwyth mainstream (all sites, whole study period)	64.3 ± 0.4	962	Tanllwyth tributaries	30.3 ± 0.5	357	$p < 0.001$
Tanllwyth mainstream (all sites, whole study period)	64.3 ± 0.4	962	Tanllwyth forest ditches	17.0 ± 0.2	831	$p < 0.001$
Tanllwyth tributaries (all sites, whole study period)	30.3 ± 0.5	357	Tanllwyth forest ditches	17.0 ± 0.2	831	$p < 0.01$

ns = Not significant at the $p < 0.05$ level.

rate at each site shows a general increase in erosion in the downstream direction, as reported by Lawler (1994).

Multiple Regression Analysis of Erosion

Bank erosion proceeds by a complex interplay of processes which involve sediment "preparation" and its subsequent removal by fluvial entrainment (Hooke, 1990). Both

Hooke (1979) and Lawler (1986) have used stepwise multiple regression in an attempt to identify the more important influences on bank erosion. In this chapter, results from these and other previous studies, previous fieldwork and relevant theory were used to select variables to help explain and predict erosion rates in this upland forested environment. In addition to mean erosion rate (MEAN ER) for each period, the following measures of erosion were computed as dependent variables: MEANMAX ER (the mean of the three highest eroding pins in each period) was felt to be a reliable measure of maximum erosion rate; and % ERODING (the percentage of erosion pins showing erosion) was a measure used by Hooke (1979) and Lawler (1986) to give an indication of the area of bank being eroded. Mean erosion rates were separately computed for tributary sites (MEAN TRIB ER) and for forest ditch sites (MEAN DITCH ER).

Independent variables deemed likely to affect erosion rates were placed into three groups: fluvial entrainment (or bank material detachment) indices, frost indices (both air and bank surface) and precipitation indices. A range of 24 independent variables were derived from hydrometeorological data sources available. The influence of frost processes on bank erosion (e.g. Wolman, 1959; Lawler, 1986, 1993a; Stott, 1997a) led to the inclusion of several indices of air and bank temperatures representing aspects of intensity, frequency and duration of freezing and thawing conditions. Of the three groups of indices the frost indices correlated best with all dependent variables. These frost indices were derived from two independent sources: the Moel Cynedd automatic weather station 200 m from the Tanllwyth gauging station, and a thermistor located on a stream bank (see Figure 3.1(b)). The strongest correlations were found between MEAN ER and the number of hours when air temperature was below 0 °C (0.82, $p < 0.01$) and MEAN ER and cumulative total degees of sub-zero air temperatures (-0.77, $p < 0.01$). Using temperatures measured by the thermistor on the bank surface at site T10 to calculate the frost indices produced weaker, but still significant correlations with MEAN ER ($p < 0.05$). Since the erosion rate of the main channel was found to be highest, and therefore most significant in terms of its contribution to catchment sediment yields, the following analysis concentrates on trying to explain variation over time (12 time periods, 22 October 1994–15 February 1997) in spatially averaged erosion rates (MEAN ER) on the Tanllwyth main channel only. However, the technique could be applied to MEAN TRIB ER and MEAN DITCH ER equally well since the correlation coefficient between MEAN DITCH ER and the number of hours when air temperature was below 0 °C was 0.91, $p > 0.001$ and a further highly significant correlation was found between MEAN TRIB ER and cumulative total degrees of sub-zero bank surface temperatures (0.85, $p < 0.001$). However, none of the fluvial entrainment indices computed from the IH Tanllwyth flume records (see Figure 3.1 for location) were significant in the ditches or tributaries.

Although multiple regression analysis was conducted successfully for individual sites, only the results for MEAN ER, the mean erosion for all sites on the Tanllwyth main channel, are presented here. Owing to the relatively small number of survey periods ($n = 12$) and the strong likelihood of intercorrelation which can cause computation difficulties in multiple regression analyses, it was decided to use only two independent variables (from separate groups), one preparation (frost) index and one fluvial entrainment (grain detachment) index. This way, statistical validity was retained and the possibility of colinearity between independent variables in the same group, or even between flow and precipitation indices was avoided. The independent variables selected to predict MEAN ER on the basis of the correlations were the number of hours when air temperature was below $0\,°C$ (AWS HRS < 0), as the highest correlating variable from the frost group, and maximum discharge (MAXQ), as the highest correlating variable from either of the other two groups.

The following model was then used to predict MEAN ER:

$$\text{MEAN ER} = (-14.61 \times \text{MAXQ}) + (0.17 \times \text{AWS HRS} < 0) + 9.29$$
$$R^2 = 0.71, \quad p < 0.01$$

The results show broad agreement with the observed pattern in mean erosion rate although with over- and under-predictions in most periods (Figure 3.3(a)). When the model was derived from the first nine periods only and used to predict MEAN ER in the last two periods, autumn 1996–winter 1997 (Figure 3.3(b)), the autumn 1996 period is over-predicted while the winter 1997 period is substantially under-predicted. Further refinement of this crude model will be an aim in the longer term as this study continues.

Photo-electronic Erosion Pin (PEEP): Increased Temporal Resolution

The photo-electronic erosion pin (PEEP) has been developed (Lawler, 1991, 1992) to address the problem of quantifying the temporal distribution of change in bank erosion between erosion pin surveys which inevitably reveal only net changes in the bank surface since the last survey. The design, calibration, installation and a variety of applications of the PEEP are reported elsewhere (Lawler, 1991, 1992, 1993b).

The PEEP was installed in Forest Ditch A (see Figure 3.1(b)) between July 1995 and July 1996. Erosion rates were generally low although one storm event in February 1996 resulted in ditch erosion. Figure 3.4(a) shows the analysis of this event. The mean length of exposed PEEP for 10–11 February was 44.1 mm. The first light rain occurred during 10 February but had little effect on stream discharge. Heavier rain continued during 11 February, with a peak rate of $6\,\text{mm}\,\text{h}^{-1}$ at 0500 and further peaks of $5\,\text{mm}\,\text{h}^{-1}$ at 1000 and $4\,\text{mm}\,\text{h}^{-1}$ at 1400. The temperature climbed slowly from around $-2.0\,°C$ during the night of 8/9 February to fluctuate around $0\,°C$ during 9 and 10 February, and then there was a sudden rise in temperature of around $3\,°C$ between 1500 and 1600 on 11 February. The first rise in discharge at the Tanllwyth gauging station 200 m away began during 10 February and peaked at 0200 on 12 February 1996 at $0.42\,\text{m}^3\,\text{s}^{-1}$. It was this relatively small flow event which appears to cause about 4.5 mm of erosion on the ditch side as the mean length of PEEP exposed changes from 44.1 to 48.6 mm overnight. During calibration, Lawler (1992) estimated the standard error of PEEP length exposed (i.e. the resolution) to be 1.26 mm. Thus, any changes in PEEP length greater than 1.26 mm, as observed here, can be deemed to be real. The much larger flood event which peaked at $1.33\,\text{m}^3\,\text{s}^{-1}$ at 2400 on 12/13 February appeared to cause no further erosion. This observation points to the likelihood of bank material being "prepared" while frozen and then whilst the interstitial ice melts a minor flow event is capable of removing it, while a subsequent, much larger

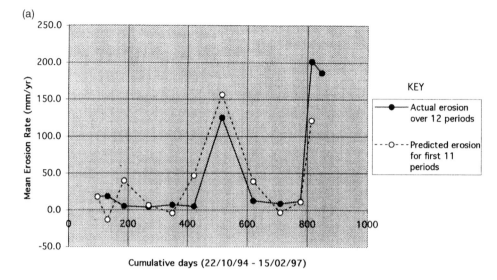

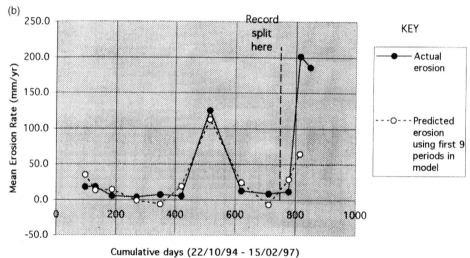

Figure 3.3 (a) Mean erosion rates on the Afon Tanllwyth observed and predicted by the multiple regression model. (b) Split data test of the multiple regression model of mean bank erosion rate on Afon Tanllwyth

event caused no further erosion. This observation is in agreement with previous studies which have identified suspended sediment exhaustion and hysteresis effects in the suspended sediment concentration versus discharge rating relationship (e.g. Walling, 1974; Klein, 1984).

In July 1996 the PEEP was moved to Site T10 on the main Tanllwyth channel in an area which had been clearfelled in March–April 1996 (see Figure 3.1(b)). The PEEP was installed some 100 mm above low flow water level in a bank which was near vertical. Figure 3.4(b) shows a moderate flow event which peaked at $0.48\,\text{m}^3\,\text{s}^{-1}$ in the late evening of 23 August and resulted in the mean length of PEEP exposed changing from 43.6 to 47.8 mm. Again, over 4 mm of material was removed and this is well in excess of the measurement error. Throughout the period 21–28 August the temperature remained between 9 and 13 °C. Figure 3.4(c) illustrates the PEEP response during the period 21–31 December 1996 at Site T10 on the Tanllwyth, during which time the discharge remained low and constant and did not affect the PEEP at all. However, bank temperature fluctuated around freezing point, with peaks during the day and troughs at night. The PEEP length appeared to increase by about 10 mm on 25 December, and remained that way for three days, before the length exposed decreased again. This oscillatory pattern, it is suggested, reflects the freezing and heaving of bank material (Lawler, 1993a). A site visit on 13 January 1997 found the PEEP partially covered with fine bank sediment which had fallen onto it from above. During this site visit all pins were re-surveyed and at most sites layers of bank sediment, 10–30 mm in thickness, were seen to have been heaved off the bank. Needle ice formation and the presence of ice lenses up to 30 mm thick were also observed.

Preliminary Effects of Clearfelling on Channel Erosion, Bank Temperatures and Frost Incidence

The influence of forest on air temperatures (Fritts, 1961; Kittredge, 1962; Smith, 1970) and stream temperatures (Greene, 1950) is well recognised. Gray and Eddington (1969) reported a marked rise in summer temperature of a woodland stream which was clearfelled. Hurst (1966) reported that shading in the forest of Thetford Chase gave considerable protection from frost. Given the now well established link between frost action (freeze–thaw) and stream bank erosion (e.g. Wolman, 1959; Hill, 1973; Blacknell, 1981; Gardiner, 1983; Curr, 1984; Lawler, 1986, 1987, 1993a; Stott, 1997a), the potential impact of clearfelling upland forest plantations on stream bank temperatures forms the subject of further investigation. Having identified a statistically significant difference in mainstream bank erosion rates before and after clearfelling an area of the Afon Tanllwyth, as compared with the Afon Cyff over the same period, the changes in the temperature regime of the stream banks which have occurred during the clearfelling phase are now investigated further.

Figure 3.5(a) and (b) shows the temporal variation in both stream bank temperature (measured by a thermistor fixed on the bank) and air temperature (measured by an automatic weather station (AWS) in a nearby forest clearing) over 28-day periods in early winter (a) before clearfelling (8 December 1995–5 January 1996) and (b) after clearfelling (8 December 1996–5 January 1997). The complex nature of the relationship is apparent and the greater variability in the AWS air temperature is clear. An analysis of these data is presented in Table 3.3 and it is the relationship between bank and AWS temperatures which can be used to assess the impact of clearfelling on bank temperatures since the AWS records remained unaffected by the clearfelling. Bank temperature indices expressed as a percentage of the AWS temperature indices are used to assess the changes which have occurred following clearfelling. Five of the six bank temperature indices chosen show an increase as compared to the AWS indices. Rating relationships between bank and air temperature for (i) pre-clearfelling (20 July 1995–8 September

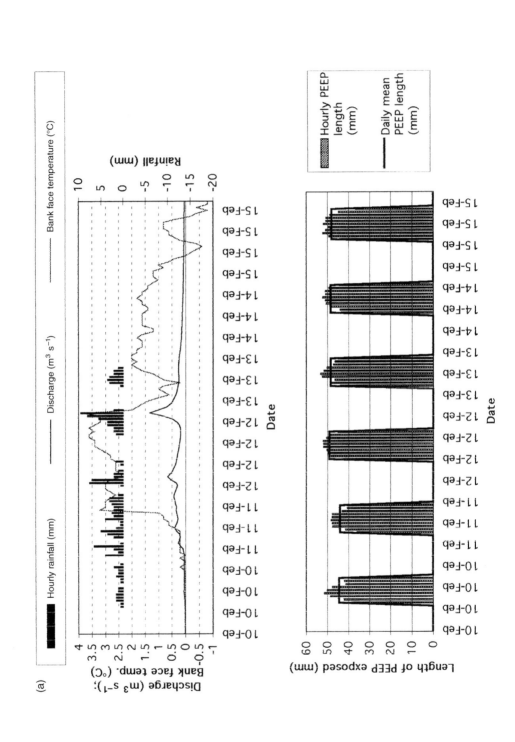

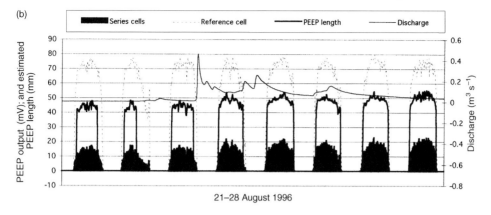

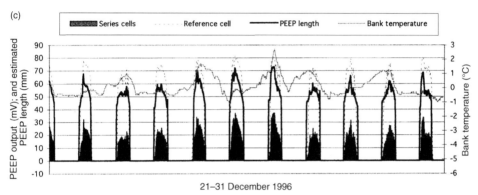

Figure 3.4 (a) Relation of rainfall, discharge, bank temperature and bank erosion (length of PEEP exposed) in Forest Ditch A, 9–15 February 1996. The minor event early on 12 February caused an estimated 4.5 mm of erosion while the much larger event during the night of 12–13 February resulted in no further erosion of the ditch side. (b) A moderate flow event on 28 August 1996 at Site T10 (in the clearfelled area on Afon Tanllwyth main channel) produced 4 mm of bank retreat, as detected by the sudden increase in the diurnal trends in the PEEP cell series outputs and derived estimated PEEP length. Bank temperature during this period remained in the range 9–13 °C. (c) PEEP record for 21–31 December 1996 at Site T10 in the clearfelled area on Afon Tanllwyth showing oscillations in bank surface (length of PEEP exposed) associated with freeze–thaw cycles as measured by a thermistor on the bank 200 mm away

1995) and (ii) post-clearfelling (20 July 1996–8 September 1996) periods are presented in Figure 3.5(c). Although rating relationships have been established for each season, the relationships in Figure 3.5(c) compare the summer ratings for (i) before and (ii) after clearfelling. There is an obvious steepening of the relationship in the post-clearfelling phase suggesting that banks have become more sensitive to changes in air temperature as measured by the AWS. This may well have implications for bank "preparation (freeze–thaw) processes" as well as for stream temperatures as reported by Gray and Eddington (1969).

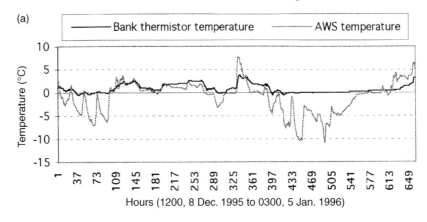

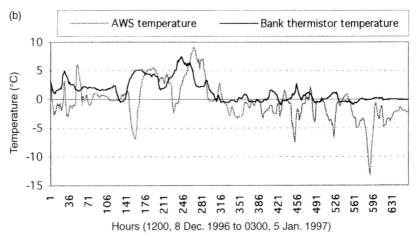

DISCUSSION

Mean bank erosion rates on the Tanllwyth and Cyff broadly compare with those reported for other similar-sized British catchments summarised in Table 3.4 (e.g. Hill, 1973; Lewin et al., 1974; Lawler, 1986, 1987; Stott, 1997a). Tributary and forest ditch erosion rates are significantly lower and this may be attributable to infrequent and lower magnitude fluvial entrainment events. The forest ditches show low, and apparently stable, erosion rates over this study period although this does not appear to have been the case over the whole of the forest cycle. Newson (1980) reported increases in ditch cross-sectional areas of up to 75% in some Tanllwyth ditches 26 years after they were first established. Neither of the ditch systems studied here were subjected to clearfelling but this will be examined as part of ongoing studies.

During field visits to the study sites, freeze–thaw processes, the growth of needle ice

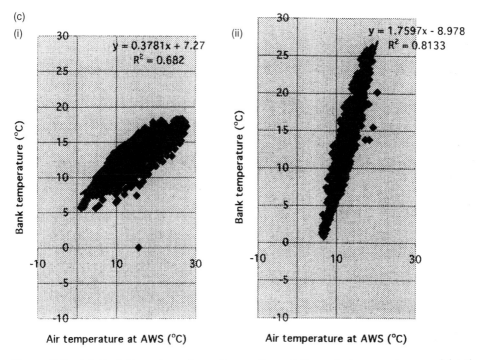

Figure 3.5 (a) Variations in automatic weather station air temperature and bank thermistor temperature under plantation forest canopy (8 December 1995–5 January 1996) in the Afon Tanllwyth catchment. (b) Variations in automatic weather station air temperature and bank thermistor temperature at clearfelled site T10 (8 December 1995–5 January 1996) in the Afon Tanllwyth catchment. (c) Relationship of bank temperature to automatic weather station air temperature in the Afon Tanllwyth (i) before clearfelling (20 July 1995–8 September 1995) and (ii) after clearfelling (20 July 1996–8 September 1996)

(Lawler, 1993a), and at a larger scale ice lenses which heave apart soil peds and blocks destroying inter-ped cohesion (Thorne, 1990) have been observed. Freezing of soil moisture can reduce erosion resistance in the cohesive banks in this study by heaving material from the bank surface in layers. This reduces cohesion when the ice melts, which allows sediment to slump and be entrained by the next rise in stage. Thorne (1990, p. 127) states that "The processes responsible for loosening aggregates are mostly driven by the dynamics and physical state of soil moisture close to the bank face". If the bank is poorly drained, positive pore water pressure acts to reduce the effective cohesion and weaken the soil. High pore pressures occur in saturated banks after heavy/prolonged precipitation, snowmelt or rapid drawdown in the channel. Although bank drainage was outside the scope of this study, it must form a focus for future research in such upland afforested areas. Vegetated and forested banks are drier because (i) the tree canopy prevents 15–30% of precipitation from ever reaching the soil; (ii) plants draw water from the soil and transpire it to the atmosphere; and (iii) suction pressures in the soil are increased by water abstraction at the roots, so that the height of the capillary fringe is increased and water is drawn towards the surface from

Table 3.3 Comparison of air temperatures at Moel Cynedd Automatic Weather Station (AWS) with thermistor on stream bank (Bank) before clearfelling (1200, 8 December 1995–0300, 5 January 1996) and after clearfelling (1200, 8 December 1996–0300, 5 January 1997)

	Before clearfelling			After clearfelling		
	AWS	Bank	Bank as % of AWS	AWS	Bank	Bank as % of AWS
Maximum (°C)	7.7	3.8	49	9.2	7.5	81
Minimum (°C)	−11.0	−0.6	6	−13.3	−1	8
Range (°C)	18.7	4.4	24	22.4	8.5	38
Mean (°C)	−1.0	0.8	80	−0.5	1.4	280
No. of frost cycles	13	7	54	8	4	50
No. of hours < 0 °C	376	63	17	434	198	46

Table 3.4 British bank erosion rates

Reference	Location	Catchment area (km^2)	Erosion rate (mm year^{-1})
Cummins and Potter (1972)	Bradgate Brook, Leicestershire	< 20	25
Hill (1972)	Clady and Crawfordsburn, Northern Ireland	3.4	30–66
Knighton (1973)	R. Bollin-Dean, Cheshire	∼ 260	230
Lewin *et al.* (1974)	Maesnant, mid-Wales	0.54	30
McGreal and Gardiner (1977)	R. Lagan, Northern Ireland	85	80–10
Hooke (1979)	Various rivers, Devon	9.6–620	80–1180
Murgatroyd and Ternan (1983)	Narrator Brook	4.75	5.2 in forest
Gardiner (1983)	R. Lagan, Northern Ireland	85	76–139
Lawler (1984, 1986)	R. Ilston, South Wales	6.75 and 13.18	38–310
Ashbridge (1995)	R. Culm, Devon	276	227–329
Stott (1997a)	Kirkton (forest)	6.85	47
	Monachyle (moorland)	7.7	59
This study	Tanllwyth (forested)	0.89	35
	Tanllwyth (clearfelled)		95
	Cyff (upland grassland)	3.1	31–65

greater depths than in an unvegetated bank, which results in increased evaporative loss (Thorne, 1990). This accepted, the effect of clearfelling on the bank moisture regime should be to increase the amount of precipitation which reaches the bank and reduce evaporative losses until vegetation regenerates and raises the height of the capillary fringe. Results from this study show a steepening of the bank versus air temperature relationship, a greater diurnal temperature range which results in a greater duration of sub-zero temperatures in winter, probably more freeze–thaw cycles in winter and higher maximum bank temperatures. Such higher maxima should increase evaporative losses and result in more extreme wetting and drying cycles. According to Thorne (1990), this generates a ped fabric with desiccation cracks between peds and a crumb structure to the soil. Cohesion between peds and crumbs is much weaker than within them, so a heavily desiccated soil may have little erosion resistance. Soil and bank stability may be further reduced by the decay of root systems from the first crop.

Management Implications of Clearfelling

The findings of this study add further support to forest management proposals and guidelines which advocate the inclusion of "buffer strips" in the management plan (e.g. Brazier and Brown, 1973; Mills, 1980; Forestry Commission, 1988; Maitland *et al.*, 1990). Mills (1980) stipulated reserve areas 10 times the width of the stream, up to a maximum of 30 m, to protect stream channels from excessive shade (temperature stability) and the inflow of sediments and toxins. Some vegetation (e.g. birch, alder, willow, rowan) is seen as desirable to give shade for fish and provide nutrients via leaf-fall. In the Kielder Forest, for example, extensive riparian corridors of broadleaf trees feature in the second planting and in some areas riparian plantation forest is even being prematurely felled in order to carry out such replanting, partly in the interest of landscape aesthetics, but also in the hope that the stream ecology will benefit. Newson (in Maitland *et al.*, 1990) points out that the whole policy is proceeding without research confirmation. This study confirms the preliminary impacts of clearfelling at Plynlimon on stream bank temperatures and tentatively links this to increased bank erosion rates observed over the timescale of this study.

Further research needs include investigation of the effects of broadleaf planting in riparian buffer strips on bank erosion processes; monitoring the effects of clearfelling and replanting on bank moisture regimes and frost activity; and the gathering of longer-term data to confirm these preliminary effects of clearfelling on bank and forest ditch erosion rates.

CONCLUSIONS

(a) Contemporary channel erosion rates in the forested Tanllwyth catchment are significantly higher than in adjacent tributaries, which in turn are significantly higher that on 40 to 50-year-old forest ditches, although rates are very similar to those measured on the nearby Afon Cyff main channel (prior to clearfelling).

(b) Clearfelling of part of the Tanllwyth catchment has resulted in a statistically

significant increase in main channel bank erosion rates as compared with the nearby Afon Cyff.
(c) Mean erosion rates can be crudely predicted by a multiple linear regression model using two independent variables (one frost-derived "bank preparation" index and one "detachment" or fluvial entrainment index).
(d) Photo-electronic erosion pin data allow examination of erosion processes at a much increased temporal resolution and revealed a distinct example of where a "frost-prepared" bank yields to erosion by a very minor fluvial entrainment event when a much larger event hours later caused no erosion. Oscillations in the bank surface appear to be associated with bank temperatures fluctuating around freezing point.
(e) Clearfelling resulted in a steepening of the bank versus air temperature relationship although detailed examination reveals the complex nature of this relationship. Clearfelling increases maxima, minima, mean, range and duration of sub-zero temperatures on stream banks. Whether or not the number of freeze–thaw cycles increases is not yet conclusive.
(f) This research has significant implications for forest managers concerned with clearfelling and the establishment of riparian buffer zones.

ACKNOWLEDGEMENTS

Liverpool John Moores University School of Education and Community Studies provided funding for datalogging equipment and travel. The Institute of Hydrology (Plynlimon Research Station) provided stream flow and automatic weather station records and Sean Crane is thanked particularly for assembling these. Steve Marks and all staff at Plynlimon supported the fieldwork. Field assistance was also provided by Adam Sawyer and Kath Stott. Prof. Rob Ferguson, Prof. Dave Huddart, Dr Greg Sambrook Smith and an anonymous referee provided many useful comments on earlier drafts of the manuscript.

REFERENCES

Ashbridge, D. 1995. Processes of river bank erosion and their contribution to the suspended sediment load of the River Culm, Devon. In Foster, I. D. L., Gurnell, A. M. and Webb, B. W. (Eds) *Sediment and Water Quality in River Catchment Systems*. John Wiley, Chichester, 229–245.

Austin, R. and Brown, D. 1982. Solids contamination resulting from drainage works in an upland catchment, and its removal by floatation. *Journal of the Institute of Water Engineers and Scientists*, **36**(4), 1–10.

Bello, A., Day, D., Douglas, J., Field, J., Lam, K. and Soh, Z. B. H. A. 1978. Field experiments to analyse runoff, sediment and solute production in the New England region of Australia. *Zeitschrift für Geomorphologie* (Supplement), **29**, 180–190.

Blacknell, C. 1981. River erosion in an upland catchment. *Area*, **13**(1), 39–44.

Brazier, J. R. and Brown, G. W. 1973. Buffer strips for stream temperature control. Oregon State University Forest Research Laboratory Research Paper 15, Corvallis.

Burt, T. P., Donahoe, M. A. and Vann, A. R. 1984. A comparison of suspended sediment yield from two small upland catchments following open ditching for forestry drainage. *Zeitschrift für Geomorphologie* (Supplement), **51**, 51–62.

Carson, M. A., Taylor, C. H. and Grey, B. J. 1973. Sediment production in a small Appalachian watershed during spring runoff: The Eaton Basin 1970–1972. *Canadian Journal of Earth Sciences*, **10**, 1707–1734.

Church, M. and Slaymaker, O. 1989. Disequilibrium of Holocene sediment yield in glaciated British Columbia. *Nature*, **337**, 452–454.

Coldwell, A. E. 1957. Importance of channel erosion as a source of sediment. *Transactions of the American Geophysical Union*, **38**, 1908–1912.

Cummins, W. A. and Potter, H. R. 1972. Rates of erosion in the catchment area of Cropston Reservoir, Charnwood Forest Leicestershire. *Mercian Geologist*, **4**, 149–157.

Curr, R. H. 1984. The sediment dynamics of Corston Brook. Unpublished PhD thesis, University of Exeter.

Davis, R. J. and Gregory, K. J. 1994. A new distinct mechanism of river bank erosion in a forested catchment. *Journal of Hydrology*, **157**, 1–11.

Dickinson, W. T. and Scott, A. M. 1979. Analysis of streambank erosion variables. *Canadian Agricultural Engineering*, **21**(1), 19–25.

Duijsings, J. J. H. M. 1986. The sediment supply by streambank erosion in a forested catchment. *Zeitschrift für Geomorphologie* (Supplement: Erosion Budgets and Their Hydrological Basis), **60**, 233–244.

Duijsings, J. J. H. M. 1987. A sediment budget for a forested catchment in Luxembourg and its implications for channel development. *Earth Surface Processes and Landforms*, **12**, 173–195.

Ferguson, R. I. and Stott, T. A. 1987. Forestry effects on suspended sediment and bedload yields in the Balquhidder catchments, central Scotland. *Transactions of the Royal Society of Edinburgh: Earth Sciences*, **78**, 379–384.

Ferguson, R. I., Grieve, I. C. and Harrison, D. J. 1991. Disentangling land use effects on sediment yield from year to year climatic variation. In *Sediment and Stream Water Quality in a Changing Environment: Trends and Explanation*, Proceedings of the Vienna Symposium, August 1991. International Association of Hydrological Sciences Publication 203, 13–20.

Forestry Commission 1988. *Forests and Water Guidelines*. HMSO, London.

Francis, I. S. and Taylor, J. A. 1989. The effect of forestry drainage operations on upland sediment yields: a study of two peat-covered catchments. *Earth Surface Processes and Landforms*, **14**, 73–83.

Fritts, H. C. 1961. An analysis of maximum summer temperature inside and outside a forest. *Ecology*, **42**, 436–440.

Gardiner, T. 1983. Some factors promoting channel bank erosion, River Lagan, County Down. *Journal of Earth Science Royal Dublin Society*, **5**, 231–239.

Graesser, N. W. C. 1979. How land improvement can damage Scottish salmon fisheries. *Salmon and Trout Magazine*, **215**, 39–43.

Gray, J. R. A. and Eddington, J. M. 1969. Effect of woodland clearance on stream temperature. *Journal of the Fisheries Research Board, Canada*, **26**, 399–403.

Greene, G. E. 1950. Land use and trout streams. *Journal of Soil Conservation*, **5**, 125–126.

Greene, L. A. 1987. The effects of catchment afforestation on public water supplies in Strathclyde Region, Scotland. *Transactions of the Royal Society of Edinburgh: Earth Sciences*, **78**, 335–340.

Gregory, K. J., Gurnell, A. M. and Hill, C. T. 1985. The permanence of debris dams related to river channel processes. *Hydrological Sciences Journal*, **30**(3), 371–380.

Grimshaw, D. L. and Lewin, J. 1980. Source identification for suspended sediments. *Journal of Hydrology*, **47**, 151–162.

Hagerty, D. J., Sharifounnasab, M. and Spoor, M. F. 1983. Riverbank erosion – a case study. *Bulletin of the Association of Engineering Geologists*, **20**(4), 411–437.

Hedin, L. O., Mayer, M. S. and Likens, G. E. 1988. The effect of deforestation on organic debris dams. *Verh. Internat. Verein. Limnol.*, **23**, 1135–1141.

Hill, A. R. 1973. Erosion of river banks composed of glacial till near Belfast, Northern Ireland. *Zeitschrift für Geomorphologie*, **17**, 428–442.

Hooke, J. M. 1979. An analysis of the processes of river bank erosion. *Journal of Hydrology*, **42**, 39–62.

Hooke, J. M. 1980. Magnitude and distribution of rates of river bank erosion. *Earth Surface*

Processes and Landforms, **5**, 143–157.

Hooke, J. M. 1990. The linkages between bank erosion and meander behaviour in gravel bed rivers. Portsmouth Polytechnic Department of Geography Working Paper 14.

Hudson, J. A. and Gilman, K. 1993. Long-term variability in the water balances of the Plynlimon catchments. *Journal of Hydrology*, **143**, 355–380.

Hurst, G. W. 1966. Temperatures in the forest of Thetford Chase. *Meteorological Magazine, London*, **95**, 123.

Imeson, A. C. and Jungerius, P. D. 1974. The widening of valley incisions by soil fall in a forested Keuper area, Luxembourg. *Earth Surface Processes and Landforms*, **2**, 141–152.

Johnson, R. C. 1993. Effects of forestry on suspended solids and bedload yields in the Balquhidder catchments. *Journal of Hydrology*, **145**, 403–417.

Kirby, C., Newson, M. D. and Gilman, K. 1991. Plynlimon research: the first two decades. *Institute of Hydrology Report* 109, 188pp.

Kirkby, M. J. 1967. Measurement and theory of soil creep. *Journal of Geology*, **75**, 359–378.

Kittredge, J. 1962. The influence of forest on weather and other environmental factors. *Forest Influences*. FAO, Rome.

Klein, M. 1984. Anti clockwise hysteresis in suspended sediment concentration during individual storms: Holbeck Catchment; Yorkshire, England. *Catena*, **11**, 251–257.

Knighton, A.D. 1973. Riverbank erosion in relation to streamflow conditions, River Bollin-Dean, Cheshire. *East Midlands Geographer*, **5(8)**, 416–426.

Law, F. 1956. The effect of afforestation upon the yield of water catchment areas. *Journal of the British Waterworks Association*, **38**, 484–494.

Lawler, D. M. 1978. The use of erosion pins in river banks. *Swansea Geographer*, **16**, 9–18.

Lawler, D. M. 1982. Processes of river bank erosion: River Ilston, South Wales, UK. Unpublished PhD thesis, University of Wales.

Lawler, D.M. 1984. Processes of river bank erosion: the River Ilston, South Wales, UK, unpublished PhD thesis, University of Wales, 518pp.

Lawler, D. M. 1986. River bank erosion and the influence of frost: a statistical examination. *Transactions of the Institute of British Geographers*, **11**, 227–242.

Lawler, D. M. 1987. Bank erosion and frost action: an example from South Wales. In Gardiner, V. (Ed.) *International Geomorphology 1986, Part 1*. John Wiley, Chichester, 575–590.

Lawler, D. M. 1991. A new technique for the automatic monitoring of erosion and deposition rates. *Water Resources Research*, **27**(8), 2125–2128.

Lawler, D. M. 1992. Design and installation of a novel automatic erosion monitoring system. *Earth Surface Processes and Landforms*, **17**, 455–463.

Lawler, D. M. 1993a. Needle ice processes and sediment mobilization on river banks: the River Ilston, West Glamorgan, UK. *Journal of Hydrology*, **150**, 81–114.

Lawler, D. M. 1993b. The measurement of river bank erosion and lateral channel change: a review. *Earth Surface Processes and Landforms, Technical Software Bulletin* 18, 777–821.

Lawler, D. M. 1994. Temporal variability in streambank response to individual flow events: the River Arrow, Warwickshire, UK. In *Variability in Stream Erosion and Sediment Transport*, Proceedings of the Canberra Symposium, December 1994. International Association of Hydrological Sciences Publication 224, 171–180.

Lawler, D. M. 1995. The impact of scale on the processes of channel-side sediment supply: a conceptual model. In *Effects of Scale on Interpretation and Management of Sediment and Water Quality*, Proceedings of a Boulder Symposium, July 1995. International Association of Hydrological Sciences Publication 226, 175–184.

Lawler, D. M. and Leeks, G. J. L. 1992. River bank erosion events on the upper Severn detected by the Photo-Electric Erosion Pin (PEEPS) system. In *Erosion and Sediment Transport Monitoring Programmes in River Basins*. International Association of Hydrological Sciences Publication 210, 95–105.

Leeks, G. J. L. 1992. Impact of plantation forestry on sediment transport processes. In Billi, P., Hey, R. D., Thorne, C. R. and Tacconi, P. (Eds) *Dynamics of Gravel Bed Rivers*. John Wiley, Chichester, 651–670.

Leeks, G. J. L. and Roberts, G. 1987. The effects of forestry on upland streams with special reference to water quality and sediment transport. In Good, J. E. G. and Institute of Terrestrial Ecology (Eds) *Environmental Aspects of Plantation Forestry in Wales*, Institute of Terrestrial Ecology, Merelwood Research Station, Grange-over-Sands, No. 22, 9–24.

Lewin, J., Cryer, R. and Harrison, D. I. 1974. Sources for sediments and solutes in mid-Wales. In Gregory, K. J. and Walling, D. E. (Eds) *Fluvial Processes in Instrumented Watersheds*, Institute of British Geographers Special Publication No. 6, 73–84.

Madej, M. A., Weaver, W. and Hagans, D. K. 1994. Analysis of bank erosion on the Merced River, Yosemite Valley, Yosemite National Park, California, USA. *Environmental Management*, **18**, 235–250.

Maitland, P. S., Newson, M. D. and Best, G. A. 1990. The impact of afforestation and forestry practice on freshwater habitats. *Focus on Nature Conservation Report* 23, Nature Conservancy Council, Peterborough.

McGreal, W. S. and Gardiner, T. 1977. Short-term measurements of erosion from a marine and fluvial environment in County Down, Northern Ireland. *Area*, **9**, 285–289.

Megahan, W. F. 1982. Channel sediment storage behind obstructions in forested drainage basins draining the granitic bedrock of the Idaho Batholith. In Swanson, F. J., Janda, R. J., Dunne, T. and Swanson, D. N. (Eds) *Sediment Budgets and Routing in Forested Drainage Basins*. USDA Forest Service Research Publication PNW-141, Pacific Northwest Forest and Range Experimental Station, Portland, OR, 114–121.

Miller, E. L., Beasley, R. S. and Lawson, E. R. 1988. Forest harvest and site preparation effects on erosion and sedimentation in the Ouachita Mountains. *Journal of Environmental Quality*, **17**, 219–225.

Mills, D. H. 1980. *The Management of Forest Streams*. Forestry Commission Leaflet No. 78, HMSO, London.

Moore, R. J. and Newson, M. D. 1986. Production, storage and output of coarse upland sediments: natural and artificial influences as revealed by research catchment studies. *Journal of the Geological Society*, **143**, 1–6.

Mosley, M. P. 1981. The influence of organic debris on channel morphology and bedload transport in a New Zealand forest stream. *Earth Surface Processes and Landforms*, **6**, 571–579.

Murgatroyd, A. L. and Ternan, J. L. 1983. The impact of afforestation on stream bank erosion and channel form. *Earth Surface Processes and Landforms*, **8**, 357–369.

Nanson, G. C. and Hickin, E. J. 1986. A statistical analysis of bank erosion and channel migration in Western Canada. *Bulletin of the Geological Society of America*, **97**, 49–54.

Newson, M. D. 1976. *The physiography, deposits and vegetation of the Plynlimon experimental catchments*. Institute of Hydrology Report 30.

Newson, M. D. 1979. The results of ten years' experimental study on Plynlimon, mid-Wales and their importance for the Water Industry. *Journal of the Institution of Water Engineers*, **33**, 321–333.

Newson, M. D. 1980. The erosion of drainage ditches and its effect on bedload yields in Mid-Wales: reconnaissance case studies. *Earth Surface Processes*, **5**, 275–290.

Newson, M. D. 1984. Table of data quoted in 'Effects of upland development on river runoff and sediment yields in the UK', reported by B. Webb. *Circulation*, **2**, 6–8.

Newson, M. D. 1986. River basin engineering-fluvial geomorphology. *Journal of the Institution of Water Engineers and Scientists*, **40**(4), 307–324.

Newson, M. D. and Leeks, G. J. L. 1985. Transport processes at the catchment scale. In Thorne, C. R., Bathurst, C. R. and Hey, R. D. (Eds) *Sediment Transport in Gravel-bed Rivers*. John Wiley, Chichester, 187–223.

Omerod, S. J. and Edwards, R. W. 1985. Stream acidity in some areas of Wales in relation to historical trends in afforestation and the usage of agricultural limestone. *Journal of Environmental Management*, **20**, 189–197.

Painter, R. B., Blyth, K., Mosedale, J. C. and Kelly, M. 1974. The effect of afforestation on erosion processes and sediment yield. In *Effects of Man on the Interface of the Hydrological Cycle with the Physical Environment*. Proceedings of the Paris Symposium, September 1974, International Association of Hydrological Sciences Publication 113, 62–67.

Robinson, M. and Blyth, K. 1982. The effects of forestry drainage operations on upland

sediment yields: a case study. *Earth Surface Processes and Landforms*, **7**, 85–90.
Slaymaker, H. O. 1972. Patterns of present sub-aerial erosion and landforms in Mid-Wales. *Transactions of the Institution of British Geographers*, **55**, 47–68.
Smith, K. 1970. The effects of a small upland plantation on air and soil temperatures. *Meteorological Magazine*, **99**, 45–48.
Soutar, R. G. 1989. Afforestation, soil erosion and sediment yields in British freshwaters. *Soil Use Management*, **5**(2), 82–86.
Stott, A. P. 1987. Medium-term effects of afforestation on sediment dynamics in a water supply catchment: a mineral magnetic interpretation of reservoir deposits in the Macclesfield forest, N. W. England. *Earth Surface Processes and Landforms*, **12**, 619–630.
Stott, T. A. 1984. Vegetation as a factor affecting river bank erosion on three rivers in mid-Wales. *Nature in Wales*, **3**(1/2), 65–68.
Stott, T. A. 1987. Forestry effects on sediment sources, dynamics and yields in the Balquhidder catchments, central Scotland. Unpublished PhD thesis, University of Stirling.
Stott, T. A. 1989. Upland afforestation – Does it increase erosion and sedimentation? *Geography Review*, **2**(5), 30–32.
Stott, T. A. 1997a. A comparison of stream bank erosion processes on forested and moorland streams in the Balquhidder Catchments, central Scotland. *Earth Surface Processes and Landforms*, **22**, 383–399.
Stott, T. A. 1997b. Forestry effects on bedload yields in mountain streams. *Journal of Applied Geography*, **17**(1), 55–78.
Stott, T. A., Ferguson, R. I., Johnson, R. C. and Newson, M. D. 1986. Sediment budgets in forested and unforested basins in upland Scotland. In Hadley, R. F. (Ed.) *Drainage Basin Sediment Delivery*. International Association of Hydrological Sciences Publication 159, 57–68.
Stretton, C. 1984. Water supply and forestry – a conflict of interests: Cray Reservoir, a case study. *Journal of the Institution of Water Engineers and Scientists*, **38**, 323–330.
Thorne, C. R. 1981. Field measurements of rates of bank erosion and bank material strength. In *Erosion and Sediment Transport Measurement*, Proceedings of the Florence Symposium, June 1981. International Association of Hydrological Sciences Publication 133, 503–512.
Thorne, C. R. 1982. Processes and mechanisms of river bank erosion. In Hey, R. D., Bathurst, J. C. and Thorne, C. R. (Eds) *Gravel Bed Rivers*. John Wiley, Chichester, 227–259.
Thorne, C. R. 1990. Effects of vegetation on riverbank erosion and stability. In Thornes, J. B. (Ed.) *Vegetation and Erosion*. John Wiley, Chichester, 125–144.
Thorne, C. R. and Lewin, J. 1979. Bank processes, bed material movement and plantform development in a meandering river. In Rhodes, D. D. and Williams, G. P. (Eds) *Adjustments of the Fluvial System*. Kendall/Hunt Publishing, Dubuque, Iowa, 117–137.
Walling, D. E. 1974. Suspended sediment and solute yields from a small catchment prior to urbanisation. In Gregory, K. J. and Walling, D. E. (Eds) *Fluvial Processes in Instrumented Watersheds*. Institute of British Geographers Special Publication 6, 169–192.
Walling, D. E. and Woodward, J. C. 1992. Use of radiometric fingerprints to derive information on suspended sediment sources. In Bogen, J., Walling, D. E. and Day, T. (Eds) *Erosion and Sediment Transport Monitoring in River Basins*. Proceedings of the Oslo Symposium 24–28 August 1992. International Association of Hydrological Sciences 210, 163–164.
Wolman, M. G. 1959. Factors influencing erosion of a cohesive river bank. *American Journal of Science*, **257**, 204–216.

4 Slope and Gully Response to Agricultural Activity in the Rolling Loess Plateau, China

TIMOTHY A. QUINE,[1] DES WALLING[1] and XINBAO ZHANG[2]
[1] *Department of Geography, University of Exeter, UK*
[2] *Institute of Mountain Disasters and Environment, Chinese Academy of Sciences, People's Republic of China*

GEOMORPHOLOGICAL PROCESSES AND ENVIRONMENTAL CHANGE: THE LOESS PLATEAU

The landscape of the Loess Plateau of China provides clear visual evidence of the dynamism of its geomorphological environment (Figure 4.1), and the sediment load transported to the lower Huang He of 1.6 Gt year^{-1} (Zhu, 1986) provides quantitative evidence for the gross magnitude of the geomorphological changes under way. The area, therefore, meets the first key requirement for a study of the interaction between geomorphological processes and environmental change, namely dynamic geomorphological processes. Another requirement for such a study is the existence of environmental change and in the Loess Plateau this may be inferred from the changing magnitude of the human population. Figure 4.2 shows the expansion of the population of the Loess Plateau from *c.* 1 million in 1000 BC to more than 72 million in AD 1981, with over half of the increase occurring since 1957 (Wen, 1993). Given such rates of population increase, an evolving interaction between human communities and land resources amounting to environmental change is to be expected. Some indication of the results of this interaction is provided by a comparison of estimates of the temporal change in total soil loss with the data for population growth (Figure 4.2). These data suggest that at the regional scale, expansion of the human population and associated intensification of human activity in the Loess Plateau has been accompanied by increasing soil erosion amounting to an increase of 100% in 3000 years, including *c.* 50% in the last century. Although these regional data identify the potential importance of human-induced environmental change for the erosion processes in the Loess Plateau, their interpretation is apt to lead to simplification of the "environmental change–process response" relationship. In particular, the conclusion may be drawn that population increase will inevitably lead to an acceleration of erosion rates and the still more dubious inference may be made that population decline will lead to a

Fluvial Processes and Environmental Change. Edited by A. G. Brown and T. A. Quine.
© 1999 John Wiley & Sons Ltd.

Figure 4.1 The study area in the rolling hills region of the Loess Plateau near Ansai, Shaanxi Province: (a) abondoned fields on the mound slopes, displaying several parallel shallow gullies, with small gully headcuts at the junction with the major gully slopes; (b) large-scale mass movements on steep gully slopes which are reported to have occurred since field abandonment in 1985 (photograph taken in 1996)

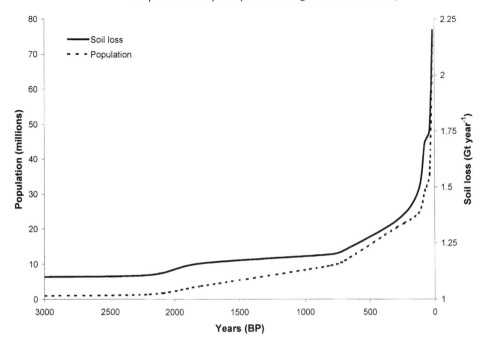

Figure 4.2 Temporal change in the population of the Loess Plateau, and in the total amount of soil lost from the area (after Wen, 1993)

reduction in erosion. There is, therefore, a need for analysis of the landscape response to human activity at a scale which permits more detailed evaluation of the complex interrelationships between land use and geomorphological processes. This study seeks to begin to address this need by examining landscape response to changing agricultural activity near Ansai in Shaanxi Province. Particular attention is paid to fluvial processes and associated geomorphological change on the slopes and to linkages between slopes and gullies (cf. Figure 4.1). Evidence derived from discussions with local farmers, from field observation and from environmental radionuclide measurements is used in the study. This evidence is examined after description of the study area and the reasons for its choice.

THE STUDY AREA

Choice of Study Area

The study area near Ansai was selected on the basis of regional and local characteristics. On the basis of macro-scale geomorphology, the Loess Plateau may be divided into two distinct regions (Figure 4.3): the Yuan area, where some of the plateau surface remains intact; and the rolling hills area, with no remaining intact plateau surface (cf. Figure 4.1). Ansai lies in the rolling hills area (Figure 4.3) which is characterised by some of the highest sediment yields in the Plateau, ranging from 10 000 to

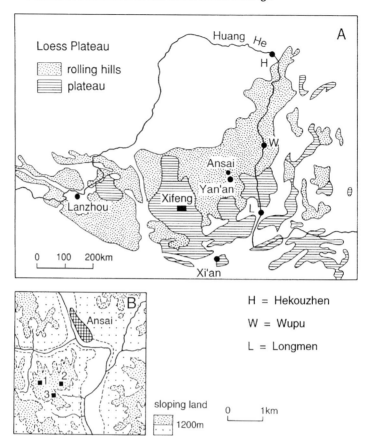

Figure 4.3 Location of the study area: (A) the major subdivisions of the Loess Plateau; (B) the location of Field A (1), the major area of landslide activity (2), and the studied abondoned fields (3)

$25\,000\,\mathrm{t\,km^{-2}\,year^{-1}}$ (cf. Gong and Xiong, 1979). Sediment load and catchment area data presented in Table 4.1 further emphasise the significance of the rolling hills area as a source of sediment to the Huang He. Data are presented for three important stations on the middle reaches of the Huang He, namely: Hekouzhen, upstream of the rolling hills; Wupu, in the middle of the rolling hills area; and Longmen, downstream of the rolling hills. The data show that the Hekouzhen–Longmen basin, dominated by the rolling hills area, is the source of 85% of the sediment load of the Huang He at Longmen even though the basin represents only 26% of the upstream catchment area at the same point. These high rates of sediment production are not only of academic interest, since off-site impacts include dam sedimentation, elevation of the channel of the Huang He with consequent increase in flood risk, and extension of the Huang He delta into valuable oilfields, etc. (cf. Tang, 1985; Douglas, 1989; Wen, 1993). Control of sediment production, therefore, remains an important goal, as reflected in the inclusion in the seventh national five-year plan of a major research project addressing the comprehensive control of soil erosion in the Loess Plateau, involving 11 experimental

Table 4.1 Sediment load and catchment area data for selected locations on the main stream of the Huang He (Zhang and Zhang, 1995)

Huang He Gauging Station	Area (10^3 km^2)	Sediment load (Gt year^{-1})
Hekouzhen	367.9	0.167
Wupu	433.5	0.673
Longmen	497.6	1.146

areas and over 600 scientists (Wen, 1993). However, control of sediment production will be possible only if the sources and mechanisms of sediment production are known. At present, these issues remain unresolved with no consensus on the relative contribution of the slopes and gullies to total sediment production (cf. Zhang and Zhang, 1995; Zhang et al., 1997a, 1997b). Consequently, there is an urgent need for improved understanding of the relationship between human activity and sediment production in the rolling hills area of the Loess Plateau and, in particular, a need to understand the impact of human activity on the slope and gully systems.

In terms of local characteristics, Ansai offers several important advantages for a study of the interrelationships between land use and geomorphological processes. Much of the accessible rolling hills area has been occupied and farmed for more than one century. However, Ansai lies close to the historic border with Mongolia in an area of frequent conflict. The population density in the area, therefore, remained low until the 19th century, and the most significant increase in population and agricultural activity has occurred within the last 60 years. Two particular advantages arise from this distinctive history. First, the critical changes occurred within "living memory" and it has been possible to obtain valuable, first-hand information concerning the population, agricultural and landscape change from senior members of the local population, and in particular Mr Chen Yuan, the "head" of the nearby Majinge village. Secondly, many of the important changes occurred within a time frame (1954 onwards) that could be subject to study using the radionuclide caesium-137 (^{137}Cs).

Ansai: Geomorphology and Environment

The most important geomorphological features of the rolling hills area near Ansai are annotated on Figure 4.1. At the most coarse level, the rolling plateau can be divided into two units with approximately equal areas; namely, the mound region between the major gullies and the gullies themselves. The loess in the area is about 200 m thick and the mound crests (*Mao*) and ridges (*Liang*) lie at an altitude of 1340–1380 m, whilst the valley floors are at 1000–1050 m. The relatively gentle mound slopes (0–30°) are usually underlain by Malan (Q_3) loess, while the very steep gully slopes (35–70°) are usually underlain by Wucheng (Q_1) and Lishi (Q_2) loess. The texture of the loess is classified as "medium", with a typical composition of 12.4% sand (0.25–0.05 mm), 71.7% silt (0.05–0.005 mm), and 15.9% clay (< 0.005 mm).

The slopes of the mounds are typically incised by linear concavities, literally translated from Chinese as "shallow gullies" (Figure 4.1(a)), and the junction between the

Table 4.2 A summary of the recent history of the study area near Ansai derived from discussions with local farmers and field observations

Date	Population/change	Mound slopes	Major gully slopes
pre-1935	Five families in Ansai county	Continuous bush and grass cover	Forested
1935	Civil war continuing; Communist party arrive with many followers		Some felling of trees for fuel
1948	Nationalists defeated by communists in north	Beginning of cultivation of slopes; no shallow gullies; no "minor" gully headcuts	Still large forested areas
1952		All slopes cultivated; introduction of contour strips	Near complete removal of forest cover; some cultivation of slopes
1952–1996	Steadily increasing population	Rill and inter-rill erosion of fields; development of shallow gullies	
1980s		Abandonment of some fields	Abandonment of cultivation
1985–1996			Massive slope failures on abandoned steep slopes
1992 and 1996	Sampling visits	Active fields; cultivation of beans and millet	Supervised grazing by goats

mound slopes and the major gully is often marked by a series of smaller gully headcuts (< 30 m wide and < 10 m high), as shown in Figure 4.1(a). The slopes of the major gullies are often characterised by vertical piping and by landslide scars and landslide-displaced slopes (Figure 4.1(b)).

The climate in Ansai is characterised by cold dry winters (January mean temperature: $-8\,°C$) and warm moist summers (July mean temperature: $22\,°C$). Annual precipitation in the area ranges from 350 to 960 mm (mean of $c.$ 550 mm) and typically 75% of it falls during the flood season from June to September. A large proportion of the flood-season precipitation is accounted for by a few heavy storms (the maximum daily precipitation of 136.5 mm with a maximum intensity of 45 mm in 30 minutes was observed on 16th July 1989). The average erosion rate in the area, based on river sediment load data, is $14\,000\,\text{t km}^{-2}\,\text{year}^{-1}$.

Ansai: Recent History and Land Use

An outline of the changes in population, agricultural activity and landscape, described by Mr Chen and others, is provided in Table 4.2. The geomorphological units referred to are discussed above and annotated in Figure 4.1. With this information and background, the study area offers the rare opportunity to examine the effect of agricultural activity on geomorphological processes in the rolling hills area over a period of $c.$ 60 years from the time of initial cultivation of the sloping land.

STUDY OBJECTIVES AND METHODS

Three major geomorphological changes which are said to have occurred since the initiation of cultivation are identified in the recent history of the study area (Table 4.2): the development of the shallow gullies; the development of gully headcuts; and the landslide activity on the steep slopes. This study, therefore, sought to investigate the relationships between the observed geomorphological changes and the environmental changes represented by the intensification of human activity. The fields on the mound slopes and the development of the shallow gullies provided the main focus of attention and in their investigation the following data were available:

(1) detailed topographic survey;
(2) field measurements of rill dimensions;
(3) ^{137}Cs-derived erosion rates;
(4) depth distributions of shear strength.

The methods used to derive items (1)–(3) are described in detail elsewhere (Quine et al., 1997; Zhang et al., 1997b). In brief, a longitudinal section of a cultivated slope (Field A), including the entire catchment of a shallow gully, was studied in detail. Topographic data were collected using a dumpy level to record elevation at 2-m intervals. In September 1992, after a flood season with a five-year return period, rill dimensions were measured at 2-m intervals downslope; all rills with a depth greater than 1 cm were recorded. 110 soil core samples were collected and analysed for ^{137}Cs by gamma spectrometry. Erosion rates were derived from the ^{137}Cs data using "calibration" procedures discussed elsewhere (Quine, 1995; Quine et al., 1997).

While investigating the vane shear strength of surface soil in eroded and uneroded areas, during October 1996, it was found that the shear strength of the cultivated loess was significantly lower than the underlying, undisturbed loess (plough soil: $3.2 \pm 1.3\,\mathrm{t\,m}^{-2}$; undisturbed loess: $7.6 \pm 0.6\,\mathrm{t\,m}^{-2}$) and that the boundary between the two could be readily identified on the basis of shear strength measurements. This serendipitous finding provided a method for establishing the depth to the undisturbed loess. Depth distributions of vane shear strength were, therefore, measured in situ at 1-m intervals along six transects (3-m intervals downslope) across the shallow gully in Field A. During the same field season, in October 1996, additional observations of the upper mound surface were undertaken, including topographic survey, rill measurements and vane shear strength measurements in abandoned fields within 500 m of Field A.

Data of relevance to the investigation of gully headcut extension and landsliding are limited to the field description of gully headcuts below both abandoned and cultivated fields and field recording of the distribution of landslide scars. However, in the absence of access to aerial photographs or detailed maps these data are qualitative and partial. Interpretation of headcut extension and landslide activity must, therefore, be based largely on inference.

CULTIVATED FIELDS

Data

The data obtained for Field A using the methods outlined above are summarised in Table 4.3. On the basis of the observation made by Mr Chen that no shallow gullies existed prior to cultivation, an estimate of the pre-cultivation surface was derived (Z_p). This was a smoothed version of the measured topography, assuming no linear concavity (shallow gully), and is clearly a very tentative estimate. Using both this estimated surface and the measured surface it was possible to estimate net rates of landform change since the initiation of cultivation, as follows:

$$LC_{48i} = (Z_{si} - Z_{pi}).1000 / T_c$$

where, LC_{48i} is the net rate of landform change at point i since 1948 – the initiation of cultivation (mm year^{-1}); Z_{pi} is the estimated elevation of point i in 1948 (m); Z_{si} is the measured elevation of point i in 1992 (m); and T_c is the time elapsed between the initiation of cultivation and the time of survey (44 years). In order to facilitate comparison with other rate estimates, the rill erosion volumes were also converted to net rates of landform change as follows:

$$LC_{ri} = - V_{ri}.1000$$

where LC_{ri} is net rate of landform change due to rill erosion at point i (mm year^{-1} – for 1992); and V_{ri} is the rill volume per unit area at point i observed in 1992 (m^3 m^{-2}). The rates derived using these approaches are illustrated in Figure 4.4, where they may be compared with the ^{137}Cs-derived rates for the period 1954–1992. Rates of landform change (mm year^{-1}) have been used in this comparison as these units closely reflect those of the source data. However, in order to permit comparison of the observed rates of denudation with other published data, spatially integrated mass rates (kg m^{-2} year^{-1}) have been calculated using the plough layer bulk density (1350 kg m^{-3}) and are presented in Table 4.4.

Information derived from the vane shear strength investigations is shown in Figures 4.5 and 4.6. The sharp boundary between the upper surface of the undisturbed loess and the overlying material is clear in the vane shear strength–depth distributions from ridge and shallow gully locations (Figure 4.5). This sharp boundary was used to identify the lateral variation in depth to undisturbed loess from ridge to ridge, through the shallow gully, which is shown in Figure 4.6. This figure shows the depth of

Table 4.3 Spatially distributed data collected at Field A study site near Ansai

Data	Units
Elevation above local datum in 1992, i.e. time of survey	m
Rill erosion volumes for 1992	m^{-3} m^{-2}
Net landform change rates for 1954–1992 derived from ^{137}Cs	mm year^{-1}
Depth to undisturbed loess	m

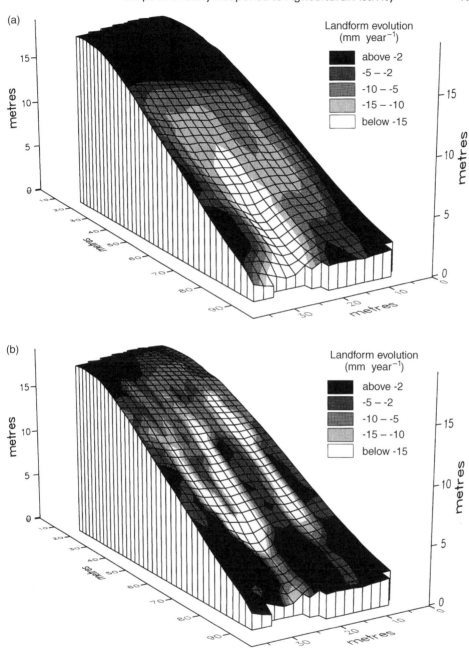

Figure 4.4 Landform evolution rates for Field A (negative rates represent erosion): (a) estimated rate for period 1948–1992; (b) rate due to rill erosion for 1992; (c) (*see over*) ^{137}Cs-derived rate for 1954–1992

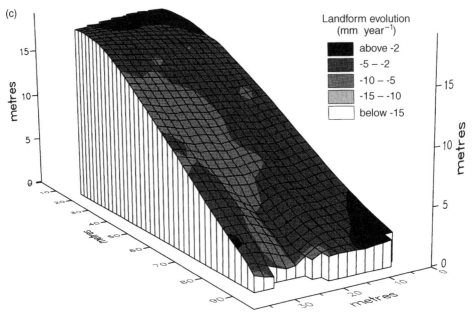

Figure 4.4c

disturbance of the loess by tillage activity on the ridges, and the greater depth to disturbed loess in the shallow gully. The significance of this observation is considered below.

Discussion: Shallow Gully Development

On the basis of the observation that the shallow gullies were not present prior to cultivation, and that the maximum rates of erosion identified using the ^{137}Cs data and the maximum rates of rill erosion lie within the shallow gully, a first explanation for their mode of formation might be based on gradual incision over the period of cultivation, as a result of fluvial erosion processes, namely, rill and inter-rill erosion.

There are three reasons for rejecting this explanation. First, the rates of landform change derived using ^{137}Cs for the period 1954–1992 (Figure 4.4(c)) are significantly lower than the rates required to develop the shallow gullies from a smooth surface since 1948 (Figure 4.4(a)). This is also seen clearly in the spatially integrated data presented in Table 4.4. Secondly, this divergence cannot be explained by suggesting that the ^{137}Cs-derived rates underestimate the true level of soil erosion, because the rill erosion data also indicate that mean annual rill erosion rates are insufficient to explain the formation of the shallow gully in the requisite period of time. Although the rill erosion rates for 1992 are close to the annual rate of erosion required to produce the shallow gully, the 1992 data do not represent mean annual rates, but instead represent an erosion season with an approximately five-year return period. (The year 1996 was also marked by a high summer rainfall, but since this occurred when the ground surface was protected by a full crop cover, little erosion ensued.) On this basis it may be

Table 4.4 Erosion rate estimates for Field A

	Whole slope (length 90 m; area 2510 m²)		Area of shallow gully (length 40–90 m; area 1560 m²)	
	kg m^{-2} year^{-1}	mm year^{-1}	kg m^{-2} year^{-1}	mm year^{-1}
^{137}Cs-derived: 1954–1992	5.1	3.8	6.1	4.5
Rill-derived: 1992	7.3	5.4	9.8	7.3
Topographicaly derived: 1948–1992	6.9	5.1	11.0	8.1

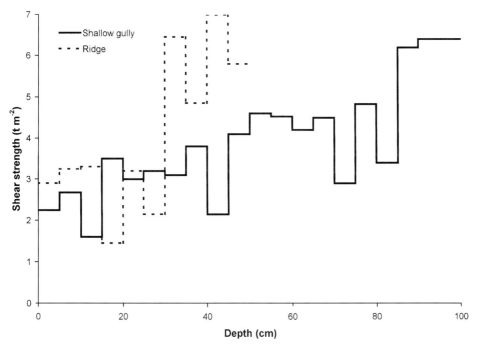

Figure 4.5 The depth distribution of vane shear strength in the shallow gully and on the adjacent ridge in field A

suggested that the ^{137}Cs-derived rates are likely to provide reasonable estimates of annual erosion by rill and inter-rill processes. Thirdly, if the shallow gully had formed by gradual incision then the depth to undisturbed loess in the gully would be expected to be the same or lower than the depth on the ridges. This is not the case (Figure 4.6). Contrary to expectations, the disturbed loess is much deeper in the shallow gully, providing evidence of greater incision in the area of the shallow gully in the past. In view of the unsatisfactory explanation provided by gradual development, other mechanisms for the formation of the shallow gullies must be sought. Two will be considered: intermittent development and combined surface and subsurface development.

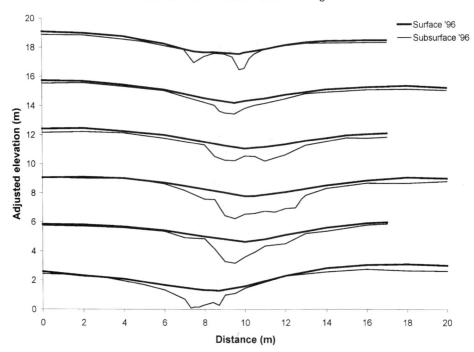

Figure 4.6 Surface of shallow gully and adjacent ridges in 1996 and depth of undisturbed loess derived from vane shear strength data for six transects at 75, 78, 81, 84, 87 and 90 m from the crest of the slope. For the purposes of graphical presentation, the elevations of the 75, 78, 81, 84 and 87 m transects have been adjusted by the addition of 12.5, 10, 7.5, 5 and 2.5 m, respectively

Intermittent Development

Re-examining the evidence presented above, an alternative explanation for the development of the shallow gully may be proposed. On the basis of the ^{137}Cs-derived soil redistribution rates, the 1954 surface may be reconstructed as follows:

$$Z_{54i} = Z_{si} - [(T_f . LC_{54i} . 1000)]$$

where Z_{54i} is the derived estimate of the elevation of point i(m) in 1954 – prior to ^{137}Cs fallout; LC_{54i} is the net landform change rate at point i derived from the ^{137}Cs data (mm year^{-1}); and T_f is time elapsed from the initiation of fallout to the time of sampling (38 years).

Figure 4.7 allows comparison of the surface elevation values for three of the six transects investigated for shear strength. For each transect, the surface elevation in 1996 and the elevation estimated for 1954 on the basis of the ^{137}Cs-derived soil redistribution rates are shown. These data suggest that by 1954 the shallow gully had already formed.

On the basis of the depth to undisturbed loess data (Figure 4.6), the maximum depth of gully incision may be estimated. If the observation that there were no shallow gullies

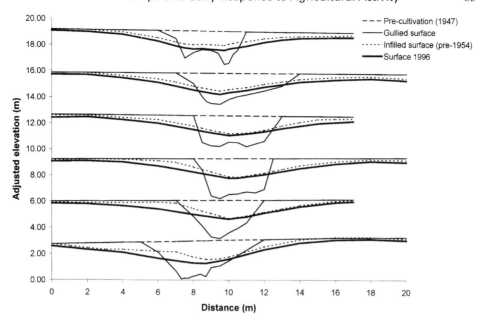

Figure 4.7 Measured elevations in 1996 and estimated surface elevations: in 1947, i.e. pre-cultivation; immediately after the postulated major episode of gully erosion between 1947 and 1954; and after gully infilling in 1954. The data shown are for six transects at 75, 78, 81, 84, 87 and 90 m from the crest of the slope. For the purposes of graphical presentation, the elevations of the 75, 78, 81, 84 and 87 m transects have been adjusted by the addition of 12.5, 10, 7.5, 5 and 2.5 m, respectively

prior to 1948 is accepted and the surface derived from 1954 is considered a reasonable estimate, then this maximum depth of incision must have occurred very soon after initial cultivation, in the period 1948–1954. This suggests that the gully formed in the initially smooth surface of the field. Both the estimated initial surface and the estimated gully form are shown in Figure 4.7. The gully form was derived by extrapolating the gully edges, inferred from the shear strength data, to the estimated initial surface (Z_p). For cultivation to continue after such gullying, it would be necessary to infill some of the gully and smooth the surface. Therefore, it is not unreasonable to suggest that the gullied surface shown in Figure 4.7 would have been rapidly transformed by human activity to that estimated for 1954. In order to establish the viability of this suggestion, the volume of soil required to infill the gully to the 1954 level was compared to the volume of erosion required to lower the remainder of the area to the 1954 level. An improbably high level of agreement was found (c. 70 m^3 in each case over the length from 75 to 90 m downslope). While it should be emphasised that there is considerable freedom to adjust this level of agreement by manipulating the gully form and the initial surface elevation, Figure 4.7 demonstrates that the noted agreement was obtained without unrealistic manipulation.

There is, therefore, some evidence to support the alternative mechanism of intermittent incision, which involves the following sequence of events:

1948 Initiation of cultivation.
1948–1954 Deep gully incision of the field in the location of the current shallow gully, followed by deliberate infilling to create a cultivable land surface.
1954–1992 Relatively little change in the form of the shallow gully.

Combined Surface and Subsurface Development

Although the preceding mechanism is consistent with the available observations and data, an alternative and consistent mechanism involving both surface and subsurface erosion processes is now considered. Pipe and tunnel erosion are widely observed in the study area, and small near-surface pipes were seen to be intimately associated with the rill network in 1992. Where pipe erosion occurs within the plough layer, it will remove ^{137}Cs and the resultant soil removal will be included in the ^{137}Cs-derived soil erosion rates. However, if pipe or tunnel erosion occurs below the plough layer, the soil removed will not be labelled with ^{137}Cs (which is deposited on the surface and mixed through the plough layer by tillage), and soil loss will not be accompanied by depletion of ^{137}Cs inventories. Therefore, soil erosion by sub-plough piping will not be represented in the ^{137}Cs-derived rates of soil redistribution. If pipe erosion occurred below the plough layer and was followed by roof collapse, then surface lowering would take place associated with the creation of a zone of disturbed loess to a depth equal to the depth of the base of the pipe. This process could account for the deep zone of disturbed loess in the shallow gully without invoking the level of gully incision shown in Figure 4.7. The sequence of events envisaged is as follows:

1948 Initiation of cultivation.
1948–1992 Phase 1: rill and ephemeral gully erosion lead to the formation of an incipient shallow gully.
Phase 2: the presence of an incipient shallow gully leads to the concentration of drainage in the shallow gully area, the initiation of piping and the continuation of rill and ephemeral gully erosion.
Phase 3: a sequence of pipe erosion and collapse leads to surface lowering; rill and ephemeral gully erosion are concentrated in the shallow gully, but counteracted by infilling by tillage. Therefore, periods of apparent landform stability are punctuated by episodic change when pipe collapse takes place.

Unfortunately this mechanism is not testable using presently available data. The only supporting evidence is negative in that local accounts of instantaneous formation of the shallow gullies (indicated by the previous mechanism) are entirely lacking. Table 4.5 summarises the erosion volumes and rates implied by the hypotheses.

Table 4.5 Process rates associated with the competing hypotheses for shallow gully formation

	Intermittent incision		Surface/subsurface	
	Total erosion ($kg\,m^{-2}$)	Erosion rate ($kg\,m^{-2}\,year^{-1}$)	Total erosion ($kg\,m^{-2}$)	Erosion rate ($kg\,m^{-2}\,year^{-1}$)
c. 1948				
Gully erosion	145			
Gully infill (redistribution)	64			
1948–1992				
Rill and inter-rill erosion	223	5.1	223	5.1
Sub-plough pipe erosion			145	3.3
Net total	368	8.1	368	8.1

ABANDONED FIELDS AND SMALL GULLY HEADCUTS

Data

Field observations were made in an area of abandoned fields ($4200\,m^2$) within 500 m of Field A. Attention focused on the surface conditions and on the evidence for surface erosion. At the time of observation in 1996, the fields had been abandoned for a 10-year period and during this time a full vegetation cover of grasses and scrub had developed. The surface of the soil was further protected by the formation of a hard, dark organo-mineral crust. The vane shear strength of this crust was greater than that of the surface of the cultivated fields (vane shear strength: cultivated surface $1.7 \pm 0.7\,t\,m^{-2}$; abandoned surface $3.0 \pm 0.4\,t\,m^{-2}$). Visible evidence of water erosion was limited to rilling and piping in the bed of the shallow gullies. There was no evidence of erosion from the inter-gully ridges.

In contrast to the absence of visible evidence of erosion on the abandoned fields, there was plentiful visual evidence for active advance of the small gully headcuts at the base of the abandoned fields (cf. Figure 4.1(a)). The walls of the headcuts were devoid of vegetation and organo-mineral crusts, and at the base of the headcuts there were accumulations of freshly eroded sediment. Furthermore, the base of many of the headcuts was marked by the mouth of large vertical pipes (1–2 m in diameter) of indeterminate depth.

In general, the headcuts below cultivated fields did not evidence the same level of activity. In most cases the faces of the headcuts were vegetated or covered in organo-mineral crusts. However, it should be noted that the cultivated slopes differed from the abandoned slopes in terms of slope aspect and slope length.

Discussion

Discussion of post-abandonment change in the fields is constrained by the absence of pre-abandonment observations. However, the limited visible evidence for erosion seems to indicate that little erosion has taken place in the abandoned fields since the cessation of cultivation. This is surprising, especially because the period of abandonment includes the summer of 1992, during which the high rates of rill erosion documented on Field A occurred (Figure 4.4(b)), and widespread erosion was evidenced in sediment yield data. The absence of erosion from the ridges may reflect the protective capacity of the vegetation cover and of the surface crust. The latter may be expected to promote runoff without particle detachment. This would lead to convergence of high volumes of runoff in the shallow gully bed and may explain the evidence for incision in the beds. However, even where incision is evident, the degree of incision is less than would be anticipated on the basis of comparison with the cultivated Field A. This may indicate that erosion of the bed was limited by the high shear strength of the basal loess, and that the runoff output from shallow gullies in the abandoned fields carried a relatively low sediment load. Such a finding would be consistent with erosion plot data from the area, which demonstrate differing patterns of runoff and sediment generation under natural vegetation, on cultivated slopes, and on abandoned field slopes (Zheng Fenli, pers. comm.):

- Natural vegetation: low runoff; low sediment production.
- Cultivated slopes: high runoff; high sediment production.
- Abandoned slopes: high runoff; low sediment production.

It is possible that these differing patterns explain the differences in headcut behaviour below cultivated and abandoned fields indicated by the, albeit limited, field observations. The erosive energy of the runoff from the cultivated fields is dissipated in detachment and transport of sediment from the fields and, therefore, makes little contribution to headcut advance. In contrast, the runoff from the abandoned fields carries little sediment and maintains a high erosive energy which is dissipated only on passing over the headcut, leading to scour and pipe erosion at the base of the headcut and thus destabilising the headcut itself. As indicated in the discussion of methods, these conclusions must rely largely on inference because of the limited nature of the data available.

If the conclusions drawn from the data are correct, then continued abandonment will probably lead to some further incision and gradual deepening of the shallow gully because the export of sediment from the shallow gully is not counteracted by infilling by tillage. However, the higher resistance to erosion of the undisturbed loess may limit this incision. The greatest geomorphological change is likely to occur at the boundary between the mound slope and the gully slope. Important erosion processes in this zone include both headcut advance and pipe erosion at the headcut base. These processes are expected to continue while high rates of runoff are produced.

SLOPE FAILURE

The final observation made by Mr Chen that appears particularly significant is that the landslide activity, evidenced by scars and resultant displaced slopes (Figure 4.1(b)), had occurred only after 1985 and had followed the abandonment of the associated fields. Detailed investigation of this phenomenon lay beyond the scope of the field programme on which this chapter is based. It is clearly a tantalising statement: is the sequence of abandonment followed by landsliding coincidental or causal? If the landsliding occurs as a result of de-stabilisation caused by undercutting at the base of the slope, and if it is assumed that the undercutting has been continuing since the initiation of cultivation, then the timing may be coincidental and reflect the passing of a threshold level of incision in the valley floor. Alternatively, increased generation of low sediment-content runoff as a result of field abandonment may have led to increased rates of incision in the valley floor and therefore contributed directly to the slope instability. Finally, Derbyshire *et al.* (1993) have demonstrated the contribution to slope instability made by drainage through pipe networks and dissolution of the calcium carbonate cement in basal loess. If the vertical piping observed below the abandoned fields is a common occurrence and was promoted by abandonment, then pipe erosion and induced basal instability may provide a process-link between field abandonment and enhanced landslide activity.

SYNTHESIS

The objective of this study was to investigate the relationships between environmental change and geomorphological processes in an area of rolling hills in the Loess Plateau. In particular, the study has examined mound-slope and gully responses to environmental change brought about by the initiation of cultivation in the area. The available data have been presented and interpretations, albeit tentative, have been discussed. Figure 4.8 represents a synthesis of the geomorphological responses on the slopes and gullies and an attempt to identify the nature of the linkages between slopes and gullies.

On the mound slopes, the shallow gullies are identified as landforms produced in response to increased runoff and erosion induced by agricultural land use. In the "intermittent incision mechanism", the shallow gullies form as an almost immediate response to the change of land use on the slopes and are maintained in dynamic equilibrium by the opposing processes of incision by water erosion and infilling by tillage. In the "surface/subsurface mechanism", the shallow gullies are believed to be in dynamic metastable equilibrium, having formed by gradual evolution in response to rill, inter-rill and sub-plough pipe erosion. Figure 4.9 summarises the relationships between landform change and time indicated by these hypotheses. This figure emphasises the stepwise landform development proposed in the "surface/subsurface mechanism" under conditions in which the processes remain essentially constant. In contrast, in the "intermittent incision mechanism" lower rates of landform change and geomorphological work are maintained after an initial rapid adjustment. Abandonment of the fields on the mound slopes is seen to result in reduced landform change on the mound slopes but increased impact on the gully slopes.

In the gully, the processes and landforms respond to both the change in fluxes from the mound slopes and the change in vegetation cover on the gully slopes. While headcut retreat at the mound–gully boundary occurs initially as a result of agricultural land use on the slopes, it is accelerated by abandonment of the slopes with the resultant increase in erosive (low sediment content) runoff. Where erosion at the mound–gully

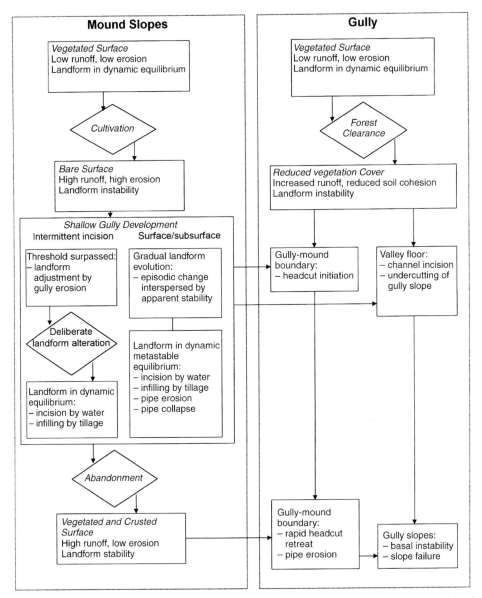

Figure 4.8 Synthesis of slope and gully responses to environmental change at Ansai

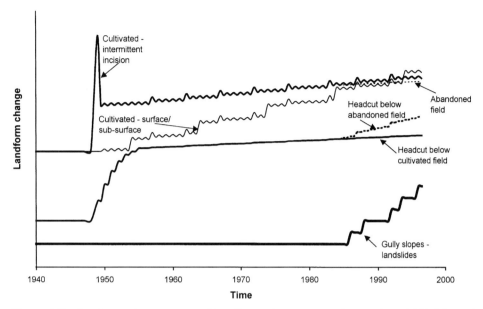

Figure 4.9 A schematic symmary of landform change over time at Ansai. "Cultivated – intermittent incision" and "Cultivated – surface/subsurface" refer to alternative explanations for the formation of shallow gullies which are discussed in the text. (Scales of landform change are different for each geomorphic unit)

boundary leads to vertical pipe development it may contribute to enhanced slope failure through de-stabilisation of the basal loess. In this case maximum geomorphological work, i.e. landsliding, may coincide with reduced human activity.

CONCLUSIONS

The study presented here emphasises the complexity of the relationship between environmental change and geomorphological processes, especially when considered at the scale of a slope–gully system. Due to the limitations of the available data, the synthesis and many of the interpretations presented here are tentative. Nevertheless, they both highlight the need for an improved understanding of the slope–gully system in the Loess Plateau and draw attention to two particular areas of current uncertainty. First, there is a need for greater understanding of the possible role of pipe erosion in the development of shallow gullies. This is not only of academic interest because, if the "surface/subsurface mechanism" is correct, erosion rates from the cultivated land are higher than those indicated by methods designed to measure surface erosion, including erosion plots, field survey and the ^{137}Cs technique. Secondly, there is a need to understand the dynamics of the slope–gully boundary zone and the influence of field abandonment in this area. Again, this is of broad interest and significance. It has been suggested that field abandonment may lead to reduced erosion on the mound slopes at the expense of enhanced landslide activity on the gully slopes. If this is the case, it must clearly be taken into account when developing catchment-wide sediment control plans.

ACKNOWLEDGEMENTS

The authors gratefully acknowledge the financial support provided by The Royal Society, the UK Overseas Development Administration, and the University of Exeter Research Fund. The authors are also grateful to Mr Meng Qingmai and the Middle Yellow River Bureau for offering logistical support for fieldwork; to Mr Wang Yukuan for assistance with sample collection and processing; to Mr Jim Grapes for overseeing the gamma spectrometry; and to Mr Rodney Fry for graphical work.

REFERENCES

Derbyshire, E., Dijkstra, T. A., Billard, A., Muxart, T., Smalley, I. J. and Li, Y.-J. 1993. Thresholds in a sensitive landscape: the Loess Region of Central China. In Thomas, D. S. G. and Allison, R. J. (Eds) *Landscape Sensitivity*. John Wiley, Chichester, 97–128.

Douglas, I. 1989. Land degradation, soil conservation and the sediment load of the Yellow River, China: review and assessment. *Land Degradation and Rehabilitation*, **1**, 141–151.

Gong, S. and Xiong, G. 1979. Source and distribution of silts in the Huanghe (Yellow River). *Remin Huanghe*, **1**, 1–17 (in Chinese).

Quine, T. A. 1995. Estimation of erosion rates from caesium-137 data: the calibration question. In Foster, I. D. L., Gurnell, A. M. and Webb, B. W. (Eds) *Sediment and Water Quality in River Catchments*. John Wiley, Chichester, 307–329.

Quine, T. A., Govers, G., Walling, D. E., Zhang, X., Desmet, P. J. J. and Zhang, Y. 1997. Erosion processes and landform evolution on agricultural land – new perspectives from caesium-137 data and topographic-based erosion modelling. *Earth Surface Processes and Landforms*, **22**, 799–816.

Tang, K. 1985. The perspectives of soil erosion and conservation in the Loess Plateau. In *The Situations, Strategies and Perspectives of Soil Sciences in China*. Chinese Association of Soil Science, Beijing, 45–48.

Wen, D. 1993. Soil erosion and conservation in China. In Pimentel, D. (Ed.) *World Soil Erosion and Conservation*. Cambridge University Press, 63–85.

Zhang, X. and Zhang, Y. 1995. Use of caesium-137 to investigate sediment sources in the Hekouzhen-Longmen Basin of the Middle Yellow River, China. In Foster, I. D. L., Gurnell, A. M. and Webb, B. W. (Eds) *Sediment and Water Quality in River Catchments*. John Wiley, Chichester, 353–362.

Zhang, X., Walling, D. E., Quine, T. A. and Wen, A. 1997a. Use of reservoir deposits and caesium-137 measurements to investigate the erosional response of a small drainage basin in the rolling Loess Plateau region of China. *Land Degradation and Development*, **8**, 1–16.

Zhang, X., Quine, T. A. and Walling, D. E. 1997b. Soil erosion rates on sloping cultivated land on the Loess Plateau near Ansai, Shaanxi Province, China: an investigation using ^{137}Cs and rill measurements. *Hydrological Processes*, **12**, 171–189.

Zhu, X. (Ed.) 1986. *Land Resources in the Loess Plateau of China*. The Northwest Institute of Soil and Water Conservation, Academia Sinica, Shaanxi Science and Technique Press.

5 River Activity in Small Catchments over the Last 140 ka, North-east Mallorca, Spain

JAMES ROSE and XINGMIN MENG
Department of Geography, Royal Holloway, University of London, UK

INTRODUCTION

The publication of *Mediterranean Quaternary River Environments*, edited by Lewin, Macklin and Woodward (1995), has drawn attention to the importance of river systems in changing the landscape of a region that is highly responsive to changes of climate and vegetation, tectonic movements and human activities. To a large extent, this responsiveness is a product of the latitudinal position of the Mediterranean region, the boundary position between major tectonic plates, the characteristic rock types and steep relief of the land area around the Mediterranean Sea, and the long history of human civilisation. A summary and assessment of the significance of these factors is set out in Macklin *et al.* (1995) and Woodward (1995), and examples of each of the controlling factors, either separately or in combination with others, are included in these volumes.

The role of climatic change and human interference on river activity is given in a number of papers in Lewin *et al.* (1995) (cf. Hunt and Gilbertson, 1995; Macklin and Passmore, 1995; Roberts, 1995) and elsewhere (White *et al.*, 1996), but with limited exceptions (Fuller *et al.*, 1996), detailed studies have been restricted to the last 50 ka and usually to the period from the Last Glacial Maximum to the present. Consequently, there is a need for more information from sites where repetitive sequences of climatic change enable process–response relationships to be tested by replication, preferably from small catchments where complex response is reduced and the effects of climatic and vegetational variations within the catchment are minimised.

In the north-eastern part of Mallorca, stacked sequences of fluvial, beach and aeolian sediments and intervening soils provide evidence for fluvial activity in part of the western Mediterranean, over a period spanning the last 140 ka. This evidence is associated with a chronostratigraphy, palaeogeography and estimates of palaeoclimate which form a basis for reconstructing river processes and associated landscapes over a number of major climatic changes from the Last Interglacial through to the Holocene. This chapter aims to describe this evidence and seeks to examine (i) the

Fluvial Processes and Environmental Change. Edited by A. G. Brown and T. A. Quine.
© 1999 John Wiley & Sons Ltd.

timing and scale of river activity in small Mediterranean catchments; (ii) the role of rivers in changing the landscapes of such small catchments; and (iii) the factors responsible for causing the variations in river activity in such regions.

LOCATION, CATCHMENT DIMENSIONS, ROCK TYPES AND GEOMORPHOLOGICAL CONTEXT

The sites are located in the north-eastern peninsula of Mallorca along the south-eastern coast of the Bay of Alcudia (Figure 5.1(a)) and are associated with two small catchments that drain north-westwards from the Sierra de Farruch which reaches an elevation of 400 m (Figure 5.1(b)). The Torrente d'es Coco and the Torrente de sa Telaia Freda respectively have an area of 3.14 and 1.89 km^2, a length of 2.4 and 2.0 km, and a range of relief (r/l, where l = length and r = range of elevation) of 163 and 252 m km^{-1} (Figure 5.1(b)). The upper part of these catchments is a broad, steep valley-head, but the lower sections (1.25–1.5 km) consist of a broad plain composed of alluvial fans and aeolian dunes with a gradient in the order of 67–80 m km^{-1}. The present river channel is narrow and incised into the valley bottom. At the present time, river activity takes the form of storm-generated flood events, capable of transporting gravels with individual clasts reaching boulder size.

The upper slopes of the catchments are composed of hard, jointed limestones, while the lower slopes consist of lithified Quaternary sands, gravels, silts and partially

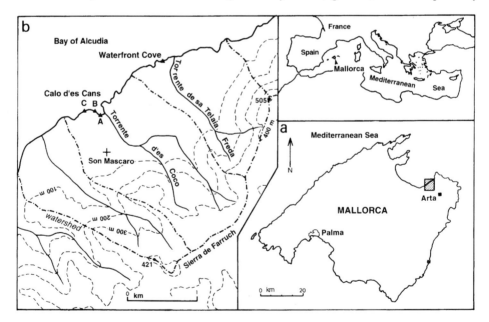

Figure 5.1 (a) Location of study area. (b) Location of sites and configuration of catchments

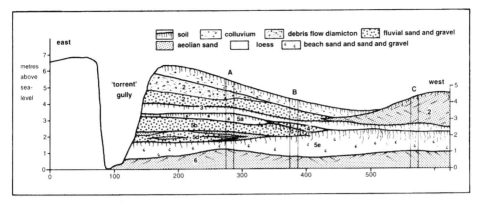

Figure 5.2 Schematic representation of sites within the alluvial fan sediment stack of Torrente d'es Coco (Cala d'es Cans)

lithified soils. There is no evidence for glaciation or periglacial activity having occurred within the catchments (Butzer, 1964). The natural vegetation is evergreen oak woodland with pine forest on the lowland plain. At the present time the upper parts of the catchments are either bare rock or patchy grass and garrigue, while the lower slopes are cultivated. The mean annual temperature of the region at sea-level is 17 °C, with a January mean of about 10 °C and a July mean of 24.5 °C. A lapse-rate of 0.8 °C per 100 m gives January temperatures of about 4 °C in the upper parts of the catchment. Mean annual precipitation is in the order of 500 mm, of which 90% falls between September and May.

The sites investigated are actively eroding sea cliffs cut into the Quaternary sediments and are predominantly aeolian sands and silts, and fluvial sands and gravels with intervening soils. Fortuitously, these sites are located where aggradation has been dominant and sediment bodies have been stacked one-upon-another with a sedimentology and thickness dependent upon the position relative to the axis of the fluvial fan (Figure 5.2). Although small channels and soils are common within the succession, the only evidence for a major erosional event is the shore platform and cliffline formed during the Last Interglacial (LIG) (Oxygen Isotope Stage (OIS) 5e) and the "torrente" gully formed in the latter part of the Holocene (late OIS 1). During cooler episodes of the Quaternary these river catchments extended much further north-westward across what is now the Bahia de Alcudia (van Andel, 1989). Low sea-levels would also have made available large expanses of unlithified sand capable of entrainment by aeolian processes.

METHODOLOGY

Each site was described in detail at representative sections according to sedimentary structures, matrix colour, clast lithology and roundedness, soil structures and magnetic susceptibility (MS) (Figure 5.3). The elevation and relative position of the stratigraphic units and present land surface was measured precisely with an electronic

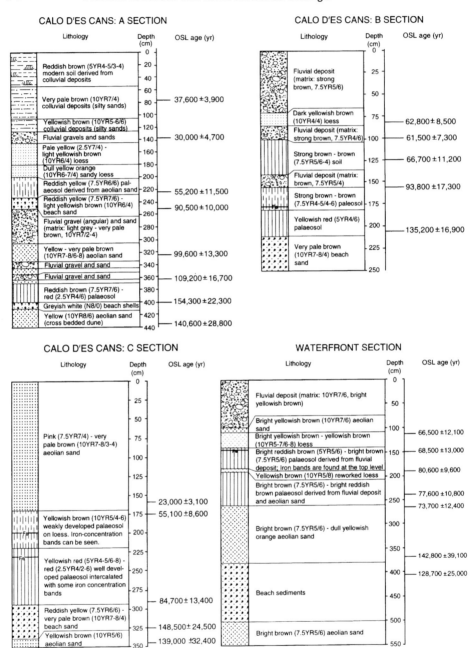

Figure 5.3 Representative sections from Calo d'es Cans and Waterfront Cove. These sections give lithological and pedological properties and OSL ages. The location of these sites within the alluvial landform is given in Figure 5.2

distance measurer (EDM), and samples were collected from measured locations for particle size, calcium carbonate, extractable iron, SEM, microfabric, and oxygen and carbon isotope analysis. Samples were also collected for optically stimulated luminescence (OSL) age estimation and terrestrial and marine molluscan faunal composition.

The geochronology of the stratigraphic units was derived from biostratigraphy of the marine and terrestrial faunas, mean amino acid racemisation (AAR) ratios of 20 individuals of *Bittium* sp. and 25 OSL determinations carried out in the Luminescence Laboratory in the Department of Geography, RHUL. The OSL results and locations of sample points are given in Figure 5.3. Quantitative estimates of mean annual temperature were derived from oxygen isotopic values representing fractionation during soil cementation. Qualitative estimates of biomass cover are indicated by carbon isotope values from pedological cements and estimates of precipitation are derived from SEM analysis of the soil cement structures. Full details of all these procedures are given in Rose (1998) and Rose *et al.* (1999).

SITES, STRATIGRAPHY AND ALLUVIAL ARCHITECTURE

Torrente d'es Coco–Calo d'es Cans Sections

The catchment is represented by three sections from Calo d'es Cans (Figures 5.2 and 5.3). The river sediments are part of a fan which rests on a Last Interglacial beach and against a cliffline that was eroded into aeolian dunes deposited during earlier cold stages. Site A is closest to the apex of the fan, site B is at the margin and site C is in the adjacent terrain that was beyond the reach of fluvial sedimentation (Figure 5.2). Three sorted sand and gravel units and one diamicton unit can be recognised and attributed to fluvial and debris flow/colluvial processes respectively. Intervening materials are beach sands, wind-blown sand and silt (loess) and soils.

The fluvial sediments are composed of thin, horizontal and cross-bedded units of moderately sorted sand and moderately sorted coarse gravel with a sand matrix. The clasts are predominantly sub-angular, composed almost entirely of limestone, and show imbrication with a dip towards the land. Occasionally the clasts are of boulder size and exist in both isolated positions or as parts of clusters. The sedimentary structures indicate low-angle bars with steep, but shallow, distal margins and thin sand beds accumulated in shallow channels on the surface of, or between the gravel bars. These sedimentary structures are typical of a shallow, ephemeral braided river system producing bedforms with a range of dimensions reflecting frequently changing discharges. OSL determinations on *in situ* fluvial sands or on interbedded aeolian sands indicate that the fluvial sediments were formed during OIS 5c, 5a, 4 and 2 and that the greatest thickness was deposited during OIS 2.

The debris flow/colluvial sediments consist of a poorly sorted gravelly silty sand unit represented in Calo d'es Cans A. This unit has two parts, differentiated by the sorting, the terrestrial molluscan fauna, and proportion of soil material indicated by the magnetic susceptibility signal. The lower part of the unit is slightly better sorted with less soil material included. This material is considered to be the suspension fraction of sediment flows, and is interpreted as a debris flow deposit. In contrast, the upper part has a higher proportion of fines, a well-developed terrestrial molluscan fauna and a high transported soil component. This part of the unit is attributed to surface wash of fine soil material from locations further upslope. The relatively unabraded condition of the thin terrestrial mollusca suggests that transport distances were small. The OSL determination and mollusc fauna indicate that sedimentation of this diamicton began in OIS 2 and continued through to early OIS 1. The present "torrente" gully is located at the eastern side of the fan and cuts through all the sedimentary units described above, indicating that it was eroded during the later part of the Holocene.

Torrente de sa Telaia Freda–Waterfront Cove Section

One section is described in detail and provided the samples for laboratory analyses and OSL determinations (Figure 5.3). This section is located at the western side of a fan complex. As at Calo d'es Cans, the sediments rest on the Last Interglacial beach. The site is close to the apex of the fan but only provides evidence for one thick fluvial unit, although other thick fluvial deposits developed further west can be related to the succession. As with Calo d'es Cans, the present "torrente" gully dissects all the units and appears to have been formed in the late Holocene.

Sedimentologically the sand and gravel units are similar to those described at Calo d'es Cans and appear to have formed by the same fluvial depositional processes. However, at Waterfront Cove there is evidence that fluvial deposition was preceded by a small amount of erosion and the fluvial sediments rest in a wide shallow channel. Stratigraphic relationships and OSL determinations indicate that the gravel unit which forms the surface fan at the site was deposited during OIS 2.

HISTORY OF RIVER ACTIVITY

Evidence for river activity is described according to the separate depositional or erosional events. These are related to particular periods of time derived from biostratigraphy, AAR and OSL and are classified, for convenience, into particular oxygen isotope stages (OIS) and substages (OISst).

No evidence can be found for river activity during OIS 5e and surface processes at this time are represented by beach sediments and soil development at the beach surface.

OISst 5d

OISst 5d is represented by thin (c. 10 cm) beds of fluvial sands and gravels composed of limestone clasts and a derived marine molluscan fauna at Calo d'es Cans A. Individual beds are separated by well-sorted aeolian sand sheets. Relatively low MS values indicate that most material was derived from eroded bedrock rather than redeposited soil material, and the fact that the unit rests on a well-developed soil indicates that erosion was minimal in the lower part of the catchment. Mean annual temperatures at the time ranged from 13.6 to 8.2 °C, biomass was low and the moisture regime was "dry".

This unit represents the initiation of an alluvial fan over a Last Interglacial beach and shore platform. Composition indicates transport of material from rock outcrops in the upper part of the catchment with a minor component from adjacent unlithified beach sediments. Interbedded aeolian sands indicate that deposition was periodic and beach material was blown across the fan surface to form sand sheets (this may also be the source of some of the small marine molluscs that are included in the river deposits). Elsewhere, as at Waterfront Cove, aeolian sand deposition was the dominant process. Development took place over a period in which there was a fall in temperature, a decrease in moisture regime and a reduction in biomass from the preceding LIG.

During OIS 5c the river deposits were buried by aeolian sand and silt sedimentation and altered by soil formation. No evidence was found for either river deposition or erosion at this time.

OISst 5b

OISst 5b is represented by 0.47 m and 0.15 m of fluvial sands and gravels at Calo d'es Cans A and B respectively. The beds are composed of limestone clasts and a sand matrix with a relatively low magnetic susceptibility (MS) and Fe content. The units rest either on aeolian sands or soil, indicating minimal or no erosion at the site prior to deposition. Mean annual temperatures at the time ranged from 10.8 to 6.7 °C, and biomass was moderate to low. Evidence is not available for moisture regime.

This unit represents the further development on an aggrading alluvial fan, which extended seawards and westwards. Its composition is almost entirely from rock outcrops in the upper part of the catchment. Beyond the fan, soil development was the dominant process on low-angle slopes. Fluvial activity at this time was associated with a fall of temperature and a reduction in biomass from OIS 5c.

During OIS 5a, a beach was deposited at Calo d'es Cans A, and soil formation took place at the other sites. There is no evidence for river activity at this time.

OIS 4

River activity is represented at Calo d'es Cans B where some 0.15 m of fluvial sands and gravels, with a moderate MS signal and a derived marine molluscan fauna rest upon a soil. Elsewhere aeolian sands and soil were formed at this period of time. Mean annual temperature during this stage was 8.2–4.9 °C, biomass was very low and the moisture regime was dry, becoming moist at the later part of the substage.

Fluvial activity at this time was limited. Material transported by the rivers includes rocks from the upper parts of the catchment, beach material deposited during OISst 5a and some soil material. The existence of beach sediments and soil below Calo d'es Cans A and B suggests that fluvial erosion is negligible and that the alluvial fan continued to aggrade in this region. The absence of fluvial material of this age at Calo d'es Cans A may be due to diversion of the river around a beach ridge causing a temporary shift of the fan axis, although no evidence could be found elsewhere of a channel formed at this time.

During OIS 3 aeolian sand and loess were deposited extensively, often associated with soil formation. There is no evidence of fluvial activity at this time.

OIS 2

The greatest thicknesses of fluvial sediments were formed during OIS 2. At Calo d'es Cans A they consist of 0.22 m of gravels and 0.7 m of debris flow diamicton. At Calo d'es Cans B and Waterfront Cove there are 0.75 m and 1.0 m of fluvial gravels respectively. MS and Fe signals from the matrix of the gravels is very low although moderate values characterise the matrix of the debris flow diamicton. Mean annual temperature at this time ranged from 13.4 to 6.3 °C, the biomass was low and the moisture regime was very dry.

These sediments indicate that OIS 2 saw the most extensive fluvial activity during the Last Interglacial/Last Glacial/Holocene period with the extension and aggradation of large alluvial fans. The sedimentology associated with the period is more complicated than that of the earlier fluvial units, with debris flow processes contributing to the sediment pile close to the fan apex, and better sorted gravels and sands deposited exclusively at the fan margin. This facies variation is attributed to the transport of only the more mobile bedload to the margins of the fans. The composition of the materials suggest that the "fluvially" transported material was derived from fresh rocks in channels and the upper part of the catchment, but that the mass movement sediments were derived from surface materials closer to the fan. All these processes were associated with low temperatures, low biomass and very dry moisture regime. This latter point may be qualified as Prentice et al. (1992) has shown that although mean moisture levels may be low, seasonality of precipitation may have been increased and winter precipitation may have been relatively high.

OIS 1 (Holocene)

During OIS 1, two quite different elements contributed to river activity within the catchments: colluviation at Calo d'es Cans A, and river incision and formation of the "torrente" gully at both catchments.

The colluvium consists of a gravelly, sandy, silty diamicton with a rich terrestrial mollusc fauna, increasingly high MS values and high Fe values. A mean annual temperature of 13.8 °C has been obtained for the colluvium. The moisture regime was moist and a high biomass was present. This colluvium is interpreted as having been formed by the surface wash from actively forming soil material, probably facilitated by human disturbance of the vegetation and land surface.

Within the context of this study the "torrente" gullies are unique and indicate fluvial erosion, rather than deposition which has otherwise been dominant. It is not possible to date the formation of these channels directly as the sediments that have been eroded from them are on the seabed beyond the present coastline, or currently in the process of transportation along the channel bed, but the fact that they dissect OIS 2 fan gravels and early Holocene colluvium suggests that they formed during the later part of the Holocene when there was a relatively dense vegetation cover in the form of degraded Mediterranean forest and garrigue and relatively high winter precipitation.

DISCUSSION

The relationship between the oxygen isotope ratio for temperate latitudes (Martinson et al., 1987), the mean, maximum and minimum temperatures, the moisture characteristics, the relative biomass and the fluvial activity at the sites are summarised in Figure 5.4. It is clear from this, and the preceding summary, that fluvial activity is associated with cooler episodes (OIS 5d, 5b, 4, 2) when precipitation and biomass were relatively low. Low total precipitation values may be misleading, as intense seasonally concentrated winter precipitation (Prentice et al., 1992) may result in geomorphologically effective, infrequent, high discharges. Nevertheless, the critical factors in this relationship appear to be rapid runoff associated with limited interception and infiltration, rapid unimpeded runoff, and available sediment supply. The first two factors are likely to be explained by the breakdown of the woodland vegetation and its replacement by open, cold steppe shrub or grassland (van Andel and Tzedakis, 1996). With the climatic regimes indicated by the palaeoclimatic values in Figure 5.4 it is likely that sediment supply was enhanced by frost processes on the jointed bedrock in the upper parts of the catchments. In this respect the results from north-east Mallorca are similar to those from non-glaciated catchments elsewhere in the Mediterranean region (Roberts, 1995).

The fact that eroded soil material contributes little to the sediment yield other than in the debris flows of OIS 2 and colluvium of OIS 1, suggests that surface wash had limited effect in most of the cooler episodes, and that channel erosion in the headwater section of the catchment was more effective. Alternatively, the soil-rich debris flow

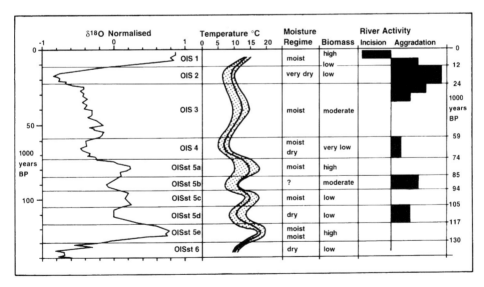

Figure 5.4 Oxygen isotope ratio (Martinson et al., 1987), mean, maximum and minimum temperatures, moisture characteristics, biomass and river activity in small catchments over the last 140 ka in north-eastern Mallorca, Spain. Also given are Oxygen Isotope Stages and Substages, after Imbrie et al. (1984). The methods of calculating temperature, and estimating moisture regime and biomass are given in Rose (1998) and Rose et al. (1999).

deposits formed during OIS 2 indicate effective surface mass movement. It is suggested that this may be due to the disturbance and mobilisation of soil material at this time by the winter growth and spring melt of ground-ice segregations. Mean annual temperatures of $c.$ 8 °C at this time indicates that this is a likely winter phenomenon, and is in accord with the observations of Butzer (1964) on cold climate landforms and sediments in the Sierra Norte of Mallorca and GCM model reconstructions for the LGM (OIS 2) (5–10 °C lower in winter; Kutzbach and Guetter, 1986). This factor may also explain the most extensive development of fluvial sediments during OIS 2.

In contrast to the explanation given above, colluviation during the earlier part of the Holocene is associated with a period of Mediterranean climate, denser natural vegetation cover and higher precipitation levels which would maintain high interception, high infiltration, throughflow and retarded runoff. In this case it is likely that human activity in the form of agricultural disturbance and grazing would be responsible for breaking down the vegetation, increasing runoff, and making available soil material (Butzer, 1976). Likewise the absence of frost processes at this time would reduce the availability of coarse clasts. The result would be that channel and bedrock erosion would be diminished, surface wash would be dominant and colluvial material would be moved from steeper to lower angle slopes.

The formation of the "torrente" gully is unique in the river history recorded, and therefore needs a unique explanation. Conceptually, this process needs either (i) a steeper gradient caused by sea-level fall or tectonic activity; or (ii) high runoff with effective channel flow, but limited sediment availability. Low sea-level is not the cause as this erosional episode took place during Holocene high sea-level, and earlier periods of low sea-level coincide with maximum river aggradation. There is no evidence of tectonic controls. However, high runoff can be generated by anthropogenically devegetated slopes. Combined with limited sediment supply due to the absence of frost shattering processes, river channel erosion is likely to be stimulated by sediment starvation after colluviation had stripped the soil from steep cultivated and grazed slopes (Macklin et al., 1995).

The absence of evidence of river activity during the intervening stages (5e, 5c, 5a and most of 3) may reflect negligible runoff or the failure to discover the evidence during this research. However, the fact that soils and aeolian sediments have been recorded for these periods, and that the soils suffered only minimal or no erosion during the subsequent fluvial episode, suggests that river activity was indeed minimal. As these periods of minimal fluvial activity are associated with higher temperatures, higher precipitation and higher biomass values, it would appear that interception and infiltration, possibly compounded by high evapotranspiration, are the dominant factors in controlling river runoff, and hence fluvial activity. It is suggested that at these times the small catchments in north-east Mallorca, and probably elsewhere in the Mediterranean region, existed as "dry valleys".

CONCLUSIONS

(i) Stacked sequences of Quaternary sediments in lowland parts of small catchments in north-east Mallorca preserve a record of fluvial activity over the last 140 ka, covering the Last Interglacial, Last Glacial and Holocene. The position of the sites within the catchment is such that depositional events are recorded and erosion has been minimal. At the sites investigated, the only significant hiatus has been due to coastal erosion by high interglacial sea-levels (OIS 5e and 1 (Holocene)). Holocene coastal erosion has created the long sections that are the basis of this study.

(ii) The small catchments indicate that changes in river activity have been brought about by changes in climate and vegetation. The sites have not been affected by glaciation, and river activity has not been affected by sea-level change. River sedimentation occurred during OIS 5d, 5b, 4, 2 and 1, and was absent or minimal during OIS 5e, 5c, 5a and most of 3 when the catchments were "dry valleys".

(iii) River activity in the form of erosion in the upper parts of the catchment and deposition in the lowland, occurred at periods of cooler climate, low precipitation and minimal biomass, although it is likely that precipitation took the form of high intensity events. The critical factors controlling river activity appear to be the breakdown of vegetation leading to reduced interception and infiltration and increased runoff. Additionally sediment supply appears to have been controlled by frost action of the jointed limestone bedrock, and segregation ice growth and melt in finer grained soil materials.

(iv) The depositional landforms consist of alluvial fans composed predominantly of sand and gravel with increments of debris flow diamicton deposited around the fan apex during OIS 2. The finer portion of the suspended load associated with the deposition of the sands and gravels was transported beyond the study site.

(v) Colluviation developed as a result of surface runoff on slopes and soils cleared or disturbed by human activity or grazing during the earlier part of the Holocene and forms a thin veneer over the fan surface.

(vi) Sediment starvation followed the early Holocene colluviation. Along with higher precipitation this resulted in late Holocene incision and the formation of the "torrente" gullies used by present-day ephemeral streams.

ACKNOWLEDGEMENTS

Thanks are expressed to Jamie Woodward and two anonymous referees for their very helpful comments on the original draft of this paper.

REFERENCES

Butzer, K. W. 1964. Pleistocene cold-climate phenomena of the Island of Mallorca. *Zeitschrift für Geomorphologie*, **8**, 7–31.

Fuller, I. C., Macklin, M. G., Passmore, D. G., Brewer, P. A., Lewin, J. and Wintle, A. G. 1996. Geochronologies and environmental records of Quaternary fluvial sequences in the Guadaloupe basin, northeast Spain, based on luminescence dating. In Branson, J., Brown, A. G. and Gregory, K. J. (Eds) *Global Continental Changes: The Context of Palaeohydrology*. Geological Society Special Publication No. 115, Geological Society, London, 99–120.

Hunt, C. O. and Gilbertson, D. D. 1995. Human activity, landscape change and valley alluviation in the Feccia Valley, Tuscany, Italy. In Lewin, J., Macklin, M. G. and Woodward, J. C. (Eds) *Mediterranean Quaternary River Environments*. A.A. Balkema, Rotterdam, 167–176.

Imbrie, J., Hays, J. D., Martinson, D. G., McIntyre, A. C., Mix, A. C., Morley, J. J., Pisias, N. G., Prell, W. I. and Shackleton, N. J. 1984. The orbital theory of Pleistocene climate: support from a revised chronology of the marine $\delta^{18}O$ record. In Berger, A. L., Imbrie, J., Hayes, J. D., Kukla, G. J. and Saltzman, B. (Eds) *Milankovitch and Climate*. Reidel, Dordrecht, 269–306.

Kutzbach, J. E and Guetter, P. J. 1986. The influence of changing orbital parameters and surface boundary conditions on climate simulations for the past 18,000 years. *Journal of Atmospheric Sciences*, **43**, 1726–1759.

Lewin, J., Macklin, M. G. and Woodward, J. C. (Eds) 1995. *Mediterranean Quaternary River Environments*. A.A. Balkema, Rotterdam.

Macklin, M. G. and Passmore, D. G. 1995. Pleistocene environmental change in the Guadaloupe basin, north-east Spain: fluvial and archaeological records. In Lewin, J., Macklin, M. G. and Woodward, J. C. (Eds) *Mediterranean Quaternary River Environments*. A.A. Balkema, Rotterdam, 103–113.

Macklin, M. G., Lewin, J. and Woodward, J. C. 1995. Quaternary fluvial systems in the Mediterranean basin. In Lewin, J., Macklin, M. G. and Woodward, J. C. (Eds) *Mediterranean Quaternary River Environments*. A.A. Balkema, Rotterdam, 1–25.

Martinson, D. G., Pisias, N. G., Hayes, J. D., Imbrie, J., Moore, T. C. and Shackleton, N. J. 1987. Age dating and the orbital theory of the ice ages: development of a high resolution 0–300,000-year chronostratigraphy. *Quaternary Research*, **27**, 1–29.

Prentice, I. C., Guiot, J. and Harrison, S. P. 1992. Mediterranean vegetation, lake levels and palaeoclimate at the Last Glacial Maximum. *Nature*, **360**, 658–660.

Roberts, N. 1995. Climatic forcing of alluvial fan regimes during the late Quaternary in the Konya Basin, south central Turkey. In Lewin, J., Macklin, M. G. and Woodward, J. C. (Eds) *Mediterranean Quaternary River Environments*. A.A. Balkema, Rotterdam, 207–217.

Rose, J. 1998. Longitudinal Palaeoenvironmental Gradients in Southern Europe during the Last Interglacial/Glacial Transition: Multiproxy Evidence at Four Reference Sites. Unpublished report to NERC.

Rose, J., Meng, X. and Watson, C. 1999. Palaeoclimate and palaeoenvironmental responses in the western Mediterranean over the last 140 ka: evidence from Mallorca, Spain. *Journal of the Geological Society*, **156**, in press.

Van Andel, T. H. 1989. Late Quaternary sea-level changes and archaeology. *Antiquity*, **63**, 733–745.

Van Andel, T. H. and Tzedakis, P. C. 1996. Palaeolithic landscapes of Europe and Environs. *Quaternary Science Reviews*, **15**, 481–500.

White, K., Drake, N., Millington, A. and Stokes, S. 1996. Constraining the timing of alluvial fan response to Late Quaternary climatic changes, southern Tunisia. *Geomorphology*, **17**, 295–304.

Woodward, J. C. 1995. Patterns of erosion and suspended sediment yield in Mediterranean river basins. In Foster, I. D. L., Gurnell. A. M. and Webb, B. W. (Eds) *Sediment and Water Quality in River Catchments*. John Wiley, Chichester, 365–389.

Section 2
CHANNEL RESPONSE

6 Impact of Major Climate Change on Coarse-grained River Sedimentation: A Speculative Assessment Based on Measured Flux

IAN REID,[1] **JONATHAN B. LARONNE**[2] **and D. MARK POWELL**[3]

[1] *Department of Geography, Loughborough University, UK*
[2] *Department of Geography and Environmental Development, Ben Gurion University of the Negev, Israel*
[3] *Department of Geography, University of Leicester, UK*

INTRODUCTION

Upon encountering a relatively abrupt change in sedimentary style within a fluvial succession, geomorphologists and sedimentologists have frequently been tempted to invoke climate-driven environmental change as the cause. The argument gains strength if the pattern can be shown to have regional significance, repeating from place to place. If there is independent evidence that corroborates the notion of a temporal shift in climate, the signal given by the sedimentary succession may be taken to be even more credible. So, for the now-defunct drainage of the western Sahel, Talbot (1980) generalises a pattern of environmental change from the sedimentary record of the terminal Pleistocene and early Holocene. He shows that the region emerged from extreme aridity as the West African seasonal monsoon was able to push further North than hitherto, bringing higher rainfall, increased vegetation cover and a change to meandering rivers that carried predominantly suspended loads of fine material (Figure 6.1). This was succeeded by aridification of the region towards the middle Holocene, the contemporary landscape then being dominated by multi-thread channels whose loads were coarser and included a substantial bedload component. The stratigraphic evidence is corroborated by the celebrated MegaChad lake-level curve that was reconstructed largely from dated littoral deposits by Servant and Servant-Vildary (1980). This illustrates a dramatic change of water-balance in the region, humidity peaking at around 9 ka BP and reflecting a temporary amelioration of the regional climate.

Similarly, although with less temporal resolution, Schumm (1968) draws out the

Fluvial Processes and Environmental Change. Edited by A. G. Brown and T. A. Quine.
© 1999 John Wiley & Sons Ltd.

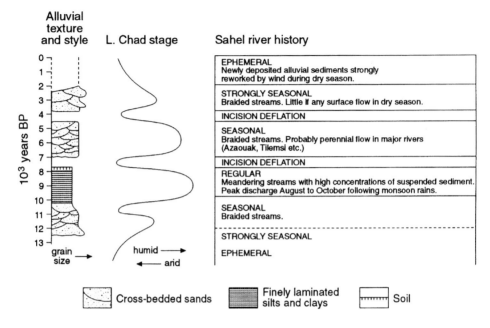

Figure 6.1 Generalised synthesis of the terminal Pleistocene and Holocene alluvial succession of the West African Sahel set against the reconstructed lake-level curve of Lake Chad (after Talbot, 1980)

distinction between the coarser-grained deposits of the arid-phase "Prior Streams" and their finer-grained successors – the "Ancestral Rivers" – in his classic study of Pleistocene drainage development on the Riverine Plain of New South Wales. Much further back in geological time, similar fairly abrupt shifts in sediment calibre from coarse to fine-grained units within a sedimentary succession have been used to infer significant changes in climate for Triassic rift basins of the northern North Sea hydrocarbon province (Frostick et al., 1988, 1992).

Notwithstanding these examples, however compelling they might be, the changes in environment and in the processes and the rates of erosion and sedimentation are inferential. To complicate matters, there are documented examples of widely differing sedimentary style that are contemporary and sit virtually side by side in the same sedimentary succession (e.g. Love, 1982).

It is only recently that *process*-based information has been available from a sufficiently diverse range of environments to test any geomorphological notions that substantial changes in environment can have a significant impact on the fluvial system. This is particularly so for the transport of coarse-grained sediment of gravel calibre, usually as bedload. We are now in a position to begin to evaluate what changes in fluvial sedimentation might be expected if the landscape were to shift in character from one that is comparatively humid to one where aridity ensures less vegetation cover and greater delivery of sediment to the river system (Figure 6.2). This newly acquired ability is timely in that there is considerable current concern about various impacts of global climate change. Just as vulnerable as any other type-environment are those sub-humid

Figure 6.2 (a) Temperate humid armoured perennial gravel-bed stream, Cowside Beck, and vegetated hillslopes, Yorkshire, England. (b) Subtropical semi-arid non-armoured ephemeral gravel-bed stream, Nahal Eshtemoa, and less vegetated hillslopes, Israel

regions that might become semi-arid should annual rainfall be reduced by only a couple of hundred millimetres.

MEASURED BEDLOAD SEDIMENT FLUX

The Information Sources

Because process-based information about bedload has been gathered painstakingly in a widening range of environmental settings, we can substitute an environmental gradient for time. However, before doing this, we have to choose our data sets with care. Because bedload flux is notoriously difficult to measure, suffering particularly from the variable efficiency of the measuring devices deployed, we have restricted our comparison to those data collected with the aid of bed-slot samplers. These devices vary in several ways, but they have one common feature: they are hydraulically "invisible", causing minimal interference with the transport process and so ensuring high levels of sampling efficiency, at least over the range of conditions for which bedload data are available. Three of the chosen data sets have been derived with vortex-tube samplers (Oak Creek: Milhous, 1973; Torlesse Stream: Hayward, 1980; Virginio Creek: Tacconi and Billi, 1987), while three come from Birkbeck-type

Table 6.1 Hydrological, hydraulic and sedimentary characteristics of gravel-bed channels in which bedload is monitored with bed-slot samplers

Characteristic	Nahal Yatir	N. Eshtemoa	Goodwin Ck	Torlesse St	Turkey Bk	Virginio Ck	Oak Ck
Annual precipitation (mm)	250	290	1400	1000	600	900	1300
Annual runoff regime	ephemeral	ephemeral	seasonal	perennial	perennial	perennial	perennial
Longitudinal slope	0.0088[a]	0.0075[b]	0.008[b]	0.067[b]	0.009[a]	0.008[b]	0.014[b]
Channel bed width (m)	3.5	6.0	3.4	3	2.9	12.0	3.7
Surface bed material D_{50S} (mm)	6[c]	16[c]	8.5	15[d]	22	28	60
Subsurface bed material D_{50SUB} (mm)	10	14	5.9	—	16	13	20
"Armour" ratio D_{50S}/D_{50SUB}	0.6	1.1	1.4	—	1.4	2.1	3
Mean bedload transport efficiency[e] (%)	17.25	13.51	2.10	0.35	0.15	0.34	0.05

[a] Average water-surface slope.
[b] Channel-bed slope.
[c] Paint-sprayed surface area sample, transformed after Kellerhals and Bray (1971) using an exponent of −1.
[d] Assumed to represent the surface layer, but not explicitly described as such.
[e] Per cent efficiency after Bagnold (1973) and defined as $100 i_b/(\omega/\tan\alpha)$, where $\tan\alpha$ is assumed constant at 0.63.

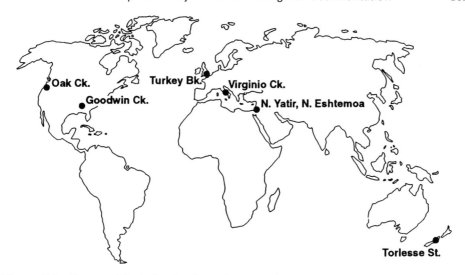

Figure 6.3 Geographical distribution of stream bedload sediment monitoring stations, the data sets of which have been used in this chapter

pressure-pillow samplers (Turkey Brook: Reid et al., 1980; Nahal Yatir, N. Eshtemoa: Laronne et al., 1992; Goodwin Creek: Kuhnle, 1992).

All of the streams have channels of similar dimensions and they all have gravel beds (Table 6.1). The East Fork River provides another potential data set (e.g. Leopold and Emmett, 1976), having been gathered with a slot sampler. It has not been included in the present analysis because its morphological and sedimentological character differs significantly from the other streams. The geographical distribution of the channels that have been chosen (Figure 6.3) gives an environmental gradient within temperate and sub-tropical zones from humid through sub-humid to semi-arid. As a result, the runoff regimes vary from perennial through seasonal to ephemeral.

Environmentally Differentiated Transport Patterns

In comparing the bedload transport data sets, several behavioural differences immediately stand out, although the contrast is strongest between the desert ephemeral streams and their perennial counterparts of more humid environments. Firstly, the relation between flow hydraulics and bedload flux is extremely complex in the perennial and seasonal streams (Figure 6.4). As a single measure of the degree of process response, correlation of bedload and stream power gives coefficients of 0.43, 0.41 and 0.61 for the perennial Torlesse Stream, Virginio Creek and Turkey Brook, respectively, and 0.52 for the seasonal Goodwin Creek. In contrast, the ephemeral gravel-bed Yatir's behaviour is much simpler, the relation between bedload flux and stream power generating a correlation coefficient of 0.93. This high degree of correlation in arid and semi-arid environments has been corroborated recently by studies of bedload transport on a neighbouring, larger gravel-bed ephemeral stream, the Nahal Eshtemoa (Reid et al., 1998). This lends confidence to the claim that desert streams behave differently (Reid and Laronne, 1995).

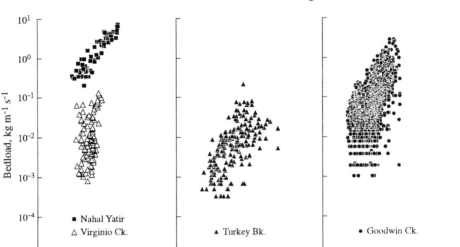

Figure 6.4 Bedload flux as a function of specific stream power in one ephemeral (Nahal Yatir), one seasonal (Goodwin Creek), and two perennial gravel-bed rivers (Turkey Brook and Virginio Creek)

The other difference to note is that bedload flux is consistently higher in the desert ephemeral stream, with the seasonal Goodwin Creek lying in an intermediate position between it and the perennial streams (Figure 6.4). Transport rates are high in the desert ephemeral stream at and just above the threshold of motion and they climb to record levels as stream power shifts through an order of magnitude, producing bedload fluxes that are two orders of magnitude higher than the perennial streams.

The reasons for these differences in transport behaviour undoubtedly lie in the supply of sediment from catchment hillslopes and the effect that this has on the channel-bed material, the immediate source of bedload. Dietrich et al. (1989) have shown through flume experiments that restricting sediment supply leads progressively to the development of a coarse armour layer and a commensurate reduction in bedload flux. Conversely, where sediment supply rates are high, as in drylands, we can expect poor or non-existent armouring (Laronne et al., 1994). Because of this, the protective surface layer that is so characteristic of perennial gravel-bed channels is absent in ephemeral streams (Figure 6.5). Of particular interest here is that, where there is a restriction of sediment supply, as is the case in an ephemeral stream that drains a catchment close to the Yatir and Eshtemoa, channel-bed armouring develops (Hassan, 1990). Although the number of case studies is not large, this range of bed conditions provides a potentially useful field corroboration of the Dietrich et al. (1989) model. It is important that the case studies are from the same topographical and lithological province.

Because armour development is weak or non-existent and the surface grain-size is commensurably small, ephemeral streams are much more efficient at conveying sediment than their perennial counterparts. A comparison of different channels is always

Figure 6.5 Vertical or near-vertical sections through surface and subsurface layers of (a) an armoured perennial gravel stream bed and (b) an unarmoured ephemeral gravel stream bed

fraught with difficulties if only because the number of variables that contribute to the permutation of process–response is large. However, bearing this in mind, we can parametrise stream behaviour using Bagnold's (1973) efficiency index, $E_b = 100i_b/(\omega/\tan\alpha)$, in which i_b is unit bedload flux, ω is specific stream power and α is the angle of internal friction. (Tan α is assumed constant at 0.63 following Bagnold, but the reservations implied in the work of Buffington *et al.* (1992) should be noted.) Mean E_b is inversely related to bed material surface size (here characterised by the median surface grain size, D_{50}; Figure 6.6). D_{50} might be taken to be a determinant of E_b and of interest here is that the Yatir and Goodwin Creek plot away from, and above, the clutch of armoured perennial streams.

It would be tempting to take this inverse relationship as representing an environmental gradient. But this can only be done by knowing that large clasts are indeed present in the bed materials of the most efficient streams and that they might well form a surface armour which would reduce efficiency were it not for the abundance of sediment of all sizes being supplied to the channels of these drier environments. This supply counteracts the winnowing effects of size-selective transport and the preferential movement of finer clasts that would otherwise lead to surface armouring. In fact, the ephemerals are transporting a capacity or near-capacity load whereas the perennials are variously inhibited by their armour layers (Reid *et al.*, 1996). We have attempted elsewhere to collapse the data of the various streams onto a single curve by plotting dimensionless transport rates against dimensionless boundary shear stress

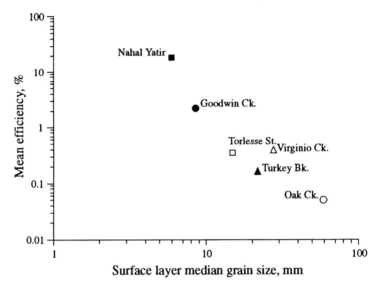

Figure 6.6 Mean bedload transport efficiency (after Bagnold, 1973) plotted against median surface bed material grain size in six gravel-bed streams that represent an environmental gradient ranging from temperate humid to subtropical semi-arid

(Reid and Laronne, 1995). With differences in grain size taken into account, the results still show the ephemeral Yatir riding above its seasonal and perennial counterparts. Furthermore, because this desert stream is working at transport capacity, its performance is more predictable and the relation between sediment transport and hydraulics is much simpler (Figure 6.4).

A corollary of this greater efficiency might be the accumulation of thicker sequences of coarse beds of fluvial sediments in semi-arid environmental settings over time. This, plus a higher preservation potential in drylands, might be a reason for the often remarkable prevalence of coarse river deposits in the sedimentary sequences of ancient deserts (Allen, 1964; Tunbridge, 1984; Frostick *et al.*, 1988).

SPECULATIVE CHANGES IN BEDLOAD YIELD

Notwithstanding the infrequent occurrence of floods in desert and semi-desert areas, high transport efficiency produces record levels of bedload sediment yield (Powell *et al.*, 1996). The simplicity of the bedload–stream power relation allows us the confidence to use a rating-curve approach to calculate yield. Any reservations about the veracity of the result will derive rather from our use of a flow duration curve which is constructed from a return period of only four years. Indeed, Schick and Lekach (1993) show the importance of record length in assessing dryland sedimentation. With this caution in mind, the mean bedload sediment yield for the 119 km^2 catchment of the ephemeral Nahal Eshtemoa is calculated to be 39 t km^{-2} year^{-1}.

For vegetated (some heavily wooded) catchments of the Ardennes, Belgium, which have approximately the same catchment area and are characterised by gravel-bed

Table 6.2 Bedload sediment yield of gravel-bed rivers

River	Location	Environment	Catchment (km^2)	Bedload sediment yield (Mg km^{-2} year^{-1})
Eshtemoa[a]	Israel	Subtropical semi-arid	119	39
Wamme[b]	Belgium	Temperate humid	139	2.21
Lhomme[b]			276	1.11
Ambieve[b]			1044	0.43
Ourthe[b]			1285	0.38

[a] After Powell et al. (1996).
[b] After Petit et al. (1996).

channels, Petit et al. (1996) provide estimates of bedload sediment yield that are only 3–6% that of the Eshtemoa (Table 6.2). These differences between environments are large. If we use them as a surrogate for environmental change in a notional time–space substitution, we can suggest that a significant drift in coarse sediment production should indeed be expected following landscape changes of the sort of magnitude known to have occurred during the Quaternary and earlier.

The lessons that we can learn from these comparisons are not only of value in interpreting ancient sedimentary sequences. Neither are they only useful in understanding how rivers as one element of the erosional system might change under future global warming scenarios. They also help us gauge the likely impact that present and future human-induced landscape changes might have on sediment supply to the river system. In the context of suspended sediment, Inbar (1992) has drawn out distinct differences in sediment yield between Old and New World Mediterranean type-climate regions. He suggests that the superior yields of the New World relate in part to the fact that human impact on the landscape is more recent and that the system is still disequilibrated under comparatively newly reduced vegetation cover. For Redwood Creek in northern California, Lisle and Madej (1992) show that clear-felling and the resulting movement of material into the river channel leads to a fining of the bed material as the armour layer is swamped. As a corollary, they report a very high bedload sediment yield of 78 tonne km^{-2} year^{-1}, albeit over only two years of observations. This is twice the average reported for Nahal Eshtemoa. Perhaps as a means of confirming the principle that changes in landscape have significant sedimentary consequences, an opposite trend is perceived for situations where the proportion of human-induced bare soil in a catchment is decreasing. Kuhnle et al. (1996) show for Goodwin Creek, Mississippi, that a decrease in agriculture and an increase in both pasture and woodland over a 9-year period has led to a significant reduction in stream sediment transport, especially of finer material.

CONCLUSIONS

A comparison of bedload transport rates in gravel-bed streams of similar size which drain landscapes that range widely in environmental character from temperate humid to subtropical semi-arid reveals significant differences in behaviour. The reasons for these differences lie in the rate of supply of sediment and the consequent degree of surface armouring that develops in the channel bed material. Dryland ephemeral streams are unarmoured and, as a result, they are characterised by high rates of bedload transport. By comparison, armoured perennials show a diminutive and disorganised response to changes in flow strength. The contrast provides a process-informed basis for invoking climate change as one of several possible factors behind shifts in sedimentary style within an ancient fluvial succession. However, the lessons we can learn are not only retrodictive. The range in sediment yield helps instruct us about present-day water resource issues such as sedimentation rates behind water impoundment structures in different environmental settings. We can also use the information to predict future changes in the efficiency of the fluvial sediment delivery system where the driving force behind evolving landscape is either climate change or human agency.

ACKNOWLEDGEMENTS

We are grateful to colleagues who gave us access to hard-won data sets of bedload flux: Paolo Billi (Virginio Creek); the late John Hayward (Torlesse Stream); Roger Kuhnle (Goodwin Creek); and Bob Milhous (Oak Creek). Their dedication to fieldwork and the information they have acquired have advanced our understanding of how the real world works.

REFERENCES

Allen, J. R. L. 1964. Studies in fluviatile sedimentation: 6. Cyclothems from the lower Old Red Sandstone, Anglo-Welsh Basin. *Sedimentology*, **3**, 89–108.

Bagnold, R. A. 1973. The nature of saltation and of 'bedload' transport in water. *Proceedings of the Royal Society of London*, **A332**, 473–504.

Buffington, J. M., Dietrich, W. E. and Kirchner, J. W. 1992. Friction angle measurements on a naturally formed gravel streambed: implications for critical boundary shear stress. *Water Resources Research*, **28**, 411–425.

Dietrich, W. E., Kirchner, J. W., Ikeda, H. and Iseya, F. 1989. Sediment supply and the development of the coarse surface layer in gravel-bedded rivers. *Nature*, **340**, 215–217.

Frostick, L. E., Reid, I., Jarvis, J. and Eardley, H. 1988. Triassic sediments of the Inner Moray Firth, Scotland: early rift deposits. *Journal of the Geological Society, London*, **145**, 235–248.

Frostick, L. E., Linsey, T. K. and Reid, I. 1992. Tectonic and climatic control of Triassic sedimentation in the Beryl Basin, northern North Sea. *Journal of the Geological Society, London*, **149**, 13–26.

Hassan, M. A. 1990. Scour, fill and burial depth of coarse material in gravel bed streams. *Earth Surface Processes and Landforms*, **15**, 341–356.

Hayward, J. A. 1980. *Hydrology and Stream Sediment from Torlesse Stream Catchment*. Tussock Grasslands and Mountain Lands Institute Special Publication, Lincoln College, Canterbury.

Inbar, M. 1992. Rates of fluvial erosion in basins with a Mediterranean type climate. *Catena*, **19**, 393–409.

Kellerhals, R. and Bray, D. I. 1971. Sampling procedures for coarse fluvial sediments. *Journal of the Hydraulic Division, Proceedings of the American Society of Civil Engineers*, **98**, 1165–1180.

Kuhnle, R. A. 1992. Bedload transport during rising and falling stages on two small streams. *Earth Surface Processes and Landforms*, **17**, 191–197.

Kuhnle, R. A., Bingner, R. L., Foster, G. R. and Grissinger, E. H. 1996. Effect of land use changes on sediment transport in Goodwin Creek. *Water Resources Research*, **32**, 3189–3196.

Laronne, J. B., Reid, I., Yitshak, Y. and Frostick, L. E. 1992. Recording bedload discharge in a semiarid channel, Nahal Yatir, Israel. In Bogen, J., Walling, D. E. and Day, T. J. (Eds) *Erosion and Sediment Transport Monitoring Programmes in River Basins*. International Association of Hydrological Sciences Publication 210, 79–86.

Laronne, J. B., Reid, I., Yitshak, Y. and Frostick, L. E. 1994. The non-layering of gravel streambeds under ephemeral flood regimes. *Journal of Hydrology*, **159**, 353–363.

Leopold, L. B. and Emmett, W. W. 1976. Bed load measurements, East Fork River, Wyoming. *Proceedings of the National Academy of Sciences*, **73**, 1000–1004.

Lisle, T. E. and Madej, M. A. 1992. Spatial variation in armouring in a channel with high sediment supply. In Billi, P., Hey, R. D., Thorne, C. R. and Tacconi, P. (Eds) *Dynamics of Gravel-bed Rivers*. John Wiley, Chichester, 277–291.

Love, D. W. 1982. Quaternary fluvial geomorphic adjustments in Chaco Canyon, New Mexico. In Rhodes, D. D. and Williams, G. P. (Eds) *Adjustments of the Fluvial System*. George, Allen and Unwin, London, 277–308.

Milhous, R. T. 1973. Sediment transport in a gravel-bottomed stream. Unpublished PhD. Thesis, Oregon State University, Corvallis.

Petit, F., Pauquet, A. and Pissart, A. 1996. Fréquence et importance du charriage dans des rivières à charge de fond caillouteuse. *Géomorphologie: Relief, Processus, Environment*, **2**, 3–12.

Powell, D. M., Reid, I., Laronne, J. B. and Frostick, L. E. 1996. Bed load as a component of sediment yield from a semiarid watershed of the northern Negev. In Walling, D. E. and Webb, B. (Eds) *Erosion and Sediment Yield: Global and Regional Perspectives*. International Association of Hydrological Sciences Publication 236, 389–397.

Reid, I. and Laronne, J. B. 1995. Bed load sediment transport in an ephemeral stream and a comparison with seasonal and perennial counterparts. *Water Resources Research*, **31**, 773–781.

Reid, I., Layman, J. T. and Frostick, L. E. 1980. The continuous measurement of bedload discharge. *Journal of Hydraulic Research*, **18**, 243–249.

Reid, I., Powell, D. M. and Laronne, J. B. 1996. Prediction of bed-load transport by desert flash floods. *Journal of Hydraulic Engineering*, **122**, 170–173.

Reid, I., Laronne, J. B. and Powell, D. M. 1998. Flashflood and bedload dynamics of desert gravel-bed streams. *Hydrological Processes*, **12**, 543–557.

Schick, A. P. and Lekach, J. 1993. An evaluation of two ten-year sediment budgets, Nahal Yael, Israel. *Physical Geography*, **14**, 225–238.

Schumm, S. A. 1968. River adjustment to altered hydrologic regime – Murrumbidgee River and paleochannels, Australia. *US Geological Survey Professional Paper* 598.

Servant, M. and Servant-Vildary, S. 1980. L'environment quaternaire de bassin du Tchad. In Williams, M. A. J. and Faure, H. (Eds) *The Sahara and the Nile*. A.A. Balkema, Rotterdam, 133–162.

Tacconi, P. and Billi, P. 1987. Bed load transport measurements by vortex-tube trap on Virginio Creek, Italy. In Thorne, C. R., Bathurst, J. C. and Hey, R. D. (Eds) *Sediment Transport in Gravel-bed Rivers*. John Wiley, Chichester, 583–606.

Talbot, M. R. 1980. Environmental responses to climatic change in the West African Sahel over the past 20 000 years. In Williams, M. A. J. and Faure, H. (Eds) *The Sahara and the Nile*. A.A. Balkema, Rotterdam, 37–62.

Tunbridge, I. P. 1984. Facies model for a sandy ephemeral stream and clay playa complex: the Middle Devonian Trentishoe Formation of north Devon, UK. *Sedimentology*, **31**, 697–715.

7 Modelling and Monitoring River Response to Environmental Change: The Impact of Dam Construction and Alluvial Gravel Extraction on Bank Erosion Rates in the Lower Alfios Basin, Greece

A. P. NICHOLAS,[1] J. C. WOODWARD,[2] G. CHRISTOPOULOS[2] and M. G. MACKLIN[2]
[1] Department of Geography, University of Exeter, UK
[2] School of Geography, University of Leeds, UK

INTRODUCTION

River channels continually adjust their morphology in response to the hydrological regime and sediment load imposed upon them by upstream drainage basin conditions. In this way the river evolves towards a form that allows the supplied sediment to be transported without major changes in channel morphology (Schumm, 1969; Hickin, 1983). The extent to which such quasi-equilibrium conditions are ever attained is currently the subject of much debate (Lewin et al., 1988). However, the related argument that a major shift in either sediment or water supply will result in a period of substantial channel adjustment is more generally accepted.

Human impacts upon water and sediment supply to rivers may promote channel change over a range of temporal and spatial scales. Channel incision and widening may be caused by increased river discharges resulting from catchment urbanisation (Park, 1977; Whitlow and Gregory, 1989) or by increased channel slopes following river channelisation (Brookes, 1985; Erskine, 1992). Erosion of channel bed and banks also occurs as part of the cycle of aggradation and degradation associated with the passage of large-scale sediment slugs in river systems (Knighton, 1989; Nicholas et al., 1995) and as a result of reduced sediment supply from upstream following reservoir construction (Petts, 1979; Rasid, 1979; Chang, 1989; Xu, 1990).

Fluvial Processes and Environmental Change. Edited by A. G. Brown and T. A. Quine.
© 1999 John Wiley & Sons Ltd.

ALLUVIAL GRAVEL EXTRACTION

Gravel extraction from channel and floodplain systems represents a profound and rapid environmental change which often promotes spectacular channel adjustment (Bull and Scott, 1974; Harvey and Schumm, 1987; Collins and Dunne, 1989; Kondolf, 1994). The geomorphological response to such intervention can be rapid and may involve channel incision, bank instability, channel widening or narrowing and changes in channel planform. Many published examples document experiences in US rivers, particularly in California, where alluvial gravel extraction is a major supplier of aggregate to the construction industry. In the south-western United States, as in the Mediterranean basin, intensive gravel mining is often a seasonal activity and takes place in summer during low stages. In-stream gravel mining has also been used as a flood management strategy to prevent channel avulsion and overbank flooding in rapidly aggrading high bedload yield systems (Kondolf, 1994). However, where rates of gravel extraction greatly exceed rates of upstream bedload input, channel bed degradation and bank erosion may proceed rapidly. The degradation process is often accelerated following destruction of the bed armour. The geomorphological responses are often exacerbated in situations where dam construction prevents upstream delivery of bedload into the mined reaches (e.g. Kondolf and Swanson, 1993). Where extraction operations take place at several sites along the channel they tend to have a cumulative geomorphological impact.

Gravel extraction is common throughout Greece and the wider Mediterranean region where low stage gravel bed rivers provide easy access to a valuable natural resource. The regulation of these operations is often ineffective and in addition to some of the geomorphological responses outlined above, excessive exploitation may lead to falling water tables, poor water quality and the deterioration of channel and riparian habitats. At present there is little published information on the response of Mediterranean river systems to in-stream gravel mining. It is important to understand the response of these systems to such recent intervention and to minimise the environmental degradation which frequently results.

THE ALFIOS RIVER BASIN

The Alfios is the largest river on the Peloponnese (with a main channel length of 112 km) and drains an area of 3600 km^2 (Argiropoulos, 1960). The basin encompasses a wide range of terrains including the high mountains of the central Peloponnese where elevations locally exceed 2400 m. The lower portion of the basin downstream of the classical site of Olympia forms the basis of the present study (Figure 7.1). The climate of the region is of marine Mediterranean type and good precipitation records are available for a number of locations within the basin and for the nearby city of Pirgos where the mean annual rainfall is 952 mm. River flows were monitored for 17 years at the Alfioussa railway bridge between 1949 and 1966, and the mean annual discharge has been estimated as 55 m^3 s^{-1} (Ministry of Environment of Greece, 1966). This flow record was used in the design of the Flokas Dam which was completed in

Monitoring River Response to Environmental Change 119

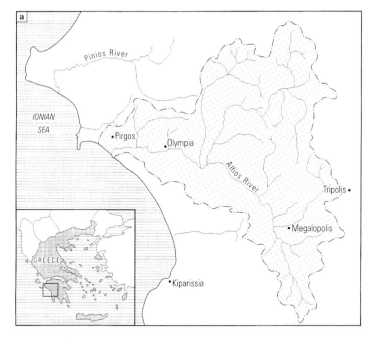

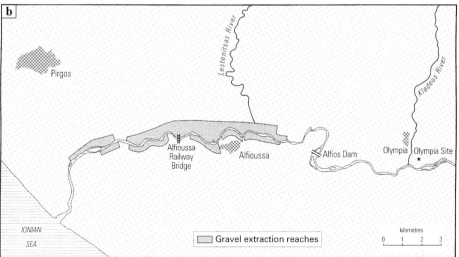

Figure 7.1 The River Alfios illustrating the Flokas (Alfios) Dam and areas of alluvial gravel extraction

1968. This concrete structure impounds the main channel approximately 5.5 km downstream of Olympia. The dam was built to provide irrigation waters from April to October. The maximum irrigation capacity is $15 \, m^3 \, s^{-1}$ and the irrigated area covers $160 \, km^2$ of the lower basin. The dam wall is about 315 m in length and its top stands 20.70 m above sea-level. The foundations of the dam lie in Quaternary-age sands and gravels which are underlain by Pliocene marine silts. The reservoir prevents bedload sediments from being transported to the reaches downstream. At present there are no data on the sedimentation rate in the reservoir.

ALLUVIAL GRAVEL EXTRACTION IN THE LOWER ALFIOS BASIN

Intensive gravel extraction began in 1967 immediately prior to the completion of the Flokas Dam. A single company operated between 1967 and 1974 and it has been estimated that they extracted around 1 million m^3 of gravel each year. In 1974 their permit was withdrawn. Since 1967 a total of eight major companies have been engaged in gravel extraction from the active channel and floodplain zone (volumetric rates of extraction for the period 1967–1995 are shown in Figure 7.2). Large amounts of gravel extraction continued until the mid-1990s but by the end of 1996 gravel extraction had ceased in most locations. Over this period it is estimated that the combined activities of these companies has resulted in the extraction of around 17.37 million m^3 of gravel. Gravel was extracted throughout the year from the bed of the main channel by cranes with "drag buckets" and from the active floodplain zone during summer low flow periods. Figure 7.2 also shows the estimated cumulative depth of degradation, within the channel reach where the extraction works have operated, over the period of interest. Degradation depths were approximated by dividing the cumulative extrac-

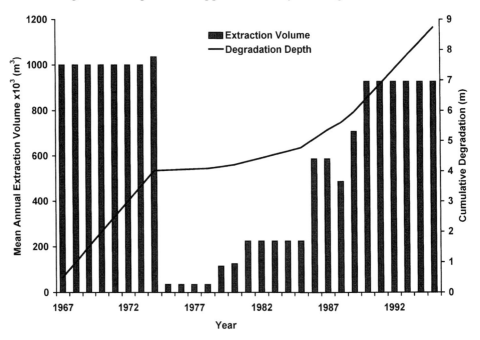

Figure 7.2 Volumetric rates of gravel extraction for the River Alfios between 1967 and 1995, and the estimated cumulative depth of channel degradation within the gravel extraction region

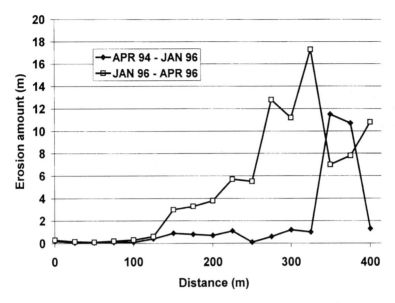

Figure 7.3 Monitored rates of bank retreat for a 400 m length of channel located 2.5 km downstream of the dam

tion volume by the width of the floodplain (250 m) and the length of channel (8 km) over which gravel was extracted. This provides a first approximation of the depth of incision due to gravel extraction which is consistent with estimates derived from channel cross-section surveys (see Figure 7.5).

Channel bank positions have been surveyed on nine occasions over the past three years (April 1994–July 1997). Figure 7.3 shows the monitored rates of bank retreat over two time periods along a 400 m length of channel located approximately 2.5 km downstream of the dam, and Figure 7.4 shows the channel bank and floodplain in the region of this monitoring. The data presented in Figure 7.3 illustrate the high degree of temporal and spatial variability exhibited by bank retreat rates throughout the study reach. Maximum rates of bank erosion of approximately 10 m year^{-1} are the result of laterally extensive bank failures which are associated with high-magnitude flow events. The exact timing of bank failures is difficult to establish because it has not been possible to make direct observations during large flood events. Average rates of bank erosion within this part of the study reach are about 3–4 m year^{-1}. The extent to which these rates of lateral erosion are typical of bank retreat over the past 30 years is difficult to determine due to the high temporal and spatial variability exhibited by the data. However, analysis of the 1988 air photograph indicates that such rates of bank retreat may be representative of the last decade.

Figure 7.5 shows channel cross-sections at three locations downstream of the dam. These sections are representative of the downstream change in morphology evident throughout the study reach and indicate a substantial and progressive increase in channel width downstream of the dam. Outflow from the dam occurs in a number of channels which quickly converge to form a single-thread river. Moving downstream the river develops alternate bars which become more extensive point bar surfaces

122　　　　Fluvial Processes and Environmental Change

Figure 7.4 Unstable bank section on the Alfios River downstream of the Flokas Dam (photograph taken in September 1994 by Jamie Woodward)

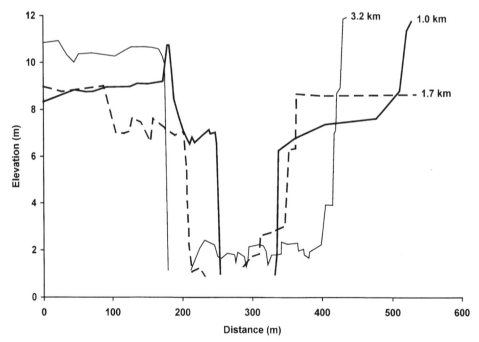

Figure 7.5 Cross-section surveys of the channel and floodplain at three locations downstream of the dam

approximately 2 km downstream. Evidence from air photographs taken in 1960 suggests that at the upstream sections (i.e. within 1 km of the dam) substantial widening of the main channel has not occurred. However, at the downstream end of the study reach (i.e. 3.2 km downstream of the dam), near the region of extensive gravel extraction, the channel may have widened by up to 100 m over the last 30 years. This suggests a mean rate of channel widening of approximately 3–4 m year^{-1} which is consistent with current rates of bank retreat monitored close to this section (discussed above). On the basis of these data it appears that while marked downstream variations in channel widening have occurred, current rates of bank retreat may be similar to average rates since 1967, thus suggesting that temporal variations in lateral erosion rates may have been relatively insignificant. The extent to which the downstream increase in channel width reflects a similar spatial trend in patterns of bed incision is difficult to establish from the cross-section surveys. However, the level of the pre-dam closure floodplain surface may be identified from the position of a flood embankment built in 1965 in response to a series of overbank flood events which occurred in previous decades. The height difference between the former floodplain and the current river channel suggests an average incision depth of 6–8 m which agrees with the estimate derived from the gravel extraction data. In an attempt to develop a better understanding of the factors controlling the evolution of the Alfios and to examine the relationships between temporal and spatial patterns of channel widening and incision, we have developed a simple one-dimensional (1D) model of flow hydraulics, sediment transport and channel change.

MODEL DEVELOPMENT

Several numerical models of river channel evolution have been developed recently which aim to simulate the processes of river bed degradation and bank erosion (e.g. Osman and Thorne, 1988; Chang, 1989; Pizzuto, 1990; Mosselman, 1995; Darby and Thorne, 1996). These models have been applied mainly to straight river channels in an attempt either to determine equilibrium channel dimensions or to investigate channel adjustment following a change in water and/or sediment supply conditions. Changes in these boundary conditions generally promote either net aggradation or degradation at a channel section. Models of river channel evolution have also been developed which simulate the migration of channels with a meandering planform where bank erosion is balanced by point bar deposition such that no net sediment imbalance results (e.g. Howard and Knutson, 1984; Johannesson and Parker, 1989; Odgaard, 1989; Howard, 1992).

In this study we present a simple 1D model of flow hydraulics, sediment transport and morphological change for meandering channels experiencing net degradation. The model is applied to a 4 km reach of channel downstream of the Alfios Dam which consists of 41 cross-sections spaced at 100 m intervals downstream. A number of simulation have been carried out to explore the sensitivity of model performance to variations in values of the important model parameters. The model has also been employed to provide further insight into the evolution of the River Alfios over the

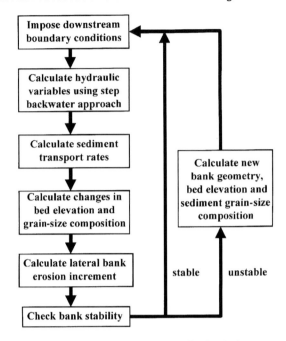

Figure 7.6 Flow chart illustrating the sequence of calculation steps carried out in the model

past three decades. A summary of the main model components is given below. Figure 7.6 illustrates the sequence of calculation steps carried out in the model.

Figure 7.7(a) shows a representation of the cross-sectional geometry within the model. The geometry of the inner and outer banks of the meandering channel reflects the processes of lateral and vertical erosion and the geometry of bank failure planes (shown in Figure 7.7(b) and described below). The channel bed has a lateral slope (S_Y) that is predicted as a simple function of lateral changes in flow depth and water surface height. Bridge (1977) presents the following equation relating lateral variations in flow depth and channel curvature

$$\frac{dh}{dR} = \frac{11h \tan \phi}{R} \tag{1}$$

where h is the flow depth, R is the radius of curvature of the bend, and $\tan \phi$ is the dynamic friction coefficient of the sediment which takes a value of approximately 0.45 (Bridge, 1976). The lateral water surface slope (S_W) resulting from super-elevation of the water surface is given by

$$S_W = \frac{v^2}{gR} \tag{2}$$

where v is the section-averaged flow velocity and g is the acceleration due to gravity.

a) Cross-sectional geometry

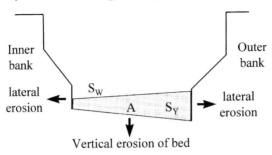

b) Bank failure block geometry

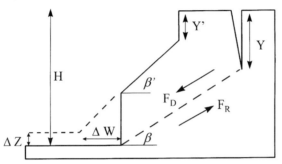

Figure 7.7 Model representations of (a) the cross-sectional geometry of the channel, and (b) the geometry of the bank failure block

The hydraulic component of the model employs a step-backwater approach to route flow through the model cross-sections (cf. Chow, 1973). Newton-Raphson iteration is used to solve the following 1D equations of mass and momentum continuity

$$Q = vA \qquad (3)$$

$$-S_f = \frac{\partial z_0}{\partial x} + \frac{\partial h_0}{\partial x} + \frac{1}{2g}\left(\frac{\partial v^2}{\partial x}\right) \qquad (4)$$

where Q is the discharge, v is the velocity, A is the cross-sectional area of the flow, z_0 is the centreline bed elevation and S_f is the centreline energy slope in the downstream (x) direction. Velocity is determined using the Darcy-Weisbach equation

$$v = \sqrt{\frac{8grS_f}{f}} \qquad (5)$$

$$\sqrt{\frac{1}{f}} = m(r/D_{50})^c \qquad (6)$$

where r is the hydraulic radius of the channel, f is the Darcy-Weisbach friction factor, m and c are empirical coefficients, and D_{50} is the 50th percentile of the bed sediment. Equation (6) is a power law approximation of a Keulegan-type resistance law (Ferguson, 1994). Studies by Griffiths (1981) and Thompson and Campbell (1979) suggest values for m and c of approximately 1.25 and 0.33, respectively.

The model employs six sediment size fractions. The proportion of sediment in each fraction for the channel bed and banks at the start of each simulation was specified using an average grain-size distribution determined from samples of point bar and bank sediments obtained using standard techniques (Church et al., 1987). The downstream sediment transport rate per unit width (qs_i) for the ith size fraction is determined using a modified form of Bagnold's (1986) relation

$$qs_i = \sigma p_i (\omega - \omega_0)^{3/2} (h_0/h_r)^{-2/3} (d_i/D_r)^{-1/2} \qquad (7)$$

$$\omega = v\tau \qquad (8)$$

$$\omega_0 = 5.75[0.04(\gamma_s - \gamma)\rho]^{3/2} (g/\rho)^{1/2} d_i^{3/2} \log(12h_0/d_i) \qquad (9)$$

where σ is an empirical parameter (discussed in the results section below), p_i is the proportion of sediment in the ith fraction, ω is the specific stream power, ω_0 is the critical specific stream power for entrainment, h_0 is the centreline flow depth, d_i is the effective sediment size, h_r and D_r are reference values of the flow depth and sediment size (assigned values of 0.1 and 0.0011 after Bagnold, 1980), and γ_s and γ are the specific gravity of sediment and water respectively. Relative size effects are incorporated within the model by determining the effective particle size (d_i) using a relation of the form of an Andrews type hiding function

$$d_i = D_{50}(D_i/D_{50})^\alpha \qquad (10)$$

where D_i is the actual size of the ith size fraction and α takes a value of between 0 and 1 (0 yields perfect equal mobility whereas 1 produces size selective transport). On the basis of evidence from field studies (e.g. Andrews, 1983; Andrews and Erman, 1986; Komar, 1987), which suggest a value for α between 0 and 0.35, α is set to 0.1 which promotes slightly size-selective transport. Rates of vertical aggradation/degradation are assumed to be uniform across the channel width and are determined from the mass continuity equation

$$-(1 - \lambda)W \frac{\partial z_0}{\partial t} = \frac{\partial Q_T}{\partial x} + B \qquad (11)$$

where λ is the sediment porosity, Q_T is the total volumetric sediment transport rate for the six size fractions, W is the channel width and B is the lateral input of sediment from the erosion and collapse of the channel banks. The model currently employs a simplified version of the sediment routing schemes proposed in recent studies (e.g. Willets et al., 1987; Hoey and Ferguson, 1994). Grain-size information is stored for a series of bed layers and exchange of sediment occurs between the topmost, active layer and the eroded/deposited material. The active layer depth (L) is determined as a function of particle size

$$L = 2D_{84} \qquad (12)$$

Following aggradation/degradation, the grain-size distribution and layer thickness are recalculated for each bed layer from the bed surface downwards.

The rate of lateral erosion of the bank toe is determined as a simple function of excess shear stress

$$\frac{\partial W}{\partial t} = E\left(\frac{\tau_b - \tau_c}{\tau_c}\right) \tag{13}$$

where E is an empirical coefficient and τ_c and τ_b are the critical and near bank shear stresses respectively

$$\tau_c = \theta(\rho_s - \rho)gD_{50} \tag{14}$$

$$\tau_b = \rho g h S_f \left(\frac{R \pm W/2}{R}\right) \tag{15}$$

where θ is Shields parameter, and ρ_s and ρ are the densities of sediment and water. Vertical and lateral erosion of the bank toe result in increases in bank height and steepness. During each time step an Osman and Thorne (1988) type stability analysis is carried out on channel banks to determine the timing of bank failures. Figure 7.7(b) shows the geometry of the channel bank failure block. The stability of the bank is determined from a factor of safety (FS)

$$FS = F_R/F_D \tag{16}$$

where F_R and F_D are the forces resisting and driving the failure

$$F_R = Cl + Wt\cos\beta\tan\phi \tag{17}$$

$$F_D = Wt\sin\beta \tag{18}$$

where C and ϕ are the bank material cohesion and angle of friction respectively, l is the length of the inclined portion of the failure block, and Wt is the weight of the failure block

$$l = \frac{H - Y}{\sin\beta} \tag{19}$$

$$Y = KH \tag{20}$$

$$Wt = \frac{\delta}{2}\left[\frac{(1 - K^2)H^2}{\tan\beta} - \frac{((H - \Delta z - \Delta W\tan\beta')^2 - Y'^2)}{\tan\beta'}\right] \tag{21}$$

where δ is the unit weight of bank material, Y is the depth of the bank tension crack, K is the ratio of the tension crack depth to the bank height (H), β is the angle of the failure plane, and Δz and ΔW are the channel depth and width increments, due to vertical and lateral erosion, since the last failure (superscript ' indicates the value of the variable at the time of the last bank failure). Progressive vertical and lateral erosion of the bank results in a reduction in FS until it falls below a critical value of one, at which point the bank fails. When this occurs sediment is supplied to the channel bed (bed elevations and grain-size distributions are recalculated) and the geometry of the bank is redefined. The bank erosion and failure component of the model includes a number of empirical parameters (e.g. E, K, C) which are likely to be characterised by significant temporal variability (both seasonal and longer term) over the course of a simulation.

Owing to a lack of field data with which to determine values for these parameters, appropriate values were taken from previous bank erosion studies (Osman and Thorne, 1988; Darby et al., 1996). Model sensitivity to variations in these parameters is considered in the next section.

SIMULATION RESULTS

The simulation model employs three boundary conditions at the upstream and downstream limits of the study reach. These are a discharge–flow duration curve derived from available flow data for the Alfios, a condition of zero sediment supply to the upstream channel section (simulating the effect of the dam) and an imposed rate of bed degradation at the downstream channel section (simulating the effect of gravel extraction in the area immediately downstream of the study reach). The amount of degradation at the downstream cross-section during each time step was determined from the estimated cumulative degradation curve shown in Figure 7.2. Values for the initial channel width and radius of curvature were determined from air photographs. Channel slope was estimated from a survey of the 1968 floodplain surface. Bed sediment grain-size distributions were determined for samples taken from modern point bar and bank sediments. Initial channel depths were estimated using the flow duration data in conjunction with the hydraulic equations outlined above.

Figure 7.8 shows the simulated changes in channel morphology over the period from 1967 to 1996 for the cross-section located 3.2 km downstream of the dam. The total amount of incision at this section over the course of the simulation is approximately 6.5 m (see Figure 7.8(a)). Two periods of rapid bed degradation occur at this section (pre-1975 and post-1987) corresponding to the times of peak gravel extraction further downstream. Figure 7.8(b) shows the temporal changes in the factor of safety (FS) of the inner and outer banks at this section. FS declines rapidly initially, then more slowly, as vertical and lateral erosion occur at the bank toe. When FS falls below one, the bank fails and FS increases in response to the creation of a new stable bank geometry. A total of 20 failures occurred over the course of the simulation. Outer bank failures are twice as frequent as inner bank failures due to the greater height and inherent instability of the former. The difference in the height of inner and outer banks is directly related to the cross-stream bed gradient which is itself a function of planform curvature (see equation (1)). For typical values of channel width and curvature for the Alfios (100 m and 1 km respectively) this yields a vertical height difference of the order of 1–2 m between the inner and outer banks. Each bank failure promotes a stepped increase in the channel width. Figure 7.8(c) shows the increase in the dimensionless channel width (width/initial width) during the simulation. Clearly the frequency of bank failures and associated increases in channel width can be related to temporal variations in the rate of bed incision. Figure 7.8(d) shows the simulated changes in the D_{84} of the active bed surface layer. During the first three years of the simulation the D_{84} increases from approximately 79 mm to 95 mm. This reflects the preferential removal of fine sediment during bed degradation promoted by the use of a value of $\alpha > 0$ in equation (10). Thereafter the D_{84} remains roughly constant although exhibiting temporal fluctuations which reflect two factors. First, each bank failure is asso-

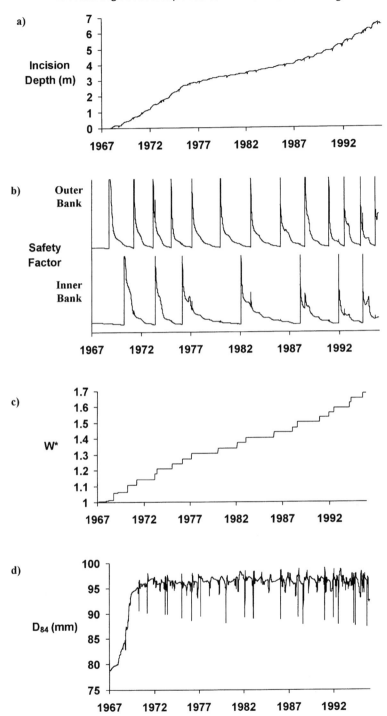

Figure 7.8 Simulated changes in channel morphology for a cross-section 3.2 km downstream of the dam over the period 1967–1996: (a) depth of incision; (b) the factor of safety of the inner and outer channel banks; (c) dimensionless channel width; and (d) the D_{84} of the active bed layer

ciated with a small negative spike in Figure 7.8(d) (the D_{84} is temporarily reduced by 5–10 mm). This results from the supply of sediment to the bed from the bank during a failure (the bank is composed of finer sediment than the channel bed). Second, in the intervening period between failures, smaller-scale fluctuations in the D_{84} occur ($\pm$ 1–5 mm), reflecting the downstream passage of pulses of fine sediment associated with upstream bank failures.

Figure 7.9 shows the downstream patterns of dimensionless channel width and incision depth over the course of a simulation. Several simple trends can be observed in the model results. The dimensionless width at the end of the simulation increases progressively in a downstream direction from approximately 1.3 at the upstream section to 1.85 at the downstream section. This may reflect a trend towards decreasing channel gradient, shear stress and lateral bank erosion rates and hence greater bank stability at upstream sections. There is also some evidence of temporal variability in patterns of widening. Immediately downstream of the dam the rate of channel widening declines throughout the simulation and by 1996 widening appears almost to have ceased. A similar temporal trend is evident in the rate of bed incision at the dam, although at the end of the simulation the incision rate is still approximately 0.2 m year^{-1} (compared to 0.5 m year^{-1} between 1968 and 1971). At the downstream end of the study reach the rate of channel widening is clearly related to the rate of bed incision and hence the rate of gravel extraction downstream (periods of rapid channel widening occur from 1968 to 1976 and from 1986 to 1996). It should be noted that the rate of incision at the downstream cross-section has been set using the estimated degradation curve in Figure 7.2. However, the effect of temporal variations in degradation rates within the gravel extraction zone is clearly felt further upstream. Periods of rapid bed degradation at the downstream cross-section promote the upstream migration of a knickpoint which ultimately drives renewed downcutting in the zone immediately downstream of the dam (cf. Kondolf, 1994). Figure 7.9(b) suggests that depths of incision will be greatest at the upstream and downstream limits of the study reach, with a minimum value somewhere in between. The location of the zone of minimum incision varies over the course of the simulation in response to changes in the rate of gravel extraction downstream. During periods when rates of gravel extraction are high the minimum moves upstream and conversely when rates of gravel extraction are low the minimum moves downstream.

A series of simulation were carried out to establish the sensitivity of the model performance to variations in the values of its main parameters. Table 7.1 shows the range of values that these parameters were assigned during the sensitivity analysis together with the default values of these parameters that were employed in the model runs, the results of which are shown in Figures 7.8 and 7.9.

For this purpose it is convenient to divide the model parameters into two groups: first, the various empirical factors controlling rates of bank erosion and failure (e.g. E, K and C), and second, the coefficient σ in equation (7). Of these two parameter sets it is σ which controls the bedload transport rate and which appears to have the greatest impact on the model output. Variations in bank stability factors (e.g. C and K) cause changes in the magnitude and frequency of bank failures. For instance, a higher value of the cohesion parameter increases the stability of the channel banks and reduces the frequency of bank failures. However, as the frequency of failures declines so their

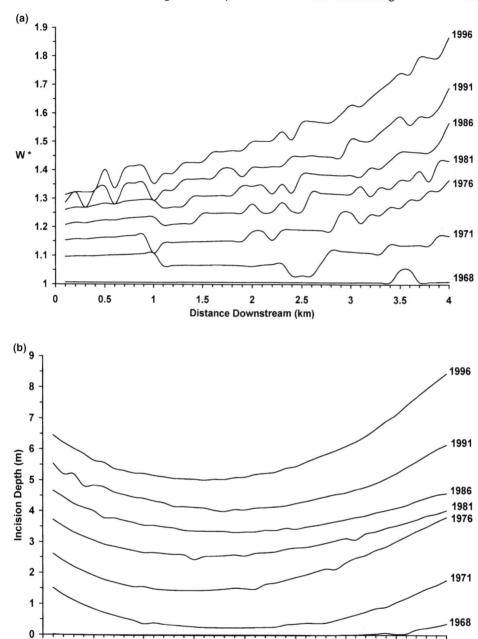

Figure 7.9 Simulated temporal changes in downstream patterns of (a) dimensionless channel width, and (b) depth of incision

Table 7.1 Parameter values employed in the model sensitivity analysis

Parameter	Range	Default
E	10^{-4}–10^{-2}	10^{-3}
K	0.3–0.7	0.5
C	2–20 kPa	10 kPa
σ	0.03–1.0	0.1

average magnitude tends to increase so that the net rate of channel widening remains to a certain extent unaffected.

Figure 7.10 shows the downstream trends in the dimensionless channel width and depth of incision at the end of four simulations in which different values of the parameter σ were employed. The total amount of channel widening during the simulation is clearly strongly controlled by σ. With $\sigma = 0.03$ the dimensionless width increases in a downstream direction from 1.2 to 2.1. In contrast, with $\sigma = 1$ the dimensionless width decreases in a downstream direction from 1.95 to 1.5. The strong dependence of the rate of channel widening on the sediment transport coefficient is clearly related to the downstream trend in patterns of bed incision shown in Figure 7.10(b). A high value of σ promotes high rates of sediment transport and removal of sediment from below the dam. The increase in σ from 0.03 to 1 results in an increase in the depth of incision at the upstream cross-section from approximately 2.5 m to 12.2 m, and this increase in incision is associated with a substantial increase in the frequency of bank failures and rate of channel widening. As is commonly the case in applications of this type, no process data are available with which to calibrate the relationship between stream power and sediment transport rate. Furthermore, available sediment transport relations calibrated for other gravel-bed rivers are unlikely to be applicable in this setting given the high degree of between-river variation that is known to exist in such relationships. Despite this it may be possible to determine a value for σ by fitting the curves shown in Figure 7.9 to measurements of downstream changes in the depth of incision determined from surveyed long profiles of the channel bed and former floodplain or terrace surfaces. Although the limited number of cross-section surveys carried out to date preclude this approach at present, the available data suggest that a value of σ of approximately 0.1 may be appropriate (this value was used in the simulation for which results are shown in Figures 7.8 and 7.9).

DISCUSSION AND SUMMARY

In river catchments where gravel extraction volumes greatly exceed the natural rate of bedload replenishment during flood events, or where impoundments prevent sediment transfer to the mined reaches, significant changes in channel geometry and floodplain morphology have been reported (e.g. Harvey and Schumm, 1987; Wyzga, 1991; Kondolf and Swanson, 1993). The magnitude of the difference between upstream supply and commercial extraction is clearly an important factor determining the nature and

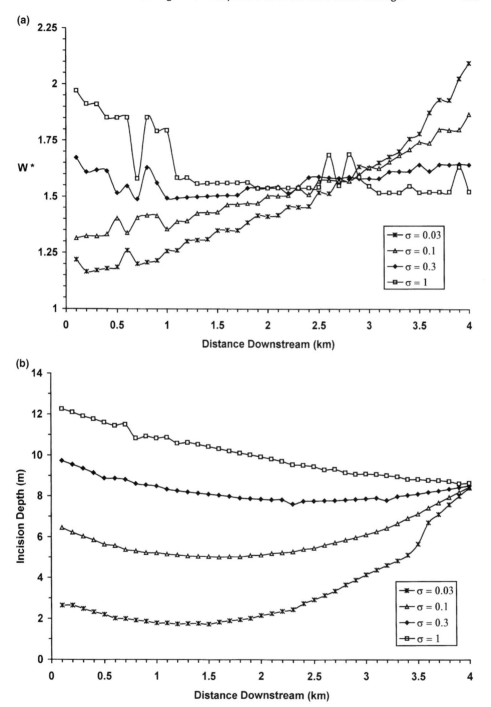

Figure 7.10 Effect of varying the sediment transport calibration coefficient (σ) on downstream patterns of (a) dimensionless channel width in 1996, and (b) depth of incision in 1996

extent of the impact on bedload sediment sources and the catchment sediment budget, but the degree of channel armouring and tributary inputs may also be important.

Gravel extraction volumes (230 000 to 580 000 m^3 year^{-1}) from reaches near the Highway 32 Bridge in Stony Creek, California, downstream of the Black Butte Dam, exceeded the pre-impoundment bedload transport rates by a factor of 2–6 between 1965 and 1990 when the area of the active channel being mined increased from 51 to 146 ha. Moreover, during this period, the extraction rates exceeded bank erosion inputs downstream of the dam by a factor of 10–30. Thus, in places, the channel bed degraded up to 5 m between 1967 and 1990 and the wide braided channel became a narrow, incised system with extensive artificially scoured surfaces (Kondolf and Swanson, 1993). Collins and Dunne (1989) attempted to establish the impact of gravel extraction activities on channel degradation in three rivers draining the Southern Olympic Mountains in Washington, USA. They compared estimates of bedload flux with data on gravel extraction volumes and concluded that the average annual input of bed material into the study reaches was actually exceeded by the extraction rates by more than ten times over a 55 year period. By comparing surveyed cross-section data at nine USGS gauging stations over this period, Collins and Dunne (1989) derived a mean degradation rate of approximately 0.03 m year^{-1} for these reaches over the last two to three decades. Harvey and Schumm (1987) have reported that the cumulative effects of gravel mining and dam construction on the Russian River and Dry Creek in California were responsible for about 3 m of base level lowering. Lateral erosion of extraction-induced incised channels is a characteristic outcome as the increasingly confined flood flows accelerate channel scour.

The lower reaches of the River Alfios have undergone substantial geomorphological adjustment over the past 30 years in response to the combined impacts of dam construction and intensive in-stream gravel extraction since the late 1960s. These two disturbances have resulted in degradation of the channel bed by approximately 7–8 m leading to reduced bank stability and channel widening. Indeed bed degradation has actually exhumed cohesive bedrock clays and this resistant surface has contributed to increased lateral erosion in some reaches. Cross-section surveys indicate that rates of channel widening increase progressively downstream of the dam, reaching maximum values of 3–4 m year^{-1} on average in the region immediately upstream of the gravel extraction works. Repeated monitoring of bank positions over the past three years suggests that current rates of bank retreat in this area are similar to these estimates of longer-term average widening rates.

The simple 1D model of channel change developed here has allowed a more detailed interpretation of temporal and spatial variations in rates of bed incision and widening downstream of the dam between 1967 and 1996. Bed degradation has proceeded in response to the removal of sediment supply below the dam and gravel extraction at the downstream end of the study site which promotes an increase in channel gradient and the upstream migration of a knickpoint. Maximum amounts of incision occur at the boundaries of the study reach where the effects of these disturbances are felt most strongly. The zone of minimum bed incision which occurs between these boundaries migrates upstream and downstream over time in response to variations in rates of gravel extraction. Marked differences are evident in temporal patterns of incision and widening at the boundaries of the study reach. Close to the downstream boundary, rates of incision and widening are highest in the periods corresponding to intensive gravel extraction in the region downstream (pre-1975 and post-1987). In contrast, immediately below the dam rates of bed degradation and bank retreat have declined progressively since the late 1960s, although downcutting is enhanced to a limited extent by the downstream reduction in base level.

Although large-scale gravel extraction downstream of the study reach has now ceased, the Alfios continues to face an uncertain future. High rates of bank retreat are likely to remain a feature of the river for many years given the height and inherent instability of the channel banks. However, rates of channel widening will decline over time as greater stability returns to the channel in the region of the former gravel extraction works. In contrast, the ongoing control on upstream sediment supply imposed by the Flokas Dam means that continued degradation of the channel bed immediately below the dam is likely to occur, although at a declining rate. In the longer-term it may be that the zone of maximum instability will switch to this part of the river as the channel downstream recovers from the effects of gravel extraction.

ACKNOWLEDGEMENTS

We are grateful to IGME for providing fieldwork permits and to the PPC of Athens for hydrological data. Thanks also to the graphics unit at Leeds University, School of Geography for providing Figure 7.1.

REFERENCES

Andrews, E. D. 1983. Entrainment of gravel from naturally sorted riverbed material. *Geological Society of America Bulletin*, **94**, 1225–1231.

Andrews, E. D. and Erman, D. C. 1986. Persistence in the size distribution of surficial bed material during an extreme snowmelt flood. *Water Resources Research*, **22**, 191–197.

Argiropoulos, P. 1960. The morphologic evolution of the rivers of the Greek realm and the influence of the transported sediments on the relief of the country. *Praktika Tis Akadimias Athinon* (Minutes of the Academy of Athens), **34**, 33–43 (in Greek).

Bagnold, R. A. 1980. An empirical correlation of bedload transport rates in flumes and natural rivers. *Proceedings of the Royal Society of London, Series A*, **372**, 453–473.

Bagnold, R. A. 1986. Transport of solids by natural water flow: evidence for a world-wide correlation. *Proceedings of the Royal Society of London, Series A*, **405**, 369–374.

Bridge, J. S. 1976. Bed topography and grain size in open channel bends. *Sedimentology*, **23**, 407–414.

Bridge, J. S. 1977. Flow, bed topography, grain size and sedimentary structure in open channel bends: a three-dimensional model. *Earth Surface Processes and Landforms*, **2**, 401–416.

Brookes, A. 1985. River channelisation; traditional engineering methods, physical consequences and alternative practices. *Progress in Physical Geography*, **9**, 44–73.

Bull, W. B. and Scott, K. M. 1974. Impact of mining gravel from urban stream beds in the Southwatern United States. *Geology*, 171–174.

Chang, H. H. 1989. Mathematical modeling of fluvial sand delivery. *Journal of Waterway, Port, Coastal and Ocean Engineering*, **115**, 311–326.

Chow, V. T. 1973. *Open Channel Hydraulics*. McGraw-Hill, Singapore.

Church, M. A., McLean, D. G. and Wolcott, R. I. 1987. River bed gravels: sampling and analysis.

In Thorne, C. R., Bathurst, J. C. and Hey, R. D. (Eds) *Sediment Transport in Gravel Bed Rivers*. John Wiley, Chichester, 43–88.

Collins, B. D. and Dunne, T. 1989. Gravel transport, gravel harvesting, and channel-bed degradation in rivers draining the Southern Olympic Mountains, Washington, USA. *Environmental Geology and Water Science*, **13**, 213–224.

Darby, S. E. and Thorne, C. R. 1996. Numerical simulation of widening and bed deformation of straight sand-bed rivers. I: Model development. *Journal of Hydraulic Engineering*, **122**, 184–193.

Darby, S. E., Thorne, C. R. and Simon, A. 1996. Numerical simulation of widening and bed deformation of straight sand-bed rivers. II: Model evaluation. *Journal of Hydraulic Engineering*, **122**, 194–202.

Erskine, W. D. 1992. Channel response to large-scale river training works: Hunter River, Australia. *Regulated Rivers: Research and Management*, **7**, 261–278.

Ferguson, R. I. 1994. Critical discharge for entrainment of poorly sorted gravel. *Earth Surface Processes and Landforms*. **19**, 179–186.

Griffiths, G. A. 1981. Flow resistance in coarse gravel bed rivers. *Journal of the Hydraulics Division, ASCE*, **107**, 899–918.

Harvey, M. D. and Schumm, S. A. 1987. Response of Dry Creek, California, to land use change, gravel mining and dam closure. *Erosion and Sedimentation in the Pacific Rim* (Proceedings of the Corvallis Symposium), IAHS Publication No. 165, 451–460.

Hickin, E. J. 1983. River channel changes: retrospect and prospect. *Special Publication of the International Association of Sedimentologists*, **6**, 61–83.

Hoey, T. B. and Ferguson, R. I. 1994. Numerical simulation of downstream fining by selective transport in gravel bed rivers: model development and illustration. *Water Resources Research*, **30**, 2251–2260.

Howard, A. D. 1992. Modeling channel migration and floodplain sedimentation in meandering streams. In Carling, P. A. and Petts, G. E. (Eds) *Lowland Floodplain Rivers: Geomorphological Perspectives*. John Wiley, Chichester, 1–41.

Howard, A. D. and Knutson, T. R. 1984. Sufficient conditions for river meandering. *Water Resources Research*, **20**, 1659–1667.

Johannesson, J. and Parker, G. 1989. Linear theory of river meanders. In Ikeda, S. and Parker, G. (Eds) *River Meandering*. Water Resources Monograph 12, American Geophysical Union, 181–214.

Knighton, A. D. 1989. River adjustment to changes in sediment load: the effects of tin mining on the Ringarooma River, Tasmania, 1875–1984. *Earth Surface Processes and Landforms*, **14**, 333–359.

Komar, P. D. 1987. Selective grain entrainment by a current from a bed of mixed sizes: a reanalysis. *Journal of Sedimentary Petrology*, **57**, 203–211.

Kondolf, G. M. 1994. Geomorphic and environmental effects of instream gravel mining. *Landscape and Urban Planning*, **28**, 225–243.

Kondolf, G. M. and Swanson, M. L. 1993. Channel adjustments to reservoir construction and gravel extraction along Stony Creek, California. *Environmental Geology and Water Science*, **21**, 256–269.

Lewin, J., Macklin, M. G. and Newson, M. D. 1988. Regime theory and environmental change – irreconcilable concepts? In White, W. R. (Ed.) *International Conference on River Regime*. Wallingford, UK, 431–445.

Ministry of Environment 1966. Preliminary report on the irrigation work planned for the Alfios Valley, Athens (in Greek).

Mosselman, E. 1995. A review of mathematical models of river planform changes. *Earth Surface Processes and Landforms*, **20**, 661–670.

Nicholas, A. P., Ashworth, P. J., Kirkby, M. J., Macklin, M. G. and Murray, T. 1995. Sediment slugs: large-scale fluctuations in fluvial sediment transport rates and storage volumes. *Progress in Physical Geography*, **19**, 500–519.

Odgaard, A. J. 1989. River meander model. I: development. *Journal of Hydraulic Engineering, ASCE*, **115**, 1451–1464.

Osman, A. M. and Thorne, C. R. 1988. Riverbank stability analysis. I: theory. *Journal of Hydraulic Engineering*, **114**, 134–150.

Park, C. C. 1977. Man-induced changes in stream channel capacity. In Gregory, K. J. (Ed.) *River Channel Changes*. John Wiley, Chichester, 121–144.

Petts, G. E. 1979. Complex response of river channel morphology subsequent to reservoir construction. *Progress in Physical Geography*, **3**, 329–362.

Pizzuto, J. E. 1990. Numerical simulation of gravel river widening. *Water Resources Research*, **26**, 1971–1980.

Rasid, H. 1979. The effects of regime regulation by the Gardiner Dam on downstream geomorphic processes in the South Saskatchewan River. *Canadian Geographer*, **23**, 140–158.

Schumm, S. A. 1969. River metamorphosis. *Journal of Hydraulics Division, ASCE*, **95**, 255–273.

Thompson, S. M. and Campbell, P. L. 1979. Hydraulics of a large channel paved with boulders. *Journal of Hydraulic Research*, **17**, 341–354.

Whitlow, J. R. and Gregory, K. J. 1989. Changes in urban stream channels in Zimbabwe. *Regulated Rivers: Research and Management*, **4**, 27–42.

Willetts, B. B., Maizels, J. K. and Florence, J. 1987. The simulation of stream bed armouring and its consequences. *Proceedings of the Institute of Civil Engineers, Part I*, **82**, 799–814.

Wyzga, B. 1991. Present-day downcutting of the Raba River channel (Western Carpathians, Poland) and its environmental effects. *Catena*, **18**, 551–566.

Xu, J. 1990. An experimental study of complex response in river channel adjustment downstream from a reservoir. *Earth Surface Processes and Landforms*, **15**, 43–53.

8 Significance of River Bank Erosion as a Sediment Source in the Alternating Flood Regimes of South-eastern Australia

WAYNE D. ERSKINE[1] and ROBIN F. WARNER[2]
[1] School of Geography, University of New South Wales, Australia
[2] Department of Geography, University of Sydney, Australia

INTRODUCTION

Rivers in central eastern Australia exhibit a distinctive hydrology characterised by significant time series changes and high variability (Erskine and Livingstone, 1999). Rainfall-driven, alternating periods of low and high flood activity have been identified on many coastal rivers in New South Wales (NSW) to the west and north of Sydney (Figure 8.1). Erskine and Warner (1988) defined flood-dominated regimes (FDRs) as time periods of several decades during which there is a marked upwards shift of the whole flood frequency curve due to significant increases in annual and summer rainfall, rainfall intensities and rainfall frequencies. They consist of multi-decadal periods of persistent flood activity. Runs of large floods occur for up to 11 years in a row separated by shorter periods of smaller floods. Drought-dominated regimes (DDRs) are time periods of several decades during which there is a marked downward shift of the whole flood frequency curve from the previous FDR due to significant decreases in rainfall. They consist of relatively long periods of low flood activity. Runs of floods occur for up to 6 years in a row separated by longer periods of little flood activity. The time periods identified by Dury (1980), Brizga et al. (1993) and Kirkup et al. (1998) refer to these shorter runs of wet and dry years of inter- and intra-decadal duration, not the alternating FDRs and DDRs defined above.

The statistical analyses of historical flood heights for three NSW coastal rivers undertaken by Erskine and Warner (1988) revealed that FDRs are characterised by a significant increase in the number of floods in various height classes (flood frequency) over DDRs. These increases in flood frequency may be accompanied by significant increases in flood heights. Furthermore, all FDRs are not the same. FDRs have been identified during the periods 1799–1820, 1857–1900 and 1949–1988 while DDRs occurred during 1821–1856 and 1901–1948. Channels usually responded to these

Fluvial Processes and Environmental Change. Edited by A. G. Brown and T. A. Quine.
© 1999 John Wiley & Sons Ltd.

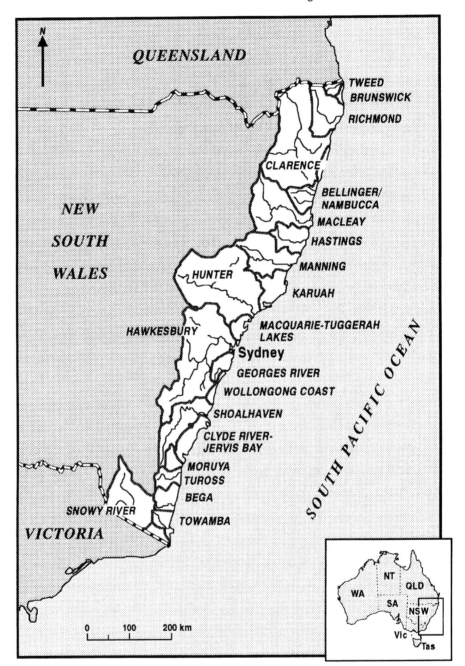

Figure 8.1 Coastal rivers of New South Wales. Those west and north of Sydney have been identified as experiencing alternating flood-dominated regimes and drought-dominated regimes

alternating flood regimes by widening and aggrading during FDRs and by contracting due to bench and floodplain formation during DDRs (Erskine, 1986a; Warner, 1987, 1993). Reddoch (1957) and Warner (1972) first noted that substantial channel erosion occurred between 1949 and 1956 on many coastal rivers north of Sydney because of the sequence of frequent, large floods. This erosion has continued during the last FDR on many rivers (Erskine, 1986a; Warner, 1994a, 1997).

The climatic causes of alternating FDRs and DDRs have not been investigated. The change in rainfall driving the alternating flood regimes is associated with large variations in mean annual rainfall (some $> 30\%$) which are, in turn, caused by large changes in summer rainfall (some $> 60\%$) and rainfall intensities (Erskine and Bell, 1982; Erskine, 1986a, 1986b). These rainfall changes have been recorded across at least the northern half of New South Wales (Cornish, 1977) and appear to be caused by cyclical, medium-term shifts in the summer rainfall belt. FDRs correspond to southerly incursions and DDRs correspond to northerly retreats of the summer rainfall belt. The most sensitive areas to record alternating flood regimes are lower rainfall areas away from pronounced orographic effects and the coast (Bell and Erskine, 1981; Erskine and Bell, 1982). Kirkup et al. (1998), in their criticism of alternating flood regimes, fail to mention the coherent spatial distribution of the synchronous changes in rainfall and runoff across more than half of New South Wales.

The variability of flood peak discharges for Australian rivers is much higher than anywhere else in the world, except southern Africa (McMahon et al., 1992). Flood variability, expressed as the standard deviation of the $\log_{10}$ of the annual maximum flood series (the Flash Flood Magnitude Index of Baker, 1977), is over twice as large for Australian and southern African rivers in comparison to that for the rest of the world (McMahon et al., 1992). Many rivers in coastal NSW and Victoria exhibit some of the highest Flash Flood Magnitude Indices recorded in the world (Erskine, 1996; Erskine and Saynor, 1996; Erskine and Livingstone, 1999). Rivers with high flood variability experience large floods relatively frequently because they have such steep flood frequency curves (Baker, 1977; Abrahams and Cull, 1979).

While catastrophic floods are truly rare events in areas characterised by low flood variability, this is not the case in areas of high flood variability (Wolman and Miller, 1960; Baker, 1977; Abrahams and Cull, 1979). For geomorphological purposes, catastrophic floods should be defined solely on the basis of magnitude (Abrahams and Cull, 1979). Empirical results indicate that floods with a peak discharge at least 10 times greater than the mean annual flood totally destroy the pre-flood channel whereas smaller floods do not cause such extensive, long-term geomorphological changes (Stevens et al., 1975, 1977). Where catastrophic floods are common, rivers are continually responding to, or recovering from, flood-induced changes (Erskine and Saynor, 1996). Therefore, these are non-equilibrium rivers, which never exhibit any form of adjustment to the mean annual flood (Stevens et al., 1975). Although Nanson and Erskine (1988) proposed the term "cyclical equilibrium" for the alternation between enlarged and contracted channels during alternating flood regimes, it has not been established that a condition of equilibrium ever exists. We believe that such behaviour should be called "cyclical disequilibrium" because the channel only becomes adjusted to the mean annual flood when the flood regime persists for long time periods in reaches with mobile channel boundaries. Therefore, these channels are also examples of non-equilibrium rivers.

Erskine and Saynor (1996) demonstrated that catastrophic floods in south-eastern Australia can generate all of the fluvial sediment yield by channel erosion. Such events have produced between at least 11 and 283 times the mean annual sediment yield by channel erosion alone. Warner (1994a) compiled a detailed record of historical channel changes on two NSW coastal rivers which showed that large volumes of in-channel sediment storage occurred during the DDR of early this century and that this material was largely reworked during the subsequent FDR. Many NSW coastal rivers have experienced the interactive effects of catastrophic floods and FDRs, a situation which should result in major channel erosion.

The purpose of this chapter is to determine the significance of sediment contributions from bank erosion to the fluvial sediment yield during alternating flood regimes. In the first part, the amounts of sediment eroded from river banks and temporarily stored in the river bed are assessed in relation to the total sediment load for the most recent FDR for a sand-bed section of the Hunter River at Singleton. In the second part, the amount of bank deposition during the DDR early this century, and the proportion of this sediment re-entrained during the subsequent FDR, are quantified for the Nepean River at Penrith. The location of these rivers is shown in Figure 8.1. Lastly the implications of this work for sustainable river management are discussed.

HUNTER RIVER AT SINGLETON

The Singleton study reach extends 35 km from the Wollombi Brook junction downstream to Scotts Flat (Figure 8.2) and is centred on the Singleton gauge where the drainage basin area is 16 400 km^2. The Hunter River in this reach has a sand-bed and a mean annual flood (1898–1978) of 609.4 m^3 s^{-1}. The channel has excavated a trench into Permian sedimentary rocks and Cenozoic terrace sediments that has been partially backfilled with Holocene alluvium (McIlveen, 1984). Bedrock is often exposed in the concave bank at bends (Figure 8.2) and partly confines channel planform. The localised expansion of the valley floor trough at Singleton occurs in the relatively erodible Mulbring Siltstone (McIlveen, 1984). A well-defined floodway has formed to the south of Singleton at Doughboy Hollow in this floodplain expansion. At the peak of the February 1955 flood, the highest event recorded since at least 1820, some 50% of the peak discharge bypassed the Hunter River at Singleton through this floodway (Cameron McNamara, 1984).

Flood History

An annual maximum flood series at the Singleton gauge was extracted for the period 1898–1978, inclusive and a log Pearson type III distribution was fitted by the method of moments (Pilgrim and Doran, 1987). Significant record losses have occurred since 1978 because of repeated recorder malfunctions. The catastrophic February 1955 flood had a peak discharge 20.5 times greater than the mean annual flood. An important aspect of the flood record is the sequence of large floods recorded between

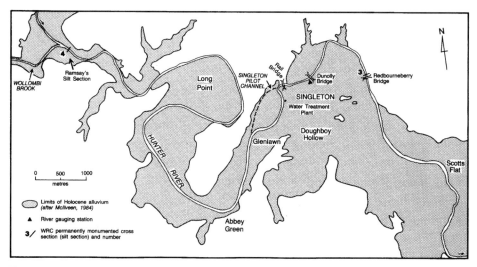

Figure 8.2 The study reach of the Hunter River at Singleton

1949 and 1956 at the start of the last FDR when 16 events greater than the mean annual flood occurred in 8 years.

Figure 8.3 shows the annual maximum series flood frequency curves (log Pearson type III distribution) at Singleton for the DDR and FDR of this century. Clearly the flood peak discharges for annual exceedance probabilities less than 95% have increased by between 48 and 142%, a result consistent with earlier findings (Bell and Erskine, 1981; Erskine and Bell, 1982). Furthermore, from statistical analysis of the flood height records at Singleton (1857–1978), Erskine and Warner (1988) found that the frequency of floods in various height classes during the FDR of the late 19th century (no discharge data available) was greater than the subsequent DDR but comparable to the most recent FDR. Furthermore, flood heights during the FDR of the late 19th century were significantly greater than the subsequent DDR but comparable to the most recent FDR.

The Flash Flood Magnitude Index for the DDR was 0.510 and increased to 0.570 for the FDR. These values are not significantly different (F test at $p < 0.05$) but are much greater than the average value for Australian streams found by McMahon et al. (1992). However, much larger values have been recorded elsewhere in the Hunter Valley (Erskine, 1986a, 1994, 1996).

Sediment Yield

The total sediment load at the Singleton gauge for the period 1949–1979 has been estimated in a variety of ways. Kriek (1970) calculated a suspended sediment rating curve from 288 depth-integrated suspended sediment samples. This was converted to a mean annual suspended sediment yield of 3.74×10^6 t year^{-1} (288 t km^{-2} year^{-1}) by the flow duration method, using Piest's (1964) duration classes (Table 8.1). The NSW Department of Land and Water Conservation has also undertaken depth-integrated suspended sediment sampling. Least-squares linear regression analysis of the log-

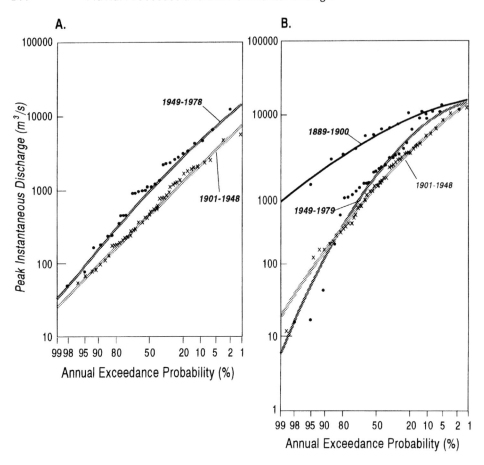

Figure 8.3 Annual maximum series flood frequency curves for (a) the Hunter River at Singleton, and (b) the Nepean River at Penrith. FDRs occurred between 1857 and 1900, and between 1949 and 1979, while a DDR occurred between 1901 and 1948

transformed data (suspended sediment concentration and load) produced equations that were significant at a probability level of at least 0.001%. Converting to a mean annual suspended sediment yield using the same method as above, produces a yield of 1.59×10^6 t year^{-1} (97 t km^{-2} year^{-1}) (Table 8.1). As expected, these two yields are within the range reported for agricultural Australian catchments by Olive and Rieger (1986). The lower yield for the Department's data relates to a wetter period than that sampled by Kriek (1970).

Standard depth-integrated suspended sediment samplers do not sample within about 0.13 m of the bed surface. Therefore, there is an unmeasured sediment discharge comprising both bedload and suspended load which passes beneath the sampler. Colby (1957) proposed a method for estimating this unmeasured sediment discharge which produces a value of 3.8×10^5 t year^{-1} (Table 8.1). This method is appropriate here because the bed material exhibits a very small proportion of gravel. Colby's (1964) method of estimating total sand discharge was used to construct a bedload rating

Table 8.1 Sediment yield estimates for the Hunter River at Singleton. Flow duration curve determined for the period 1949–1979

	Data source		
Mean annual yield	Kriek (1970)	Department of Land and Water Conservation	Calculated
1. Suspended sediment ($\times 10^6$ t)	3.74	1.59	—
2. Unmeasured sediment ($\times 10^6$ t)	—	—	0.38
3. Bed material ($\times 10^6$ t)	—	—	0.72
4. Total sediment load ($\times 10^6$ t)			
1 + 2	4.12	1.97	—
1 + 3	4.46	2.31	—

1, Determined by suspended sediment rating curve–flow duration technique.
2, Calculated using the method of Colby (1957).
3, Calculated using the method of Erskine et al. (1985) which applied the Colby (1964) sand function.

curve which has been converted into a mean annual yield by combining it with the flow duration curve (Erskine et al., 1985). Colby's (1964) method was adopted because it is one of the more accurate sand discharge functions. The result is a yield of 7.2×10^5 t year^{-1} which is nearly double that calculated by the first method. This should be expected with sand because it is transported both in suspension and in contact with the bed as migrating bedforms, depending on stream power. Therefore, the determination of the total sediment yield by suspended sediment and bedload yields (1 + 3 in Table 8.1) will contain an unknown amount of double accounting. For this reason, the total sediment yield obtained by combining suspended sediment yields with estimates of the unmeasured sediment discharge are preferred (1 + 2 in Table 8.1). However, the two suspended sediment rating curves produce markedly different results (Table 8.1) because Kriek's (1970) data were collected for a drier period than the Department's data. The Department's suspended sediment yield is preferred because the period 1949–1979 as a whole has been characterised by high rainfall. Therefore, the estimated mean annual total sediment yield at Singleton is approximately 2×10^6 t year^{-1} (122 t km^{-2} year^{-1}). It should be stressed that the use of a single suspended sediment rating curve can underestimate sediment loads (Rieger and Olive, 1988). However, these are likely to be small in the present case because of the large basin area and consequent long sediment response times. Nevertheless, the above yields are subject to many errors (Erskine et al., 1985; Rieger and Olive, 1988) and must be viewed as first estimates only.

Channel Response to Alternating Flood Regimes

Evidence presented before the Hunter River Floods Commission in 1869 and 1870 provides some indication of channel changes up to 1870. The Commission, after assessing all available information, concluded as follows:

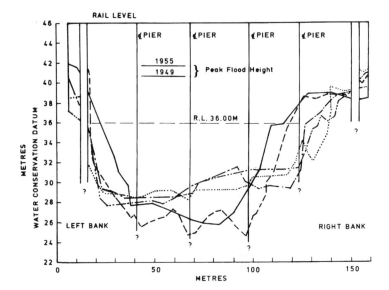

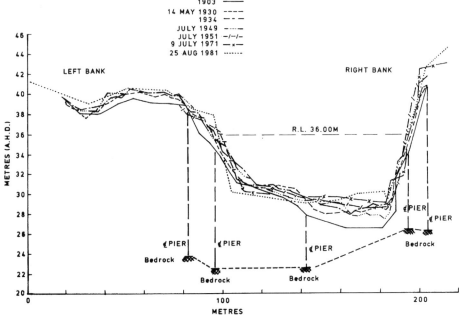

Figure 8.4 Cross-sectional changes at four sites in the Singleton study reach. For the location of sites, see Figure 8.2

River Bank Erosion as a Sediment Source   147

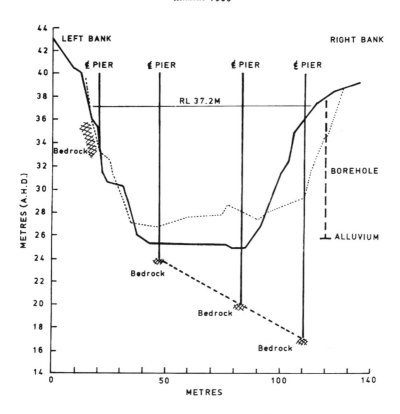

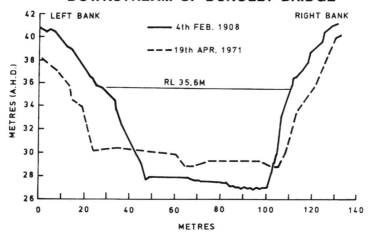

148   Fluvial Processes and Environmental Change

> Mr W.C. Leslie, who has resided at Singleton since 1841, ... is of opinion that ... the channel has doubled in width, and been cleared of many obstacles to the free discharge of flood-waters. From Mr Alexander Munro also we have some valuable evidence as to the enlargement of the channel near Singleton ... (Hunter River Floods Commission, 1870a, p.8)

The transcripts of Leslie's and Munro's evidence (Hunter River Floods Commission, 1870b, Minutes of Evidence, pp. 58–61) indicate that substantial bank erosion, channel widening and tree removal had occurred by large floods after about 1832, with the 1857 flood at the start of the FDR being the most destructive. Photographs taken in 1866 show the Hunter River as having treeless, eroding banks 10 years after the onset of the FDR. Raine and Gardiner (1995) also concluded that substantial channel erosion occurred last century. They quite rightly mentioned forest destruction in the catchment and on the floodplain as a significant factor contributing to these early channel changes. However, they fail to mention any flood effects, despite the description of massive flood damage during the onset of the FDR in the late 1850s and 1860s in Hunter River Floods Commission (1870b). The combination of initial land settlement, the change from a DDR to an FDR and the occurrence of isolated catastrophic floods resulted in substantial channel changes in the latter part of last century. No discharge data exist to quantify these catastrophic floods but contemporary descriptions of channel changes refer repeatedly to large-scale erosion (Hunter River Floods Commission, 1870b).

More recent channel changes can be determined from a better quality database. Figure 8.4 shows cross-sectional changes at four sites in the Singleton study reach. Further channel widening of up to 30% occurred after the first survey at three of the four sites. The results for the rail bridge indicate that the catastrophic February 1955 flood effected the greatest enlargement. All sites exhibit aggradation since the last century or early this century. Thalweg depths decreased by an average of 2 m. The chronology of aggradation during this century can be determined from a plot of temporal variations in gauge height for a constant discharge. Temporal trends in cease-to-flow height (i.e. the gauge height at which the river just stops flowing), corrected for any changes in gauge zero, will provide a reliable record of bed level dynamics where the gauge is unaffected by a control structure, such as at Singleton. However, cease-to-flow conditions have not been recorded at Singleton since 1958 because of flow regulation. As a result, recent rating tables have not been extrapolated to zero flow. Therefore, gauge heights corresponding to a discharge of $0.694\,\mathrm{m^3\,s^{-1}}$ as well as zero flow, when available, were extracted from all existing rating tables. Figure

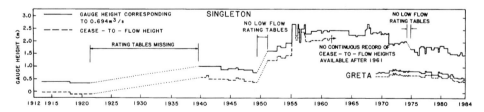

Figure 8.5 Specific gauge plots for the Hunter River at Singleton and Greta gauges. Greta is located on the Hunter River downstream of Singleton

8.5 is a plot of specific gauge for both the Singleton and Greta (located on the Hunter River downstream of Singleton) gauges and shows that trends for a discharge of $0.694\,\mathrm{m^3\,s^{-1}}$ closely parallel those for zero flow, when concurrent records are available. This means that the specific gauge plot for a discharge of $0.694\,\mathrm{m^3\,s^{-1}}$ can be used to accurately infer variations in bed level. The available data indicate that 0.65 m of aggradation occurred between 1921 and 1940. Bed level rose by a further 1.9 m between June 1949 and February 1955 and remained high until early 1957 when it dropped by 0.36 m. In January 1971 degradation commenced and since then bed levels have dropped by a further 0.85 m. In 1984 bed level was still more than 1.1 m higher than it was in 1913. Sand extraction has also contributed to this degradation (Erskine et al., 1985).

Aggradation has altered the bed-material size distribution. The descriptions of the Hunter River upstream of Singleton by John Howe (an explorer who discovered Singleton) in November 1819 refer to gravel bars and riffles as well as rare "sand shoals" (Campbell, 1928). Howe again referred to "pebbly falls" (gravel riffles) in March 1820 when he followed the Hunter River between Singleton and Maitland (Jervis, 1945). Evidence collected by the Hunter River Floods Commission (1870b) confirms these descriptions. Old photographs demonstrate that, by the early part of this century, the bed material had changed to predominantly medium sand with a small gravel fraction which is similar to today's bed material (Erskine et al., 1985).

The Department of Land and Water Conservation obtained a detailed record of post-1951 channel changes at three sites between the Wollombi Brook confluence and Elderslie (downstream of Singleton). All sites widened by over 10%, mainly during the 1955 and 1971 floods. Aggradation was also recorded at all sites, with Ramsay's, the most upstream (Figure 8.2), exhibiting the earliest change. This is to be expected because Wollombi Brook injected large amounts of sand into the Hunter River during and immediately after the catastrophic June 1949 flood (Erskine, 1996; Erskine and Saynor, 1996). This material was transported downstream as a sand slug.

To determine if the above channel changes have resulted in temporal shifts in the rate of change of mean hydraulic variables with discharge, log-linear hydraulic geometry equations were calculated by the method of least-squares for all available river gauging data collected within 100 m of the gauge (Table 8.2). The data were split into three time periods: 1924–7 June 1949, 22 June 1949–1956, and 1957–1981. These periods refer respectively, to the DDR, the sequence of large floods at the beginning of the current FDR and the latter part of the most recent FDR. Although all equations in Table 8.2 are significant at the 1% level or less, there are no significant differences in the slopes of the equations between the various time periods. This is not surprising because the data for all three time periods plot in Rhodes' (1977) channel type 8. Therefore, the rate of change of mean hydraulic variables with discharge has not altered significantly over time.

The hydraulic geometry equations have also been used to calculate bed-material load discharge rates, assuming that median bed-material size has remained constant. Colby's (1964) method was used again. For comparative purposes, the sand flux was calculated for a discharge of $609.4\,\mathrm{m^3\,s^{-1}}$ (mean annual flood for 1898–1978) for all three time periods. The largest sand flux ($64\,500\,\mathrm{t\,day^{-1}}$) was found for the latest period (FDR) and the smallest ($29\,500\,\mathrm{t\,day^{-1}}$) for the earliest period (DDR). The

Table 8.2 Hydraulic geometry equations for the Hunter River at Singleton gauging station for various time periods. All equations are significant at the 1% level or less

Time period	Hydraulic coefficients			Hydraulic exponents		
	a	c	k	b	f	m
1924–1949	12.88	0.211	0.388	0.333	0.440	0.228
1949–1956	19.19	0.173	0.306	0.261	0.481	0.256
1957–1981	18.11	0.186	0.297	0.269	0.454	0.277

$w = aQ^b$, $d = cQ^f$, $u = kQ^m$, and where w is channel width, d is mean depth, u is mean velocity, and Q is discharge.

middle period had an intermediate sand flux (38 300 t day^{-1}). These differences are caused by changes in mean velocity. Aggradation since 1949 has increased sand supply which has induced a minor hydraulic adjustment, i.e. slightly increased flow velocity. This, in turn, has meant that the channel responded to increased sand supply by increasing its sand transport capacity.

Significance of Bank Erosion and Bed Aggradation

Sediment volumes derived from bank erosion during the last FDR were calculated by superimposing channel traces of varying dates on a common base map, digitising the surface area eroded and multiplying by the bank height. Figure 8.6 shows one area of floodplain destruction by lateral migration at Scotts Flat (Figure 8.2), where the sequence of large floods between 1949 and 1956 initiated the erosion (Shattock, 1966). At the site enlarged in the inset of Figure 8.6, 19.4 ha (3.1×10^6 t) of floodplain were eroded between early 1949 and late 1964. For the total length shown in the upper part of Figure 8.6, approximately 5×10^6 t were eroded between 1949 and 1974. The Department of Land and Water Conservation carried out repeated river training works here to stabilise the channel (Shattock, 1966).

Figure 8.7 shows shifts in channel location of the Hunter River at Singleton between 1939 and 1974. Lateral migration of the bend at Glenlawn eroded 5.95×10^6 t between 1951 and 1974. There was also marked downstream lateral migration of the loop immediately upstream of the rail bridge between 1939 and 1974. This erosion was started by the June 1949 flood; the position of this loop in 1939 was essentially the same as that shown on a State Rail Authority plan dated 12 June 1866. The Singleton pilot channel (Figure 8.7) was constructed on a number of occasions by the Department of Land and Water Conservation to divert the channel to a better approach to the rail bridge. Despite these works the channel has not been relocated to the new course. However, extensive bank stabilisation works were undertaken at many locations downstream of Glenlawn (Figure 8.7). Works commenced in 1952/3 at Dunolly Bridge and included the construction of groynes, grids of cylindrical wire-mesh fences, rock sausages and stone rip-rap as well as bank battering and brush cover, tree planting and minor channel realignment (Committee of Advice of Flood Control and Mitigation, 1955; Reddoch, 1957; Shattock, 1966). These works proved successful in reducing bank erosion rates. Rankin (1982) and Erskine (1990, 1992) outline in detail the design of such works.

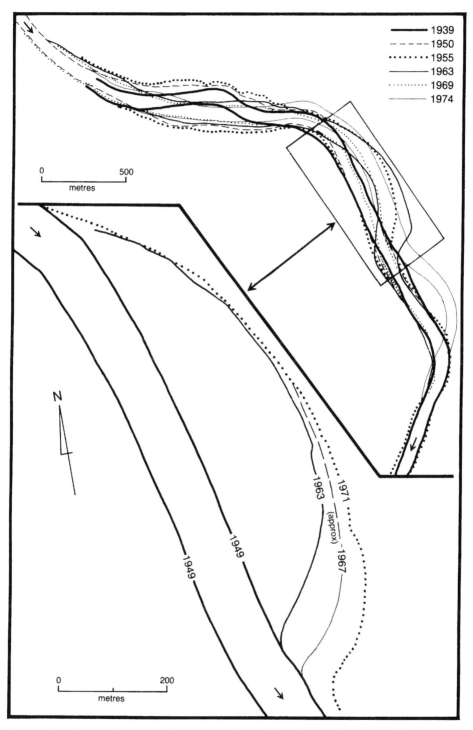

Figure 8.6 Channel pattern changes at Scotts Flat. For the location of the site, see Figure 8.2

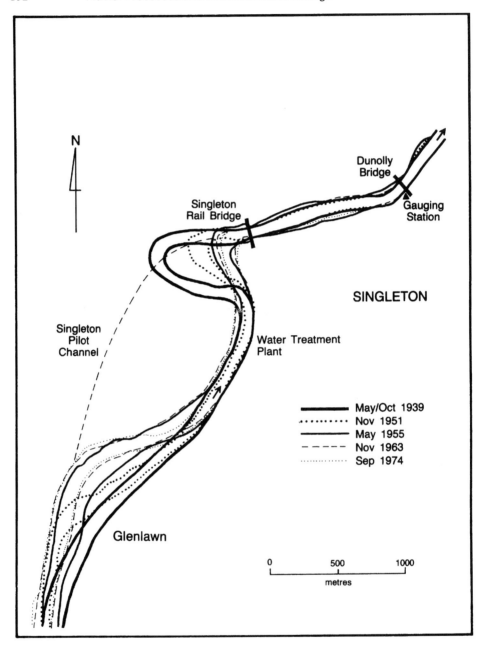

Figure 8.7 Channel pattern changes at Singleton. For the location of the site, see Figure 8.2.

For the total 35 km study reach, approximately 18×10^6 t of sediment were eroded from the channel and floodplain between 1949 and 1979. This source alone can account for 29% of the total sediment yield during the 31-year study period. However, it must be stressed that extensive, synchronous bank stabilisation works undertaken within the study reach have reduced bank erosion rates. Furthermore, if a longer study reach had been investigated, the amount of bank erosion would have increased greatly. For example, Erskine and Saynor (1996) found that 13.25×10^6 t of sediment were generated by channel enlargement on Wollombi Brook by the catastrophic June 1949 flood.

Assuming that the bed level changes at the Singleton gauge are representative of the whole study reach (Figure 8.4 indicates that this is a reasonable assumption), approximately 6.6×10^6 t of sand (more than three times the mean annual total sediment yield) were temporarily stored in the bed between June 1949 and February 1955. Since 1957, 4.19×10^6 t of sand (63% of the mass deposited between 1949 and 1955) have been re-entrained from the bed. Clearly the bulk of the sediment eroded from river banks within the study reach was transported through the reach.

NEPEAN RIVER AT PENRITH

The Penrith study reach extends 5 km from the Glenbrook Creek delta (a very coarse-grained tributary mouth bar) to Penrith Weir (Figure 8.8). The reach covers the transition of the Nepean River from the Fairlight Gorge to an alluvial plain where channel bank materials are highly variable and where channel capacity exceeds the 1% annual exceedance probability flood (Warner, 1987). The mean annual flood was $1318.6 \, m^3 \, s^{-1}$ (1889–1979) and the bed material is pebble and cobble gravel (Warner, 1983, 1987). The five sections (Figure 8.8) defined by Erskine and Warner (1988) are also used here because they are important to understanding channel changes during alternating flood regimes.

Drainage basin area at the Penrith gauging station near the Victoria rail bridge is 11 008 km^2. The Nepean River is highly regulated by two water supply schemes. First, the Upper Nepean Water Supply Scheme consists of four large dams, two small dams and two diversion weirs which were built progressively between 1880 and 1935 for water supply to Sydney (Sammut and Erskine, 1995). Secondly, the $2.057 \times 10^9 \, m^3$ capacity Warragamba Dam, which also supplies Sydney, commenced storing water in 1960. Warragamba Dam regulates 83% of the total area upstream of Penrith and traps most of the mean annual sediment yield of about 1.3×10^6 t year^{-1} (Scholer, 1974).

Flood History

An annual maximum flood series at the Penrith gauge for the period 1889–1979 was obtained from gauging records. The flood series was split into the three periods corresponding to the FDRs of the late 19th century and late 20th century, and the DDR of the early 20th century. A log Pearson type III distribution was fitted to each period by the method of moments (Pilgrim and Doran, 1987). Figure 8.3 shows the plotted curves and demonstrates that the magnitude and frequency of floods during

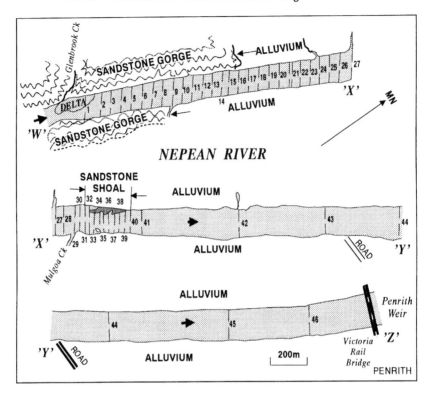

Figure 8.8 The study reach of the Nepean River at Penrith, showing the locations of the 46 cross-sections surveyed in 1900

the FDR of the late 19th century were much greater than during any subsequent period. Erskine and Warner (1988) reported a similar result at up- and downstream gauging stations. The flood frequency curve for 1949–1979 intersects the curve for 1901–1948 at a probability of 85% and, at smaller probabilities, flood peak discharges have increased by between 11 and 67%, despite significant flow regulation. The most pronounced effect of flow regulation has been to progressively increase flood variability. Between 1889 and 1900, the Flash Flood Magnitude Index was only 0.314 but increased to 0.621 (1901–1948) and then to 0.757 (1949–1979). This indicates that the channel has become more sensitive to flood-induced change.

The two largest floods at Penrith since European settlement were recorded in 1867 and 1900 during an FDR (Saynor and Erskine, 1993). Both of these events had peak discharges greater than 10 times the mean annual flood. However, palaeoflood analysis by Riley *et al.* (1989) and Saynor and Erskine (1993) in the Fairlight Gorge upstream of the reach shown in Figure 8.8 demonstrates that at least one flood approaching the magnitude of the probable maximum flood occurred in the late Holocene.

Temporal Variations in Erosion and Sedimentation

The database of Warner (1987) and Erskine and Warner (1988) is reanalysed here in terms of the volume of sediment eroded from, or deposited in, the study reach between 1900 and 1983. In 1900, 46 cross-sections were surveyed between Glenbrook Creek and Victoria Bridge (Figure 8.8). These were relocated and resurveyed in 1982 and 1983. Furthermore, analysis of 1949 and 1970 vertical air photographs were combined with the survey data.

Figure 8.9 details the time series changes in channel width. Substantial contraction occurred during the DDR, especially in the Shoal and Upper and Lower Alluvial sections. The average reduction in width was 23 m, with a maximum of 55 m being recorded in the Shoal where the sandstone ledges acted as nuclei for sedimentation. About 10.6 ha of land were deposited as in-channel benches or incipient floodplains up to 6 m above weir crest level. With the change to an FDR, widening was initiated in the same areas, particularly between 1949 and 1970. The average increase in width was 13 m, with the largest change again being recorded in the Shoal (35 m). Nearly 4 ha of land were eroded up to 1983, but the channel was still narrower than in 1900. Extensive bank protection works were carried out. Channel enlargement by the erosion of

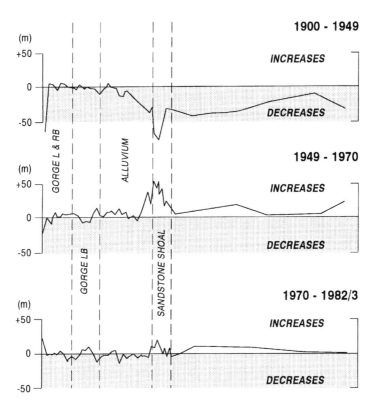

Figure 8.9 Channel width changes at Penrith for 1900–1949, 1949–1970 and 1970–1982/3

in-channel benches is shown in Figure 8.10 at a cross-section just upstream of Victoria Bridge. At the Victoria rail bridge, the channel was much narrower in 1863 at the start of the FDR last century than it was in 1976 (Figure 8.10). By 1900, channel width here had increased by 45 m following the FDR and the catastrophic floods of 1867 and 1900. At the end of the DDR in 1949, width had decreased by 40 m before widening again during the succeeding FDR. Clearly the channel synchronously responds to the time series changes in flood activity.

Changes in sediment storage were calculated for each time period after 1900. Table 8.3 outlines the results for each of the five sections defined in Figure 8.8. During the DDR, $5.5 \times 10^5 \, m^3$ were added to the banks, whereas during the first 34–35 years of the following FDR, $1.6 \times 10^5 \, m^3$ were lost from the banks. As the weir pool increased in depth by an average of 1.5 m and in capacity by $6.0 \times 10^5 \, m^3$ between 1900 and 1982/3, and the banks accreted by $3.9 \times 10^5 \, m^3$, about $1 \times 10^6 \, m^3$ of sediment were

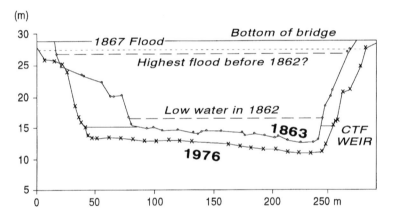

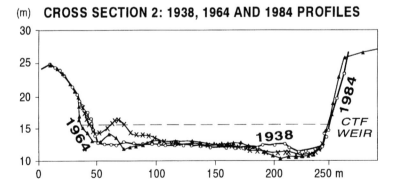

Figure 8.10 Cross-sectional changes at a site near Victoria Rail Bridge and at Victoria Bridge

Table 8.3 Changes in sediment storage volumes for each section of the Penrith study reach. For the location of sections, see Figure 8.8

Section	1900–1949 ($\times 10^3$ m^3)	1949–1982/3 ($\times 10^3$ m^3)	1900–1982/3 ($\times 10^3$ m^3)
Upper Gorge	− 5	− 11	− 16
Lower Gorge	− 2.5	− 10	− 12.5
Upper Alluvium	− 40	− 13	− 53
Shoal	− 55	+ 36	− 19
Lower Alluvium	− 450	+ 160	− 290
Total	− 552.5	+ 162	− 390.5

Note: a negative sign indicates that deposition has occurred whereas a positive sign indicates that erosion has occurred.

eroded from the bed between 1900 and 1982/3. This bed degradation resulted from the combined effects of the FDR since 1949 and bedload starvation below Warragamba Dam since 1960. However, Turner and Erskine (1997) found that essentially no further changes in bed levels occurred between 1982/3 and 1996 because the bed has armoured.

DISCUSSION AND CONCLUSIONS

Brizga et al. (1993) and Kirkup et al. (1998) challenged the existence of alternating FDRs and DDRs on the basis that no objective method was used to identify the start and end points of the flood regimes. Erskine and Bell (1982) and Erskine (1986a, 1986b) analysed rainfall records for the Hunter and Macdonald Valleys to the north of Sydney by the cusum technique to determine historical rainfall trends. They found that annual rainfall throughout these areas increased abruptly in the mid to late 1940s (the modal minimum cusum occurred in 1946) and that the rainfall increase was statistically significant, except in high rainfall areas. Subsequent work has shown that rainfall decreased in about 1900 (the modal maximum cusum occurred in 1900) and that the rainfall decrease was statistically significant (Erskine and Warner, 1998). The flood records were then split at 1900 and 1946 to determine whether synchronous changes had occurred. The results showed that there had also been marked shifts of the flood frequency curves over the same periods that were statistically significant (Erskine and Bell, 1982; Erskine and Warner, 1988). Brizga et al. (1993) and Kirkup et al. (1998) misrepresent our work by concentrating on inter- and intra-decadal periods and by ignoring the fundamental link between rainfall and floods. Furthermore, none of the data they present are directly comparable to ours. Brizga et al. (1993) use mean daily flow for their flood frequency analysis at Singleton instead of the more appropriate peak instantaneous discharge (Figure 8.3). Kirkup et al. (1998) use greatly different threshold flood heights yet make direct comparisons with our analyses. This alone explains the discrepancies.

Coastal rivers to the west and north of Sydney, NSW, respond dramatically to rainfall-driven, alternating periods of high and low flood activity. Catastrophic floods only occur during the FDRs and magnify the physical impacts of an increase in flood frequency. Brooks and Brierley (1998) criticised the FDR/DDR theory on the basis of their single case study of the Bega River in southern New South Wales (Figure 8.1). It has never been argued that alternating flood regimes are important there and our more recent work has found that no catastrophic floods have been recorded on the Bega River.

Sandy channels are the most sensitive and bedrock channels are the least sensitive to alternating flood regimes (Erskine and Warner, 1988). Substantial channel widening has been documented by bank erosion and by the destruction of in-channel benches and floodplains on sand-bed and some gravel-bed rivers. Bank erosion, not accompanied by concomitant deposition, enlarges the channel, particularly where lateral migration is active. Warner (1997) also documented the unusual case of convex bank erosion where the concave bank cannot erode because it is fixed by bedrock. He also found that channels in high energy reaches with resistant, densely vegetated boundaries are slow to adjust to alternating flood regimes, often resulting in overbank adjustments on cleared floodplains by alluvial stripping. Stripping by meander and parallel chute development has been identified thus far (Warner, 1988, 1997) and has been shown to develop rapidly during FDRs and to infill during DDRs. The disequilibrium floodplain development model of Nanson (1986), which is based on progressive vertical accretion followed by episodic but catastrophic stripping, has been shown by Warner (1997) to be parallel chute cutting during FDRs. The type example cited by Nanson (1986) was the Manning River at Charity Creek which exhibited a chute during the FDR of the late 19th century and which infilled during the subsequent DDR, only to be eroded again during the last FDR (Warner, 1997). It was only the last phase of stripping that was recognised by Nanson (1986).

Although the sediment generated by bank erosion at Singleton and Penrith signifies a significant sediment source, the above estimates must be viewed as minima because of extensive bank protection and river training works. Therefore, bank erosion rates have been retarded, to varying degrees, during most of the last FDR. The adoption of a longer study reach would have increased greatly the estimates of bank erosion (Erskine, 1992, 1994, 1996). Furthermore, large-scale extraction has also been undertaken from in-channel benches since the 1970s, thus totally removing sediment stores which could have been reworked by the river at a later time (Erskine et al., 1985). At Penrith, bank protection works have also been carried out during the current FDR to protect expensive waterfront properties. Furthermore, Warragamba Dam has suppressed downstream flood peaks since 1960 which should reduce bank erosion.

River bank erosion is a major sediment source on many NSW coastal rivers north and west of Sydney during FDRs. The occurrence of frequent, large events does not allow sufficient time between floods to enable significant channel and floodplain recovery by the re-establishment of riparian and floodplain vegetation. As a result, channel erosion and alluvial stripping are not always caused by the single largest flood but by slightly smaller events during runs of frequent large events (Warner, 1972; Erskine and Bell, 1982; Nanson, 1986; Erskine, 1992, 1996). The volume of sediment going into storage during a DDR is largely or completely eroded during the next FDR (Warner, 1994a). Rivers which experience alternating periods of high and low flood activity and large flood variability will always experience significant damage. It should also be stressed that:

1) the amount and extent of bank erosion by severe floods has often increased since European settlement because of reduced bank resistance due to vegetation clearing and stock damage; and
2) the rate of vegetation regeneration following severe floods is now much slower than in the past for the same reasons.

Therefore, the purpose of river management works should be to accelerate the rate of natural biogeomorphic processes to ensure that the post-flood unstable phases are as short as possible. (Erskine, 1993, p.42)

This clearly contradicts Brooks and Brierley's (1997, p. 292) unsubstantiated assertion that the condition of the landscape at the time of flood events has not been considered by the present authors in previous research.

Episodic but large sediment contributions from bank erosion during runs of frequent large floods exceed by many times the mean annual sediment yield. These episodic inputs are often partly stored in the river bed as bedload waves where large dams do not cause downstream sediment starvation. While bank erosion does occur by the well-known lateral migration process, channel enlargement of essentially non-migrating rivers is also very common. There is a large literature on meander migration, but there is a dearth of information on non-migrating and often bedrock-confined rivers. Nanson (1986) and Warner (1992) emphasised the significance of the bedrock valley-floor trough as a control on channel and floodplain erosion and sedimentation. Where there is little room for sediment storage in narrow bedrock- and terrace-confined sections, inset landforms are extensively reworked during FDRs. This effectively cleans out all stored material. Where there are large sediment accumulations in wide valley-floor troughs, lateral migration is a more common response which can still input massive quantities of sediment into the channel.

Warner (1993, 1994a, 1994b) concluded that in-channel sediment storage during DDRs caused substantial channel contraction by the construction of inset floodplains or benches. It is this temporarily stored sediment which is usually reworked during the subsequent FDR. River and catchment management agencies in NSW have long failed to appreciate the dynamic and alternating energy regimes of many coastal rivers. As a result, enormous amounts of money have been spent on protecting temporary sediment storages which have been repetitively cut and filled since European settlement (Warner, 1988). Kirkup et al. (1998) and Brooks and Brierley (1997, 1998) totally ignore such sediment trading on multi-decadal timescales. While this erosion does constitute the major sediment source, the zoning of marginal buffer strips of at least 50 m width (they need to be scaled according to channel width) on the major rivers, in which structures cannot be built and vegetation cannot be destroyed, would largely remove the need for costly "band aid" works during every FDR. There is no point in trying to restore a FDR channel to that which existed during the previous DDR. Yet this is exactly what has been attempted by so-called "river improvement" (Shattock, 1966). The memory of government officers and politicians is too short to manage rivers on a sustainable basis. Furthermore, the mere observation of a sand slug does not necessarily imply that it is a "safe" resource for extraction. This material is usually laterally redistributed into benches and floodplains during periods of low flood activity to be accessed again during subsequent periods of high flood activity (Erskine, 1994). Extraction of this material will cause long-term (i.e. over decades) channel enlargement as the size of the alluvial store is progressively depleted.

There are always going to be problems in communication between researchers and managers because managers do not always read scientific journals and because researchers do not always translate their results into a framework suitable for management. Nevertheless, research findings must be applied to policy and management if ecologically sustainable development is to be achieved. A simplistic interpretation of the present results is to use them as a justification for continued large-scale spending on costly river engineering works during FDRs (Erskine, 1990, 1992). However, a more enlightened view would be to frame management responses which abolish the need for such protection of ephemeral fluvial landforms. Riparian landowners and river managers need to be educated that cyclical channel changes occur in response to hydrological cycles which recur at time spans of 40–50 years. Nanson and Erskine (1988) called such river behaviour cyclical equilibrium although the development of equilibrium conditions at any time has never been established. Cyclical disequilibrium is a more accurate description of the documented channel adjustments. Innovative management responses would recognise these hydrologically driven channel responses and would reserve buffer strips for the re-establishment of riparian corridors and for the provision of public access to rivers. Non-equilibrium rivers require new management strategies.

ACKNOWLEDGEMENTS

Our research was supported by the Australian Research Council, an Australian Postgraduate Research Award and the Faculty of Applied Science, University of New South Wales.

REFERENCES

Abrahams, A. D. and Cull, R. F. 1979. The formation of alluvial landforms along New South Wales coastal streams. *Search*, **10**, 187–188.

Baker, V. R. 1977. Stream-channel response to floods, with examples from central Texas. *Geological Society of America Bulletin*, **88**, 1057–1071.

Bell, F. C. and Erskine, W. D. 1981. Effects of recent increases in rainfall on floods and runoff in the upper Hunter Valley. *Search*, **12**, 82–83.

Brizga, S. O., Finlayson, B. L. and Chiew, F. H. S. 1993. Flood dominated episodes and river management: a case study of three rivers in Gippsland, Victoria. *Hydrology and Water Resources Symposium*, Newcastle, 30 June–2 July 1993, Institution of Engineers Australia, 99–103.

Brooks, A. P. and Brierley, G. J. 1997. Geomorphic responses of lower Bega River to catchment disturbance, 1851–1926. *Geomorphology*, **18**, 291–304.

Brooks, A. P. and Brierley, G. J. 1998. The role of European disturbance in the metamorphosis of lower Bega River. In Finlayson, B. L. and Brizga, S. O. (Eds) *River Management: The Australasian Experience*. John Wiley, Chichester, in press.

Cameron McNamara Pty. Ltd. 1984. *Singleton Floodplain Management Study*. Singleton Shire Council, North Sydney.

Campbell, J. F. 1928. John Howe's exploratory journey from Windsor to the Hunter River in 1819. *Journal and Proceedings of the Royal Australian Historical Society*, **14**, 232–241.

Colby, B. R. 1957. Relationship of unmeasured sediment discharge to mean velocity. *Transactions of the American Geophysical Union*, **38**, 708–717.

Colby, B. R. 1964. Discharge of sands and mean-velocity relationships in sand-bed streams. *US Geological Survey Professional Paper 462A*.

Committee of Advice on Flood Control and Mitigation 1955. *Interim Report No. 2 Flood Control and Mitigation Proposals for the Hunter River Valley*. NSW Government Printer, Sydney.

Cornish, P. M. 1977. Changes in seasonal and annual rainfall in New South Wales. *Search*, **8**, 38–40.

Dury, G. H. 1980. Step-functional changes in precipitation at Sydney. *Australian Geographical Studies*, **18**, 62–78.

Erskine, W. D. 1986a. River metamorphosis and environmental change in the Macdonald Valley, New South Wales since 1949. *Australian Geographical Studies*, **24**, 88–107.

Erskine, W. D. 1986b. River metamorphosis and environmental change in the Hunter Valley, New South Wales. Unpublished PhD Thesis, University of New South Wales.

Erskine, W. D. 1990. Hydrogeomorphic effects of river training works: the case of the Allyn River, NSW. *Australian Geographical Studies*, **28**, 62–76.

Erskine, W. D. 1992. Channel response to large-scale river training works: Hunter River, Australia. *Regulated Rivers: Research and Management*, **7**, 261–278.

Erskine, W. D. 1993. Erosion and deposition produced by a catastrophic flood on the Genoa River, Victoria. *Australian Journal of Soil and Water Conservation*, **6**, 35–43.

Erskine, W. D. 1994. Sand slugs generated by catastrophic floods on the Goulburn River, New South Wales. *International Association of Hydrological Sciences*, **224**, 143–151.

Erskine, W. D. 1996. Response and recovery of a sand-bed stream to a catastrophic flood. *Zeitschrift für Geomorphologie*, **40**, 359–383.

Erskine, W. D. and Bell, F. C. 1982. Rainfall, floods and river channel changes in the upper Hunter. *Australian Geographical Studies*, **20**, 183–196.

Erskine, W. D. and Livingstone, E. A. 1999. In-channel benches: the role of floods in their formation and destruction on bedrock-confined rivers. In Miller, A. and Gupta, A. (Eds) *Varieties of Fluvial Forms*. John Wiley, Chichester, 445–475.

Erskine, W. D. and Saynor, M. J. 1996. Effects of catastrophic floods on sediment yields in southeastern Australia. *International Association of Hydrological Sciences*, **236**, 381–388.

Erskine, W. D. and Warner, R. F. 1988. Geomorphic effects of alternating flood- and drought-dominated regimes on NSW coastal rivers. In Warner, R. F. (Ed.) *Fluvial Geomorphology of Australia*. Academic Press, Sydney, 223–244.

Erskine, W. D. and Warner, R. F. 1998. Further assessment of flood- and drought-dominated regimes in southeastern Australia. *Australian Geographer*, **29**, 257–261.

Erskine, W. D., Geary, P. M. and Outhet, D. N. 1985. Potential impacts of sand and gravel extraction on the Hunter River, New South Wales. *Australian Geographical Studies*, **23**, 71–86.

Hunter River Floods Commission 1870a. *Progress Report of Commission Appointed to Enquire into and Report Respecting Floods in the Hunter River*. NSW Government Printer, Sydney.

Hunter River Floods Commission 1870b. *Report of Commission Appointed to Enquire into and Report Respecting Floods in the District of the Hunter River*. NSW Government Printer, Sydney.

Jervis, J. 1945. The route to the North. John Howe's journey of 1820. *Journal and Proceedings of the Royal Australian Historical Society*, **31**, 276–281.

Kirkup, H., Brierley, G., Brooks, A. and Pitman, A. 1998. Temporal variability of climate in southeastern Australia – a reassessment of flood- and drought-dominated regimes. *Australian Geographer*, **29**, 241–255.

Kriek, P. N. 1970. A report on sedimentation research in the Hunter Valley, NSW. Unpublished paper presented at 42nd ANZAAS Congress, Port Moresby, August 1970.

McIlveen, G. R. 1984. *Singleton 1:25 000 Geological Sheet 9132-IV-N*. NSW Geological Survey, Sydney.

McMahon, T. A., Finlayson, B. L., Haines, A. T. and Srikanthan, R. 1992. *Global Runoff*. Catena Paperback, Cremlingen-Destedt.

Nanson, G. C. 1986. Episodes of vertical accretion and catastrophic stripping: a model of disequilibrium floodplain development. *Geological Society of America Bulletin*, **97**, 1467–1475.

Nanson, G. C. and Erskine, W. D. 1988. Episodic changes of channels and floodplains on coastal rivers in New South Wales. In Warner, R. F. (Ed.) *Fluvial Geomorphology of Australia*. Academic Press, Sydney, 201–221.

Olive, L. J. and Rieger, W. A. 1986. Low Australian sediment yields – a question of inefficient sediment delivery? *International Association of Hydrological Sciences*, **159**, 355–364.

Piest, R. F. 1964. Long-term sediment yields from small watersheds. *International Association of Scientific Hydrology*, 65, 121–140.

Pilgrim, D. H. and Doran, D. G. 1987. Flood frequency analysis. In Pilgrim, D. H. (Ed.) *Australian Rainfall and Runoff*, Vol. 1. Institution of Engineers Australia, Barton, 195–236.

Raine, A. W. and Gardiner, J. N. 1995. *Rivercare. Guidelines for ecologically sustainable management of rivers and riparian vegetation*. Land and Water Resources Research and Development Corporation Occasional Paper Series No. 03/95.

Rankin, D. 1982. Stabilising stream channels by river training and interaction with the environment. *Civil Engineering Transactions Institution of Engineers Australia*, **CE24**, 135–142.

Reddoch, A. F. 1957. River control works in the non-tidal section of the Hunter River and its tributaries. *Journal of Institution of Engineers Australia*, **29**, 241–247.

Rhodes, D. D. 1977. The b-f-m diagram: graphical representation and interpretation of at-a-station hydraulic geometry. *American Journal of Science*, **277**, 73–96.

Rieger, W. A. and Olive, L. J. 1988. Channel sediment loads: comparisons and estimation. In Warner, R. F. (Ed.) *Fluvial Geomorphology of Australia*. Academic Press, Sydney, 69–85.

Riley, S. J., Creelman, R., Warner, R. F., Greenwood-Smith, R. and Jackson, B. R. 1989. The potential in fluvial geomorphology of a new mineral identification technology (QEM*SEM). *Hydrobiologia*, **176/177**, 509–524.

Sammut, J. and Erskine, W. D. 1995. Hydrological impacts of flow regulation associated with the Upper Nepean Water Supply Scheme, NSW. *Australian Geographer*, **26**, 71–86.

Saynor, M. J. and Erskine, W. D. 1993. Characteristics and implications of high-level slackwater deposits in the Fairlight Gorge, Nepean River, Australia. *Australian Journal of Marine and Freshwater Research*, **44**, 735–747.

Scholer, H. A. 1974. *Geomorphology of New South Wales Coastal Rivers*. The University of New South Wales Water Research Laboratory Report No. 139.

Shattock, W. H. 1966. A review of river improvement works on non-tidal streams in New South Wales. *Journal of Institution of Engineers Australia*, **38**, 275–282.

Stevens, M. A., Simons, D. B. and Richardson, E. V. 1975. Nonequilibrium river form. *Journal of Hydraulics Division, Proceedings American Society of Civil Engineers*, **101**, 557–566.

Stevens, M. A., Simons, D. B. and Richardson, E. V. 1977. Closure-nonequilibrium river form. *Journal of Hydraulics Division, Proceedings American Society of Civil Engineers*, **103**, 197–198.

Turner, L. M. and Erskine, W. D. 1997. Thermal, oxygen and salt stratification in three weir pools on the Nepean River, NSW. In *Proceedings Science and Technology in the Environmental Management of the Hawkesbury-Nepean Catchment*, 10–11 July 1997, University of Western Sydney, Nepean, Kingswood. Institution of Engineers Australia National Conference Publication NCP 97/01 and Geographical Society of New South Wales Conference Papers No. 14, 87–92.

Warner, R. F. 1972. River terrace types in coastal valleys of New South Wales. *Australian Geographer*, **12**, 1–22.

Warner, R. F. 1983. Channel changes in the sandstone and shale reaches of the Nepean River, New South Wales. In Young, R. W. and Nanson, G. C. (Eds) *Aspects of Australian Sandstone Landscapes*. Australian and New Zealand Geomorphology Group Special Publication No. 1, 106–119.

Warner, R. F. 1987. The impacts of alternating flood- and drought-dominated regimes on channel morphology at Penrith, New South Wales, Australia. *International Association of Hydrological Sciences*, **168**, 327–338.

Warner, R. F. 1988. Environmental management of coastal rivers. *Planner*, **3**(7), 32–35.

Warner, R. F. 1992. Floodplain evolution in a New South Wales coastal valley, Australia: spatial process variations. *Geomorphology*, **4**, 447–458.

Warner, R. F. 1993. Channel erosion and sedimentation in the alternating hydrologic regimes of the Hawkesbury River, NSW, Australia. In Thoms, M. C. (Ed.) *Catchments and Coasts in Eastern Australia*. Department of Geography, University of Sydney Research Monograph No. 5, 25–39.

Warner, R. F. 1994a. Temporal and spatial variations in erosion and sedimentation in alternating hydrological regimes in southeastern Australian rivers. *International Association of Hydrological Sciences*, **224**, 211–219.

Warner, R. F. 1994b. A theory of channel and floodplain responses to alternating regimes and its application to actual adjustments in the Hawkesbury River, Australia. In Kirkby, M. J. (Ed.) *Process Models and Theoretical Geomorphology*. John Wiley, Chichester, 173–200.

Warner, R. F. 1997. Floodplain stripping: another form of adjustment to secular hydrologic regime change in Southeast Australia. *Catena*, **30**, 263–282.

Wolman, M. G. and Miller, J. P. 1960. Magnitude and frequency of forces in geomorphic processes. *Journal of Geology*, **68**, 54–74.

9 Middle to Late Holocene Environments in the Middle to Lower Trent Valley

A. J. HOWARD,[1] D. N. SMITH,[2] D. GARTON,[3] J. HILLAM[4] and M. PEARCE[5]

[1] Centre for Geoarchaeological Studies, School of Geography, University of Leeds, UK
[2] Department of Ancient History and Archaeology, University of Birmingham, UK
[3] Trent & Peak Archaeological Trust, Nottingham, UK
[4] Dendrochronology Laboratory, Research School of Archaeology and Archaeological Sciences, Department of Archaeology and Prehistory, University of Sheffield, UK
[5] Department of Archaeology, University of Nottingham, UK

INTRODUCTION

Within British river systems, it is suggested that the early Holocene was a period of (relative) stability (Rose *et al.*, 1980), with incision and slow alluviation until *c.* 5200 BP (Macklin and Lewin, 1993). After 5000 BP, a significant change in river activity with respect to erosional and depositional processes and groundwater conditions occurred (Robinson and Lambrick, 1984; Macklin and Needham, 1992; Brown *et al.*, 1994). The mechanisms of change are unclear, although argument centres around an anthropogenic (e.g. Robinson and Lambrick, 1984; Lambrick, 1992) or climatically driven response (Macklin *et al.*, 1992a; Macklin and Lewin, 1993). However, as Macklin *et al.* (1992b) have suggested, these need not be seen as competing hypotheses. In addition, other authors have stressed the importance of local geomorphic conditions in explaining catchment history (e.g. Taylor and Lewin, 1997).

This chapter outlines the evidence for lowland environmental conditions of the River Trent, around *c.* 4000 BP, primarily from Langford Quarry, in the Middle to Lower Trent Valley of Nottinghamshire (Figure 9.1).

Fluvial Processes and Environmental Change. Edited by A. G. Brown and T. A. Quine.
© 1999 John Wiley & Sons Ltd.

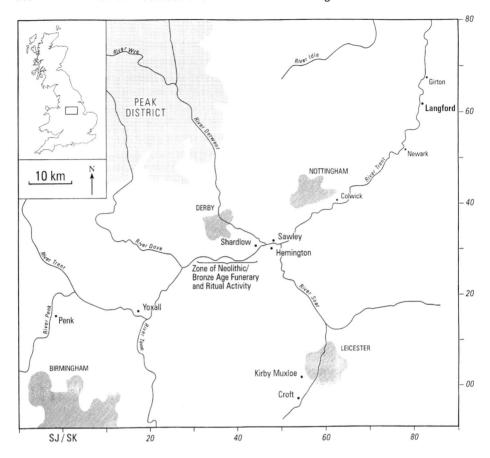

Figure 9.1 The River Trent and associated tributaries. Sites referred to in the text are marked

THE EARLY AND MIDDLE HOLOCENE ENVIRONMENTAL BACKGROUND

In contrast to other major lowland British river systems such as the Severn (Brown, 1983, 1987), the Trent has received little attention, although studies within tributary valleys have provided an important regional picture (e.g. Brown and Keough, 1992; Brown et al., 1994). In addition, palaeobotanical studies of a peat-infilled palaeochannel at Girton Quarry, c. 10 km downstream of Newark, and at the extreme eastern edge of the floodplain, indicates early Holocene vegetation colonisation by birch and pine, and subsequently by thermophilous species such as oak, elm, hazel, alder and lime, resulting in a dense closed-canopy mixed woodland by the later Mesolithic (Green, 1996; Green and Garton, 1998). The sedimentology suggests a "relatively stable" fluvial regime of single or possibly multiple anastomosing (*sensu* Smith and Smith, 1983) channels (Knight and Howard, 1994), with evidence for low energy levels supported by the insect remains (Dinnin, 1992). The basal part of the sequence is tentatively dated from the pollen spectra to the Early Boreal (Pollen Zone 5, c. 9500–8000 BP) and the middle part to the Boreal/Atlantic transition (c. 7000 BP). This fluvial regime may have continued into the Bronze Age (c. 4000 BP), since additionally recorded sections of the palaeochannel, exposed to the south during topsoil stripping, revealed peat interdigitated with sediments washed into the channel from human activity around a "burnt mound" (Garton, 1993).

Analysis of pollen and insect remains from a floodbasin close to the modern Trent at Bole Ings, c. 32 km downstream of Newark, has revealed a detailed palaeoenvironmental record between c. 8300 and 2700 BP, of a low energy, stable river landscape (Dinnin and Brayshay, 1994; Dinnin, 1997). Possible evidence of deforestation exists from c. 5000 BP with indications of intensification after c. 3570 BP. However, it appears that this clearance was more regional and had little effect on the local floodplain woodland environment. The sparing of such tracts of woodland may be explained by the marginal waterlogged nature of the soils (and hence less suitability for agriculture). Evidence from other sites in the Lower Trent and Humberhead Levels suggests that significant fine-grained alluviation of eroded topsoils has occurred since the Romano-British Period, linked to changing agricultural practices (Buckland and Sadler, 1985; Riley et al., 1995).

PALAEOENVIRONMENT: LANGFORD QUARRY

Langford Quarry is situated in the centre of the modern floodplain, adjacent to the River Trent, within the sands and gravels of the valley floor. The exposed sediments comprise Mercia Mudstone bedrock, overlain by c. 4–6 m of sand and gravel, in turn buried beneath c. 2–3 m of fine-grained, inorganic, red-brown alluvium. The sands and gravels are characterised by the inclusion of large tree remains (with intact root boles) and initial similarities may be drawn with the zone of Holocene reworked sands and gravels at Colwick, c. 25 km upstream (Salisbury et al., 1984).

Nowhere at Langford have Devensian deposits been conclusively proved within the sediment body; isolated disturbance structures have been recorded, possibly indicating periglacial activity, but caution has been exercised in the interpretation of these structures, as suggested by Worsley (1996).

The sediments at the quarry were initially studied between 1991 and 1994 following the discovery in the processing plant of a Middle Bronze Age rapier typologically datable to the 15th century BC (D. Knight, pers. comm.). Subsequent fieldwork led to the recovery of a human skull, a red deer femur, auroch bones and a dog skull (Lillie, 1992). This material, and the sedimentology of the sands and gravels, led Lillie and Grattan (1995) to suggest deposition by flooding compounded by hydrological change associated with Bronze Age deforestation of the landscape. However, the lack of secure contexts for the datable material warrants caution with this interpretation (Howard and Knight, 1995). In 1995/6, discovery during extraction at the quarry of four human crania by the drag-line operator, and an articulated human pelvis and femur (by archaeologists) during subsequent inspection of the quarry face, resulted in further work.

Sedimentology and Archaeological Remains

The recorded face (Figure 9.2) comprised massive to sub-horizontally bedded, fine to coarse, clast-supported sand and gravel, overlying and truncated by a complex sequence of planar and trough cross-bedded pebbly sands with multiple reactivation surfaces, broadly infilling a palaeochannel c. 15 m wide and c. 2 m deep (Figure 9.3). This sand-filled channel contained sizeable tree trunks (up to 8 m long), together with smaller branches of wood (including oak, ash, hazel and yew) and fine organic debris. The sands and gravels behind the recorded face were stripped vertically downward to the top of the *in situ* pelvis and femur which became the surface of excavation (Figure 9.2). Rescue excavation of a 2-m-wide box (excavated in 10-cm hand-shovel spits) recovered nearly 200 pieces of human and animal bone including horse, wild and domestic cattle, pig, possibly wild boar, red deer, roe deer, dog and sheep (J. Rackham, pers. comm.). The majority were single bones although there were some semi-articulated groups including part of a human rib-cage and pelvis. This articulation and condition of the bone material suggested little, if any reworking. Dating of a single human rib from the articulated rib-cage (Figure 9.2) yielded a radiocarbon age determination of 3780 ± 50 BP (Beta-87093) and calibrated at 2 sigma to 2350–2030 BC, implying a date for the deposition of the bone in the Late Neolithic to Early Bronze Age.

The quality and age of the remains resulted in funding for further investigation involving excavation of an additional four trenches and 100% wet sieving of the contents of a fifth trench, parallel and adjacent to the first trench. Further archaeological material collected (from beneath the level of the ^{14}C-dated rib) included five human crania, two sheep crania and two domestic cattle skulls, one deer and one dog skull, a few flint artifacts (mostly blades) and a piece of fine wickerwork (c. 30 × 40 cm).

This further excavation allowed the site to be interpreted as a sizeable organic accumulation in a river channel, here termed a log-jam, over 10 m wide, with the

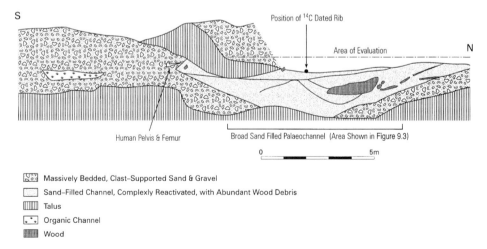

Figure 9.2 Recorded section of the quarry face. Area of archaeological evaluation and later full excavation, and position of *in situ* human pelvis and femur are annotated

Figure 9.3 Recorded section of the quarry face and area of evaluation. Pieces of the wood forming part of the log-jam are visible, tagged by white markers

human and animal remains and other organic debris accumulating within the channel and against the gravel-bank. The origin of the human remains is uncertain, although one explanation is some sort of mortuary practice, perhaps linked to the ritual focus of rivers and wetlands in later prehistory (Garton *et al.*, 1996; cf. Bradley, 1990).

Dendrochronology

Dendrochronological analysis of timbers collected from the log-jam (during excavation) and the wider quarry area (by C. Salisbury and R. Howard of Nottingham University in 1995; and J. Hillam *et al.*, University of Sheffield in 1996) revealed three datable chronologies. Five samples of wood from the log-jam and 19 (11 of which were parts of the same two or three trees) from the wider quarry environment were dated to a chronology spanning 513 years, between 2637 and 2125 BC. With specific reference to samples from the log-jam (ES 09, 11, 13, 20 and 26), the chronology spanned 281 years between 2424 and 2143 BC. In addition, two earlier chronologies were recognised from the wider quarry environment: six samples spanning 212 years between 4232 and 4021 BC; and two samples spanning 122 years between 2979 and 2858 BC.

Insect Remains

Two death assemblages were recovered from deposits of sand and silt behind the log-jam and a third from the soil in a tree bole (Tables 9.1 and 9.2). The majority of the insects recovered were beetles (Coleoptera) or caddis flies (Trichoptera).

Table 9.1 The distribution of the Elmid species from archaeological deposits in the Middle Trent Watershed

	Head water		Floodplain				Main channel back-swamps	
	Kirby Muxloe	Croft	Penk	Yoxal	Shardlow	Girton	Hemington	Langford
Helichus substriatus (Müll.)		✓						
Stenelmis canaliculata (Gyll.)	●						●	●
Elmis aenea (Müll.)		●	●				●	●
Esolus parallelepipedus (Müll.)	✓	●	✓				●	●
Oulimnius spp.	✓	●	●	✓			●	●
Limnius volckmari (Panz.)		●	✓	✓			●	●
Riolus cupreus (Müll.)		✓	✓				✓	✓
R. subviolaceus (Müll.)							●	●
Macronychus quadrituberculatus (Müll.)								●

✓, Few individuals present; ●, > 10 individuals present.
Sources: Rosseff *et al.* (in prep), Smith (1994a, 1994b, 1996, 1998, in press), Howard *et al.* (in prep).

A significant proportion of the water beetles (72–83% of all aquatic species present) are associated with fast running water conditions, with elmids ("riffle beetles") being particularly well represented. These include *Elmis aenea* (Müll), *Esolus parallelepipedus* (Müll), *Oulimnius* spp., *Limnius volckmari* (Panz.), *Riolus cupreus* (Müll), *Riolus subviolaceus* (Müll) and *Stenelmis caniculata* (Gyll.). Such species cling to the underside of stones and gravel on channel bottoms in areas of strong currents (Holland, 1972; Steffan, 1979; Friday, 1988). Other Dryopidae present, including *Helichus substriatus* (Müll) and *Macronychus quadrituberculatus* (Müll), are often associated with submerged wood in such river conditions (Holland, 1972; Steffan, 1979; Friday, 1988), perhaps indicating that the log-jam may have remained underwater.

Comparison of the elmid assemblage from Langford with those from other archaeological sites in the middle Trent catchment indicate that the (Langford) samples contain a wider and more numerous range of elmids than the headwater and floodplain back-swamp sites (Figure 9.1 and Table 9.1). The only relatively comparable site is the 11th century AD bridge at Hemington, Leicestershire, which is clearly associated with a main channel of the Trent. These species were probably widespread in the main channels of many Neolithic and Bronze age lowland rivers (e.g. the Upper Thames: Robinson, 1991, 1993), and the Warwickshire Avon (Osborne, 1988), but are absent from such river systems today (Holland, 1972; Shirt, 1987; Osborne, 1988; Robinson, 1991). Osborne (1988) has suggested that this decline is probably due to the in-channel deposition of alluvial silts after the Bronze Age. However, recent work at the Hemington bridges site suggests that some lowland river sections may not have been effected by fine-grained sedimentation until much later and that local factors have to be considered (Smith, in press).

A large range of insects associated with woodland were also recovered, accounting for between 20 and 30% of terrestrial insects present (Table 9.2). Robinson (1983, 1993) has suggested that such numbers of woodland taxa indicate a dense tree canopy. The range of trees associated with these species suggests the presence of oak, beech, lime, elm, ash, hazel and alder. In addition, the insects present the full range of ecological niches found in mature woodland (i.e. feeders of foliage, sap wood, dry tree moulds and bracket fungi, well-decayed wet and rotten wood, and predators). The woodland faunas, particularly from the tree bole, are directly derived from the woodland floor, rather than as accidental captures in water-bodies and back-swamps generally assumed to be part of a "flight fauna" (i.e. those species, such as the Scolytids, which are very active flyers and will distribute themselves widely across the landscape).

Despite the breadth of this woodland fauna, only one species present is considered to be rare, the lime feeder *Ernoporus caucasicus*. It is only found in small numbers at a few scattered sites in the British Isles today (Hyman and Parsons, 1992), although fossil finds suggest that it was more widespread, and its decline may be due to the reduction of lime since the Bronze Age (Girling, 1982, 1985; Robinson, 1991, 1993; Dinnin, 1997). None of the 35 *Urwaldrelikt* ("old woodland") species, which are now extinct in the British Isles but have been found at a number of Neolithic and Bronze Age sites (Osborne, 1972; Buckland, 1979; Girling, 1985; Buckland and Dinnin, 1993; Whitehouse, 1997), are recorded. This may be either because in the lowland woods of the Midlands such *Urwaldrelikt* species were not as abundant in the insect fauna as they

Table 9.2 The woodland insects from Langford Quarry (taxonomy follows Lucht, 1978)

Species	Associated tree
Carabidae	
Cychrus caraboides (L.)	—
Pterostichus oblongopuntatus (F.)	—
Abax parallepipedus (Pill. Mitt.)	—
Histeridae	
Abraeus globosus (Hoffm.)	Mainly moist rotten wood of beech and ash
Catopidae	
Nemadus colonoides (Kr.)	Birds' nests and in the dry interior of oak, beech, elm and lime
Rhizophagidae	
Rhizophagus picipes (Ol.)	Predator under bark of pine, beech and oak
Mycetophagidae	
Litargus connexus (Fourcr.)	On fungus under the bark of beech, ash and oak
Mycetophagus quadriguttatus Müll	On bracket fungus and fungus under the bark of oak and beech
Colydiidae	
Cercylon histeroides (F.)	Predating on bark beetles in oak, lime and beech
Cisidae	
Ennearthron cornutum (Gyll.)	In tree fungi and under bark
Anobiidae	
Hedobia imperialis (L.)	In the brittle or rotten wood of deciduous trees
Grynobius planus (F.)	In the brittle or rotten wood of deciduous trees
Ochina ptinoides (Marsh.)	In old and dead ivy stems
Xestobium rufovillosum (Geer)	In red or white rots in decaying oak and willows
Dorcatoma flavicornis (F.)	In the dead wood of mainly red rot in oak
Scarabaeidae	
Aphodius zenkeri Germ.	Usually in shaded deer dung
Lucanidae	
Dorcus parallelopipedus (L.)	In the rotten and crumbling wood of deciduous trees
Cerambycidae	
Grammoptera ruficornis (F.)	Larvae develop in the twigs of deciduous trees
Scolytidae	
Scolytus mali (Bechst.)	Under the bark of fruit and elm trees
S. scolytus (F.)	Under the bark of elm
Hylesinus oleiperda (F.)	In twigs of ash
Leperisinus varius (F.)	Under the bark of ash
Pteleobius vittatus (F.)	Under the bark of elm and occasionally ash
Dryocoetes villosus (F.)	Under the bark of oak and beech
Dryocoetes alni (Georg.) Notable B	Under the bark of alder, beech and hazel
Ernoporus caucasicus Lindem. RDB 1	Under the bark of lime, mainly small-leafed lime
Xyleborus saxeseni (Ratz.)	In oak, elm and pine
Platypodidae	
Platypus cylindrus (F.)	In the dead wood of oak, ash and beech
Curculionidae	
Phylobius maculicornis Germ.	Feeds on foliage of a range of trees and bushes
Polydrusus pterygomalis Boh.	Feeds on foliage of a range of trees and bushes
Stereocorynes truncorum (Germ.)	Damaged and decaying wood of a range of deciduous trees
Curculio venosus (Grav.)	In oak acorns
Curculio salicivorus Payk.	On willow
Trachodes hispidus (L.)	In dry dead wood of a range of deciduous trees
Acalles turbatus (Bohem.)	In dead wood of a range of deciduous trees
Rhynchaenus stigma (Germ.)	On the leaves

were in undisturbed carr woodlands (e.g. those of the Somerset levels: Girling, 1982) and the burnt pine woods of Thorne Moors (Whitehouse, 1997) at this time; or because the faunas found at Langford are so localised (in one case they are from a single tree bole) that by chance these species are just not present.

The death assemblages contain some evidence for grazed clearings in these woodlands. Around 15% of the terrestrial species are associated with plants in grassland or, more often, with the dung of animals lying in the open (including *Geotrupes* and *Aphodius* dung beetles). The insects alone do not allow distinction of glades resulting from the action of domesticated or wild animals (e.g. deer: Buckland and Edwards, 1984), although the presence *of Aphodius zenkeri* Germ may be of interest since this normally feeds on deer dung (Jessop, 1986).

The fauna from the tree bole includes numbers of species which are often seen as typical of human settlement in the archaeological record (i.e. Hall and Kenward, 1990; Kenward and Hall, 1995). These include *Cryptophagus* (probably *C. scanicus* or *C. dentatus*), several individuals of the spider beetle *Ptinus fur*, the darkling beetle *Tenebrio molitor* and the carpet beetle *Attagenus pellio*. However, these species have also been found in dead wood (Harde, 1984). Their occurrence here would appear to partially confirm the suggestion of Kenward and Allison (1994a, 1994b) that dead wood may have provided the original home for these species before they started to utilise the environments offered by human settlement. The find of *Tenebrio molitor* is also of note since this is the first individual of this species recovered from the U.K. palaeoentomological record (P. C. Buckland, pers. comm.).

DISCUSSION

It is suggested that the organic-rich sediment body with associated archaeological remains at Langford accumulated as a log-jam within a small, shallow, high energy channel. The form of the channel, sedimentological characteristics (including multiple reactivation surfaces) and lack of cohesive bank materials suggests that it could have been part of a multiple, unstable system with a variable flow regime. The identification of three dendrochronologies within the sediments at Langford suggests reworking of the sands and gravels and incorporation of tree remains since $c.$ 6000 BP and supports the concept of long-term reworking of sands and gravels of the floodplain (Howard and Knight, 1995), in contrast to the hypothesis of their emplacement by Bronze Age flooding (Lillie and Grattan, 1995). The origin of the log-jam is uncertain and deliberate impounding of the water course cannot be ruled out. Some of the smaller wood fragments and three of the trunks had deliberately cut ends (Figure 9.4) and were the result of human action as opposed to activity of animals such as beavers (Coles, 1992). This could suggest woodland management (Brown and Keough, 1992) or wider-scale forest clearance associated with agricultural development and/or ritual use of the landscape, of which there is significant evidence in the Middle Trent Valley around the Trent–Derwent confluence, $c.$ 30 km upstream (Knight and Howard, 1994). However, recent work at Bole Ings (Dinnin and Brayshay, 1994) indicates that although regional deforestation was occurring possibly from $c.$ 5000 BP, and certainly with intensity from $c.$ 3570 BP, the floodplain woodland was left largely undisturbed, with clearance

Figure 9.4 Large tree trunk with deliberately cut end. Scale-bar divisions are 0.10 m

of the drier soils of the higher terraces, although the authors do recognise that this is in contrast to other prehistoric sites in the Trent basin, e.g. Fisherwick. Investigations at Girton, c. 5 km downstream, implies a significantly different fluvial landscape of a "relatively stable" single or possibly multiple (anastomosing) system with abundant tree cover, although this site, which is not securely dated, is at the extreme eastern edge of the floodplain and may not be representative of channel and floodplain conditions in the active zone. The incorporation of large trees with intact root boles indicates uprooting of floodplain woodland and thus energetic (flood) conditions – an argument used for the incorporation of variably aged large tree remains within the sand and gravel body at Colwick, c. 25 km upstream (Salisbury et al., 1984). The radiocarbon and dendrochronological results indicate an upper age limit for the sediments recorded within the log-jam at Langford of between c. 4300 and 3980 BP, fitting approximately within the period of hydroclimatological variability between 4800 and 4200 BP, which included significant gravel deposition (Macklin and Needham, 1992; Macklin and Lewin, 1993). However, the direct evidence for woodland clearance supports the argument for consideration of human impact. What is clear is that once large organic debris has been introduced into the channel, this in turn can become a causal factor influencing channel planform and flow regime at both local and reach scales (Keller and Swanson, 1979; Mosley, 1981; Wallerstein and Thorne, 1994).

This study therefore questions the validity of single-site models and reinforces the necessity to reconstruct riverine landscapes at a multi-site reach scale. It seems likely that these contrasting environments at Langford, Girton and Bole Ings existed contemporaneously with the differing styles explained by site location on the floodplain with respect to the river. This model is supported by data collected upstream in the Middle Trent. Around Shardlow, close to the floodplain edge, radiocarbon dating of three separate peat-infilled palaeochannels indicates stability of the floodplain between 2500 and 4000 BP (Beta-100926, Beta-100927: Trent & Peak Archaeological Trust; Beta-099237, Beta-099239: Knight and Malone, 1997). In contrast, at Colwick and Hemington, adjacent to the river, sedimentological investigations by Salisbury *et al.* (1984) and Salisbury (1992) indicate higher energy conditions similar to those observed at Langford.

CONCLUSIONS

This study illustrates the importance of fluvial environments in preserving evidence of human activity and landscape change. It also illustrates the need for palaeoenvironmental interpretation of river systems at a multi-site, reach scale. At Langford, the identification of coarse sediment deposition, combined with direct evidence of human impact on the landscape, might result in speculation of changing catchment hydrology linked to causal mechanisms involving anthropogenic destabilisation. However, the sediments at Girton and Bole Ings indicate contemporary stability of the floodplain. Significant palaeoenvironmental and palaeoentomological information at Langford was gained from coarse sands and gravels and indicates that if only "classic peat sections" are sampled in river valleys, a great quantity of valuable information could be lost.

ACKNOWLEDGEMENTS

This research could not have taken place without the permission, patience and generous logistical support of Tarmac Quarry Products (Eastern) Ltd, in particular Mr Neil Beards (Estates Manager) and Mr Richard Hunter (Langford Quarry Manager). Funding was jointly provided by Tarmac and Nottinghamshire County Council and the support and advice of the County Archaeologist, Mr Mike Bishop is gratefully appreciated. Alongside T&PAT staff, the efforts of volunteer (student) excavators from the Department of Archaeology, Nottingham, the Retford and Hunter Archaeological Society and Artemus fieldwork group is acknowledged. The dendrochronology was funded through English Heritage Research Project AT116. Insect remains were processed by Jayne Barrett. The wood was identified by Dr C. Salisbury who also provided invaluable general advice. Thanks to the Graphics Unit, School of Geography, Leeds, for drawing the figures.

REFERENCES

Bradley, R. 1990. *The Passage of Arms: An Archaeological Analysis of Prehistoric Hoards and Votive Deposits.* Cambridge University Press.

Brown, A. G. 1983. Floodplain deposits and accelerated sedimentation in the lower Severn basin. In Gregory, K. J. (Ed.) *Background to Palaeohydrology.* John Wiley, Chichester, 375–397.

Brown, A. G. 1987. Holocene floodplain sedimentation and channel response of the lower Severn, United Kingdom. *Zeitschrift für Geomorphologie* **31**, 293–310.

Brown, A. G. and Keough, M. 1992. Palaeochannels, paleolandsurfaces and three dimensional reconstruction of floodplain environmental change. In Carling, P. A. and Petts, G. E. (Eds) *Lowland Floodplain Rivers: Geomorphological Perspectives.* John Wiley, Chichester, 185–202.

Brown, A. G., Keough, M. and Rice, R. J. 1994. Floodplain evolution in the East Midlands, United Kingdom: the Lateglacial and Flandrian alluvial record from the Soar and Nene valleys. *Philosophical Transactions of the Royal Society of London,* **A348**, 261–293.

Buckland, P. C. 1979. *Thorne Moors: A Palaeoecological Study of a Bronze Age Site (A Contribution to the History of the British Insect Fauna).* Department of Geography Occasional Publication 8, University of Birmingham.

Buckland, P. C. and Dinnin, M. J. 1993. Holocene woodlands: the fossil insect evidence. In Kirby, K. and Drake, C. M. (Eds) *Dead Wood Matters: the Ecology and Conservation of Saproxylic Invertebrates in Britain.* English Nature Science 7, Peterborough, 6–20.

Buckland, P. C. and Edwards, K. J. 1984. The longevity of pastoral episodes of clearance activity in pollen diagrams: the role of post-occupation grazing. *Journal of Biogeography,* **11**, 243–249.

Buckland, P. C. and Sadler, J. P. 1985. The nature of late Flandrian alluviation in the Humberhead Levels. *East Midland Geographer,* **11**, 239–251.

Coles, B. 1992. Further thoughts on the impact of beaver in temperate landscapes. In Needham, S. and Macklin, M. G. (Eds) *Alluvial Archaeology in Britain.* Oxbow Monograph 27, Oxbow Books, Oxford, 93–99.

Dinnin, M. 1992. Islands within islands: the development of the British entomofana during the Holocene and the implications for conservation. Unpublished PhD Thesis, University of Sheffield.

Dinnin, M. 1997. Holocene beetle assemblages from the Lower Trent floodplain at Bole Ings, Nottinghamshire, UK. *Quaternary Proceedings,* **5**, 83–104.

Dinnin, M. and Brayshay, B. 1994. Palaeoenvironmental evidence for the Holocene development of the River Trent floodplain at Bole Ings, Nottinghamshire. Unpublished ARCUS Report 161, University of Sheffield.

Friday L. E. 1988. A key to the adults of the British water beetles. *Field Studies,* **7**, 1–152

Garton, D. 1993. A burnt mound at Waycar Pasture, Girton, Nottinghamshire: an interim report. *Transactions of the Thoroton Society of Nottinghamshire,* **97**, 148–149.

Garton, D., Howard, A. J. and Pearce, M. 1996. Neolithic riverside ritual? Excavations at Langford Lowfields, Nottinghamshire. In Wilson, R. J. A. (Ed.) *From River Trent to Raqqa. Nottingham University Archaeological Fieldwork in Britain, Europe and the Middle East, 1991–1995.* Nottingham Studies in Archaeology Volume 1, Oxford, 9–11.

Girling, M. A. 1982. Fossil insect faunas from forest sites. In Limbrey, S. and Bell, M. (Eds) *Archaeological Aspects of Woodland Ecology.* British Archaeological Report S146, Oxford, 129–146.

Girling, M. A. 1985. An "Old-Forest" beetle fauna from a Neolithic and Bronze age peat deposit at Stileway. *Somerset Levels Papers,* **11**, 80–83.

Green, F. 1996. Pollen analysis from an infilled meander of the River Trent, Girton Quarry, Nottinghamshire. Unpublished report, Trent & Peak Archaeological Trust, Nottingham.

Green, F. and Garton, D. 1998. *Evidence for Early and Middle Holocene palaeoenvironments of the River Trent from Girton, Nottinghamshire.* Unpub. Report, Trent and Peak Archaeological Trust, Nottingham.

Hall, A. R. and Kenward, H. K. 1990. *Environmental Evidence from the Collonia.* The Archaeology of York, 14/6, Council for British Archaeology, London.

Harde, K. W. 1984. *A Field Guide in Colour to British Beetles.* Octopus Books, London.

Holland, D. G. 1972. *A Key to the Larvae, Pupae and Adults of the British Species of Elminthidae.* Freshwater Biological Association Scientific Publication No. 26.

Howard, A. J. and Knight, D. 1995. Geomorphological and palaeoenvironmental investigations in the Lower Trent Valley: Bronze Age "Flood Events" or long term fluvial reworking? *East Midland Geographer,* **18**, 89–91.

Hyman, P. S. and Parsons, M. S. 1992. *A Review of the Scarce and Threatened Coleoptera of Great Britain,* No. 3. UK Nature Conservation, Peterborough.

Jessop, L. 1986. *Handbooks for the Identification of British Insects. V. Part 11. Dung Beetles and*

Chafers. Coleoptera: Scarabaeoidea. Royal Entomological Society of London.
Keller, E. A. and Swanson, F. J. 1979. Effects of large organic material on channel form and fluvial processes. *Earth Surface Processes*, **4**, 361–380.
Kenward, H. K. and Allison, E. P. 1994a. A preliminary view of the insect assemblages from the Early Christian Rath Site at Deer Park Farms, Northern Ireland. In Rackham, J. (Ed.) *Environment and Economy in Anglo-Saxon England.* Council for British Archaeology Research Report 89, London, 89–103.
Kenward, H. K. and Allison, E. P. 1994b. Rural origins of the urban insect fauna. In Hall, A. R. and Kenward, H. K. (Eds) *Urban–Rural Connexions: Perspectives from Environmental Archaeology.* Oxbow Monograph 47, Oxbow Books, Oxford, 55–57.
Kenward, H. K. and Hall, A. R. 1995. *Biological Evidence from Anglo-Scandinavian Deposits at 16–22 Coppergate.* The Archaeology of York, 14/7, Council for British Archaeology, London.
Knight, D. and Howard, A. J. 1994. Archaeology and alluvium in the Trent Valley: An archaeological assessment of the floodplain and gravel terraces. Unpublished report, Trent & Peak Archaeological Trust, Nottingham.
Knight, D. and Malone, S. 1997. Evaluation of late Iron Age and Romano-British settlement and palaeochannels of the Trent at Chapel Farm, Shardlow and Great Wilne, Derbyshire. Unpublished report, Trent & Peak Archaeological Trust, Nottingham.
Lambrick, G. 1992. Alluvial archaeology of the Holocene in the Upper Thames Basin 1971–1991: a review. In Needham, S. and Macklin, M. G. (Eds) *Alluvial Archaeology in Britain.* Oxbow Monograph 27, Oxford, 209–226.
Lillie, M. C. 1992. Contributions towards an understanding of Holocene alluvial sequences in the Lower Trent Valley. Unpublished MSc Thesis, University of Sheffield.
Lillie, M. C and Grattan, J. P. 1995. Geomorphological and palaeoenvironmental investigations in the Lower Trent Valley. *East Midland Geographer*, **18**, 12–24.
Lucht, W. H. 1987. *Die Kafer Mitteleuropus Katalog.* Goecke & Evers Velag, Krefeld.
Macklin, M. G. and Lewin, J. 1993. Holocene river alluviation in Britain. In Douglas, I. and Hagedorn, J. (Eds) *Geomorphology and Geoecology, Fluvial Geomorphology. Zeitscrift für Geomorphologie*, supplement 85.
Macklin, M. G. and Needham, S. 1992. Studies in British alluvial archaeology: potential and prospect. In Needham, S. and Macklin, M. G. (Eds) *Alluvial Archaeology in Britain.* Oxbow Monograph 27, Oxford, 9–23.
Macklin, M. G., Rumsby, B. T. and Heap, T. 1992a. Flood alluviation and entrenchment: Holocene valley floor development and transformation in the British uplands. *Geological Society of America Bulletin*, **104**, 631–643.
Macklin, M. G., Passmore, D. G. and Rumsby, B. T. 1992b. Climatic and cultural signals in Holocene alluvial sequences: the Tyne basin, northern England. In Needham, S. and Macklin, M. G. (Eds) *Alluvial Archaeology in Britain.* Oxbow Monograph 27, Oxbow Books, Oxford, 123–139.
Mosley, M. P. 1981. The influence of organic debris on channel morphology and bedload transport in a New Zealand forest stream. *Earth Surface Processes*, **6**, 571–579.
Osborne, P. J. 1972. Insect faunas of Late Devensian and Flandrian age from Church Stretton, Shropshire. *Philosophical Transactions of the Royal Society of London*, **B263**, 327–367.
Osborne, P. J. 1988. A Late Bronze age insect fauna from the River Avon, Warwickshire, England: its implications for the terrestrial and fluvial environment and for climate. *Journal of Archaeological Science*, **15**, 715–727.
Riley, D. N., Buckland, P. C. and Wade, J. S. 1995. Aerial reconnaissance and excavation at Littleborough-on-Trent, Notts. *Britannia*, **26**, 253–284.
Robinson, M. A. 1983. Arable/pastoral ratios from insects? In Jones, M. (Ed.) *Integrating the Subsistance Economy.* British Archaeological Report S181, Oxford, 19–47.
Robinson, M. A. 1991. Neolithic and Late Bronze Age insect assemblages. In Needham, S. (Ed.) *Excavation and Salvage at Runnymede Bridge 1978.* British Museum Press and English Heritage, London, 277–325.
Robinson, M. A. 1993. The scientific evidence. In Allen, T. G. and Robinson, M. A. (Eds) *The*

Prehistoric Landscape and Iron Age Enclosed Settlement at Mingies Ditch, Hardwick-with-Yelford, Oxon. Thames Valley Landscapes: The Windrush Valley Volume 2. The Oxford Archaeological Unit, 101–141.

Robinson, M. A. and Lambrick, G. H. 1984. Holocene alluviation and hydrology in the Upper Thames Basin. *Nature,* **308**, 809–814.

Rose, J., Turner, C., Coope, G. R. and Bryan, M. D. 1980. Channel changes in a lowland river catchment over the last 13,000 years. In Cullingford, R. A., Davidson, D. A. and Lewin, J. (Eds) *Timescales in Geomorphology.* John Wiley, Chichester, 159–176.

Roseff, R., Brown, A. G., Butler, S., Hughes, G., Monkton, A., Moss, A. and Smith, D. N. 1998. Archaeological and environmental investigation of a Late Glacial and Flandrian river sequence on the River Soar at Croft, Leicestershire, in prep., Unpublished Report, Birmingham University Field Archaeology Unit.

Salisbury, C. R. 1992. The archaeological evidence for palaeochannels in the Trent Valley. In Needham, S. and Macklin, M. G. (Eds) *Alluvial Archaeology in Britain.* Oxbow Monograph 27, Oxbow Books, Oxford, 155–162.

Salisbury, C. R., Whitley, P. J, Litton, C. D. and Fox, J. L. 1984. Flandrian courses of the River Trent at Colwick, Nottingham. *Mercian Geologist,* **9**, 189–207.

Shirt, D. B. 1987. *British Red Data Books: 2. Insects.* Nature Conservancy Council.

Smith, D. G. and Smith, N. D. 1983. Sedimentation in anastomosed river systems: examples from alluvial valleys near Banff, Alberta. *Journal of Sedimentary Petrology,* **50**, 157–164.

Smith, D. N. 1994a. An assessment of the Coleoptera from the burnt mound at Girton Quarry, Nottinghamshire. Unpublished report to Trent & Peak Archaeological Trust.

Smith, D. N. 1994b. The insect remains from Yoxal, Staffordshire. In Roseff, R. (Ed.) *An Archaeological Evaluation at Yoxal, Staffordshire.* Birmingham University Field Archaeology Unit, Report No. 313, Birmingham.

Smith, D. N. 1996. Shardlow Stocking Area. Shardlow Quarry, Derbyshire. The coleopterous faunas from the relict channel deposits. Unpublished report, Trent and Peak Archaeological Trust, Notingham.

Smith, D. N. 1998. The Coleoptera from the environmental sequences at Kirby Muxloe, Leicestershire. In Beamish, M. and Cooper, L. (Eds) *Excavations of Bronze Age to Iron Age Sites on the Leicester Western By-Pass.* Leicestershire Museums Publications.

Smith, D. N. in press. Disappearance of Elmid "riffle beetles" from lowland river systems – the impact of alluviation. In O'Connor, T. and Nicholson, R. (Eds) People as an Agent of Environmental Change. Oxbow Monograph, Oxbow Books, Oxford.

Steffan, A. W. 1979. Familie: Dryopidae. In Freude, H., Harde, K. W. and Lohse, G. A. (Eds) *Die Käfer Mitteleuropas,* Band 6. Goecke & Evers Verlag, Krefeld.

Taylor, M. P. and Lewin, J. 1997. Non-synchronous response of adjacent floodplain systems to Holocene environmental change. *Geomorphology,* **18**, 251–264.

Wallerstein, N. and Thorne, C. R. 1994. *Impact of In-channel Organic Debris on Fluvial Process and Channel Morphology, Yazoo Basin, Mississippi.* Occasional Paper 24, Department of Geography, University of Nottingham, Nottingham.

Whitehouse, N. J. 1997. Insect remains associated with *Pinus sylvestris* L. from the mid-Holocene of the Humberhead Levels, Yorkshire, UK. *Quaternary Proceedings,* **5**, 293–303.

Worsley, P. 1996. On ice-wedge casts and soft sediment deformations. *Quaternary Newsletter,* **78**, 1–7.

Section 3

FLOODPLAIN PROCESSES

10 Alluvial Microfabrics, Anisotropy of Magnetic Susceptibility and Overbank Processes

C. ELLIS[1] and A. G. BROWN[2]
[1] *AOC (Archaeology Group) Ltd, Edinburgh, UK*
[2] *Department of Geography, University of Exeter, UK*

INTRODUCTION

An essential part of any environmental interpretation of alluvial sediments is the determination of alluvial facies and processes of deposition (Brown, 1997). There are a wide variety of potential depositional processes on floodplains including in-channel processes such as bar formation and bank deposition, and overbank processes such as settling due to velocity reductions, impact sedimentation (Brown and Brookes, 1997), draw-down sedimentation, colluviation and subaerial processes such as aeolian deposition. In floodplains with relatively stable channels key processes are levee formation and sedimentation in floodplain hollows, secondary channels and palaeochannels. Channel infilling can form a major component of the floodplain sediment body and yet little is known of the processes of sedimentation and controlling conditions. A particular problem facing the reconstruction of alluvial response to environmental change is the determination of the number of functioning channels. Although the processes of avulsion have been studied in detail (Richards *et al.*, 1993) channel abandonment has received far less attention partly because it is often a partial and gradual process.

Many sediment characteristics can be used to infer depositional processes, including grain-size distribution (Bagnold and Barndorf-Neilson, 1980; Erskine *et al.*, 1982; Brown, 1985; Marriott, 1996), organic matter content and type, sedimentary structures and sedimentary fabric. Fabric measurements of coarse fluvial sediments have long been used to determine hydrodynamic conditions of deposition, e.g. bed conditions, velocity and flow direction (Todd, 1996). However, fabric analyses of fine-grained fluvial sediments, silt and clay-sized material, have received far less attention. This is primarily because of the tedious nature of conventional microscopic measurements and secondly because interpretation is complicated by an array of post-depositional factors. Alluvial microfabric is the end result of long-term reorganisation of soil/sediment constituents by weathering, bioturbation, stress and translocation (Greene-

Fluvial Processes and Environmental Change. Edited by A. G. Brown and T. A. Quine.
© 1999 John Wiley & Sons Ltd.

Kelly and Mackney, 1970). The reorganisation of microfabrics can be recognised in parallel-striated and cross-striated b-fabrics suggesting past vertic behaviour. Cross-striated b-fabrics have been regarded as indicative of shearing stresses created by shrink–swell, dry–waterlogged conditions, partial gleying, or load pressure in clayey soils or sediments (Brewer, 1964; Burnham, 1970; Greene-Kelly and Mackney, 1970; Miedema and Oort, 1990; Sole-Benet *et al.*, 1990). Speckled or crystallitic b-fabrics are porous with an elastic porphyric microfabric (Miedema and Oort, 1990). The intensity of the b-fabric can be quantified using point counting (Bullock *et al.*, 1985). The extinction of the birefringence streaks gives the orientation of the clay mineral grains, their arrangement and spatial patterning; this can be used to determine the degree and type of reorganisation of alluvial microfabrics (Courty *et al.*, 1989). Burnham (1970) suggests that the main controlling factors of fabric type in calcareous silts and claystones (including fine alluvial sediments) are the environment and conditions of deposition, and within this the chemical content of the water, electrolyte content, clay mineralogy, proportion of coarse sand, organic matter and degree of loading may play a role. The type of fabric in clay-rich sediments is related to the form in which the clay particles are deposited. Flocculated clay occurs when the particles are suspended in a solution rich in electrolyte; the concentration of electrolyte being determined by clay mineralogy, hydrogen ion content and the form of other ions present. The form adopted by flocculated clay is that of domains which are composed of randomly oriented clay particles, and the reorientation of clay (granostriation) will occur around mineral grains, voids and occasionally silt particles, due to stress imposed by a number of processes including compaction, de-watering or gleying. Clay coatings may also result from a number of processes, common in floodplains where floodwaters slake the soil, releasing fine clay/silt-sized particles which are drawn-down by the falling alluvial watertable and deposited in pores and voids. In gleyed soils the fluctuating groundwater slakes the soil releasing silt/clay into suspension, and Theocharopoulos and Dalrymple (1987) found that a build up of laminations can take place when the previous coating is still in a wet form, resulting in a gradual build up of micro-laminae. Abundant micro-laminated and orientated, speckled and pure clay coatings in pores, vugs, channels and chambers indicate clay translocation under at least periodically subaerial conditions. Alluvial microfabric can therefore indicate the initial conditions of deposition (subaerial or permanently submerged) and post-depositional disturbance by a variety of processes, so providing a sedimentary history for a segment of floodplain (Figure 10.1). Measurement of microfabric based on the variations in magnetic susceptibility of samples has yielded results comparable to those obtained optically (Hamilton and Rees, 1970).

Anisotropy of magnetic susceptibility (AMS) is a method of three-dimensional petrofabric analysis and is used in the estimation of preferred orientation of grains in sediments (Rees, 1961; Hamilton, 1967; Hamilton and Rees, 1970; Ellwood, 1984). The purpose of studying AMS is to obtain information on the depositional processes of all the grains, although the magnetic content is generally a small proportion of the whole sample (Hrouda, 1982; Johns and Jackson, 1991). The magnetic susceptibility method assumes that for non-spherical, mineral grains the easy direction of magnetism is parallel to the long dimensions of the grains. The correlation between magnetic susceptibility and grain shape enables directional information to be ex-

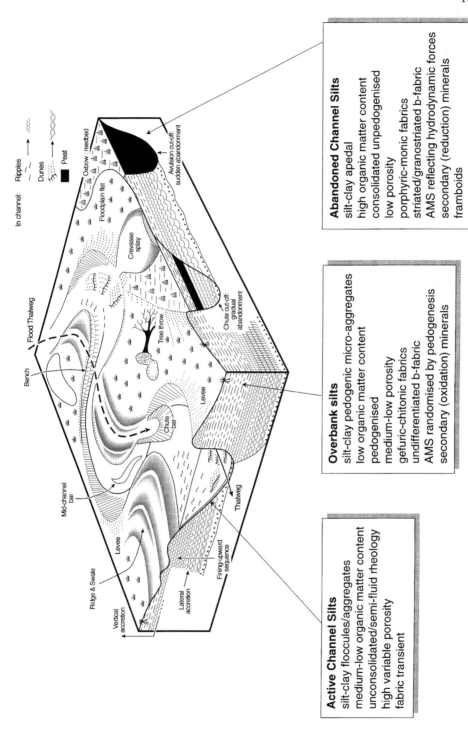

Figure 10.1 A diagrammatic representation of the floodplain environment, sediments and some sediment properties

pressed by a triaxial ellipsoid whose axes represent maximum, intermediate and minimum susceptibilities (Figure 10.5). The resultant susceptibility ellipsoid represents the sum of all the individual magnetic grains within a specific specimen (Hrouda, 1982), and so differs from conventional microfabric methods which directly measure the orientation of individual mineral grains.

The original purpose of the research was pragmatic, fuelled by a desire to use palaeomagnetic dating techniques to provide absolute dates for alluvial chronologies from eight sites at Hemington Fields, Leicestershire (Ellis and Brown, 1998). However, in order to determine the reliability of the natural remanent magnetisation directions it was necessary to differentiate primary from secondary matrix microfabrics. The differentiation of primary and secondary microfabrics necessitated an understanding of the theoretical questions as to the depositional and post-depositional processes of fine sediments in a specific fluvial environment, and in particular palaeochannels. This chapter presents examples of microfabric data from three key alluvial units found at Hemington Fields, north Leicestershire (SK 461 307). The study utilised three types of fabric measurement: anisotropy of magnetic susceptibility, micromorphology and scanning electron microscopy.

SITE DESCRIPTIONS

The site of Hemington Fields is a working gravel quarry, adjacent to the confluence of the present-day River Trent and River Derwent (Figure 10.2). The floodplain is some 4 km wide; on the east side of the floodplain lies the confluence of the River Soar and River Derwent. The simplified stratigraphy at Hemington Fields comprises Mercia Mudstone bedrock with thin localised gypsum beds. Lying unconformably above the irregular bedrock surface are approximately 2–2.5 m of Late Glacial stadial (Younger Dryas) tabular and shallow cross-bedded gravels (Brown and Salisbury, 1998). Capping these are approximately 2–2.5 m of medieval gravel. Cut into the medieval gravel are a complex series of palaeochannels (Figure 10.2). The Holocene floodplain is covered by approximately 1–2 m of silt/clay overbank sediment.

During the medieval period the river gradually evolved through a braided/anastomosing phase into a meandering, single channel form through a process of avulsion (Brown, 1996; Brown and Salisbury, 1998). The palaeochannels are a consequence of this change in the system. The palaeochannels are characterised by a basal infill of black silty sand and silt (10YR 1.7/1) referred to as unit 4 (Figure 10.3). This unit typically contains a small amount of detrital organic material. When freshly exposed the sediment is black, homogeneous and very moist. Contact with air causes a rapid colour change to brownish black (10YR 2/3) and the formation of coarse prismatic peds. Unit 3, generally stratigraphically higher than unit 4, comprises dark grey laminated clayey silt with rootlets. Unit 2 is a dark grey/brown silt with indistinct laminae, diffuse mottles, some iron concretion and rootlets. Fossil rootlets occur in both the overbank and palaeochannel sediments. For the stratigraphy of the sections discussed in this chapter and the sedimentary unit numbers, see Figure 10.3.

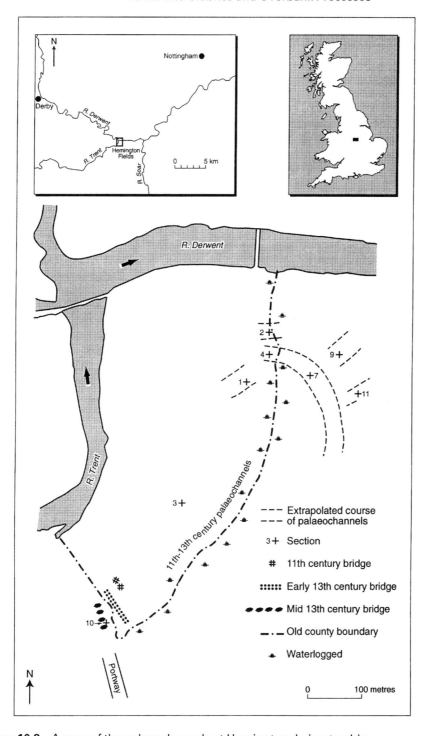

Figure 10.2 A map of the palaeochannels at Hemington, Leicestershire

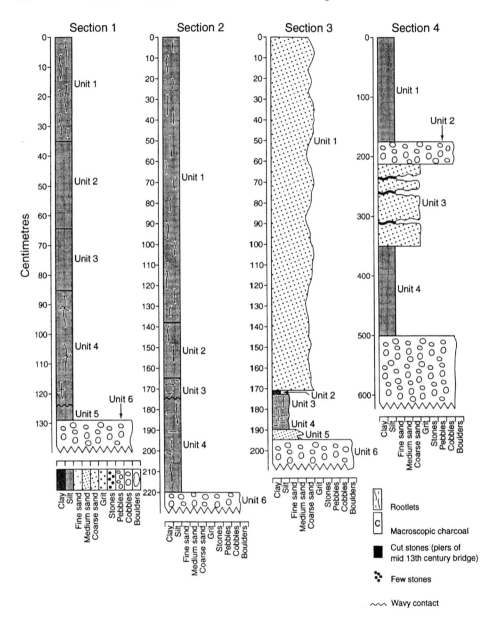

Figure 10.3 The sedimentary logs from the palaeochannels at Hemington, Leicestershire

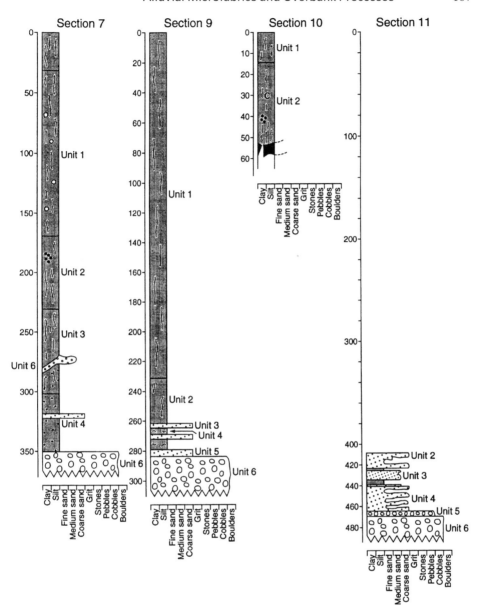

FIELD AND LABORATORY PROCEDURES

Vertical palaeochannel sections were created during gravel extraction; the sampled sediment was fine grained and *in situ*. The sections were cleaned, described and recorded in the form of standard sedimentary logs (Figure 10.3). Orientated, undisturbed samples were taken using aluminium monolith tins (54 × 11 × 9 cm); the tops of the tins were horizontal and the length vertical. The monolith tins were orientated to true north prior to removal. The samples were wrapped in foil and plastic sheet to avoid desiccation and inhibit bioturbation.

Using a stainless steel blade and knife, orientated cubic specimens were cut from the monolith samples, transferred to perspex cubes, capped and sealed. Volume susceptibility (κ) was measured using an a/c bridge circuit. Anisotropy of susceptibility was measured using a computer-linked Molspin Minisep anisotropy delineator instrument. Each specimen was measured in three different orientations. The three principal axes of susceptibility – maximum, intermediate and minimum – were calculated to an accuracy of $\pm 5\%$.

Two specimens for thin-section slides were taken from section 1 and one specimen from section 4. The specimens were made following the methods of Murphy (1986) and analysed using a standard petrographic microscope. The thin-section descriptions are based upon standard terminology following Bullock et al. (1985). For scanning electron microscope analysis, specimens were cut from undisturbed monolith samples. These specimens were gently broken in half to reveal a natural surface. The specimens were freeze-dried, following the method of Gillot (1969). The specimens were attached to aluminium stubs with the direction of north engraved to provide a reference point for microfabric analysis. Particle size analysis was performed on all units used for palaeomagnetic sampling. The wet sieve method was used down to 63 μm. The < 63 μm fraction was analysed in suspension (with 0.1% Calgon) using a Microtechnics model 5000ET SediGraph after dispersal in an ultrasound bath for 10 minutes. All the units measured were dominated by silt-size particles with variable amounts of medium–fine sand and little clay. A typical set of analyses is given in Figure 10.4. This sediment texture reflects the origin of the sediments which, given the location of the palaeochannel and proximity of the active channel (Brown, 1998), was from relatively high velocity, out of channel flow with deposition via both turbulent diffusion and convective processes typical of near-channel locations (Marriott, 1996). Further information on the conditions of deposition was sought from the magnetic parameters.

THEORY OF ANISOTROPY OF SUSCEPTIBILITY

The magnetic susceptibility of sediment when measured in a weak magnetic field is a second-rank tensor coefficient relating an induced magnetic moment to the inducing magnetic field (Bathal, 1971; Ellwood, 1984). This tensor is represented geometrically by a triaxial ellipsoid known as the susceptibility ellipsoid (Hrouda, 1978). The susceptibility ellipsoid represents the sum of all the individual magnetic grains within a specimen, the different types of magnetic minerals weighted according to their susceptibilities, particle anisotropies and preferred orientations (Hrouda, 1982). The shape of the susceptibility ellipsoid is also dependent on particle size and shape. The ellipsoid comprises three principal susceptibilities and their directions are the principal directions: these are the maximum (K_{max}), the intermediate (K_{int}) and the minimum (K_{min}) susceptibilities. When all three principal susceptibilities are equal then the shape anisotropy resembles a sphere (Figure 10.5(a)a) and the sample would be isotropic. In anisotropic fabrics three possible generalised grain forms can be identified: prolate, oblate and triaxial (Figure 10.5(a)b, c, d respectively). Oblate grains are disc-shaped

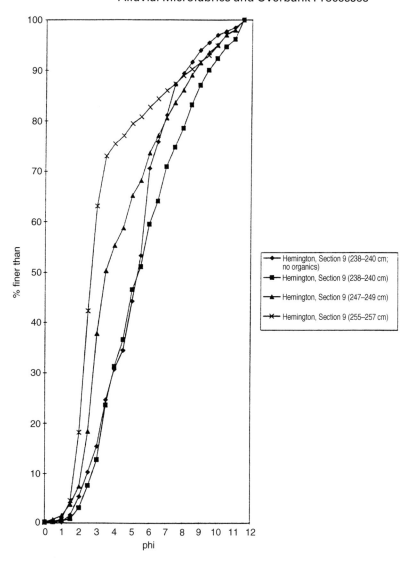

Figure 10.4 Particle size distributions from section 9

and are characterised by axes $K_{max} \approx K_{int} > K_{min}$. Prolate grains are elongate and cigar-shaped, and are characterised by $K_{max} > K_{int} \approx K_{min}$. Triaxial grains are characterised by $K_{max} > K_{int} > K_{min}$.

From the above, many parameters have been formulated, but meaningful values for the magnitude of the fabric and shape of the ellipsoid are difficult to determine and so the simplest parameters available have been used. These are the degree of anisotropy (A), magnetic lineation (L) and magnetic foliation (F) (Table 10.1). The parameter q relates the relative strength of the magnetic lineation to that of foliation (the quotient of L/F); it is a lineation parameter. The value q is independent of the initial magnetic susceptibility of a given sediment as well as being independent of the proportion of the

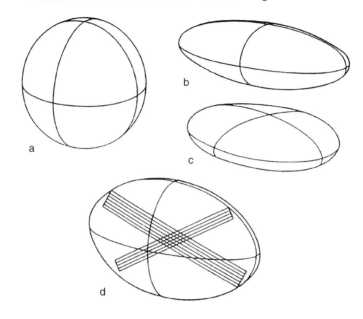

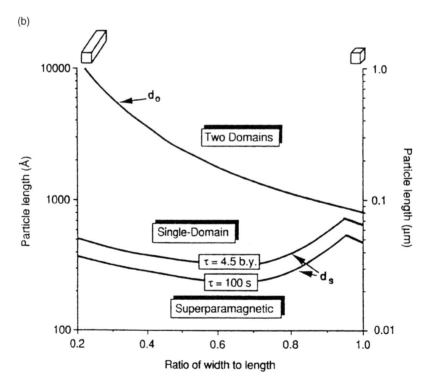

Figure 10.5 (a) Ellipsoid shapes where a = a sphere, b = prolate, c = oblate, d = triascial and (b) the relationship between magnetic grain shape and magnetic properties (adapted from Butler, 1992)

Table 10.1 Definitions for magnetic fabric analysis parameters

---
Mean intensity of susceptibility
$\mathbf{K} = (K_{max} + K_{int} + K_{min})/3$
Magnetic lineation $\mathbf{L} = (K_{max} - K_{int})/\text{mean } K$
Magnetic foliation $\mathbf{F} = (K_{int} - K_{min})/\text{mean } K$
Degree of anisotropy $\mathbf{A} = (K_{max} - K_{min})K_{int}$

---

magnetic material within a sediment, and can therefore be used to make comparisons between the fabrics of different samples and sediments. The value of q ranges from 0 to 2, indicating pure foliation to pure lineation. The critical change from foliation to lineation occurs at 0.67 (Hamilton and Rees, 1970). When the force of the magnetic field is not significant then the value of q ranges between 0.06 and 0.42 (Hamilton and Rees, 1970).

Magnetic fabric data are observations made in three dimensions and so measures of dispersion and distribution are displayed on spheres (Fisher, 1953; Figure 10.6). There are two types of microfabric elements: the first is a planar structure (foliation), often due to the effects of gravity; and the second is linear (lineation), often due to the action of fluid forces, e.g. hydrodynamic forces (Hamilton, 1967). The direction of magnetic lineation (often produced by forces tangential to the bed) is identical to that of the K_{max}, and the magnetic foliation (produced by the action of gravity in settling elongated and flattened particles with their axes as low as possible) is perpendicular to the K_{min} direction. The minimum susceptibility direction tends to be perpendicular to the bedding and the intermediate is usually near the horizontal.

Foliation may indicate a variety of processes such as gravitational fallout and compaction while lineation is sometimes indicative of flowing water, deposition on a slope or other similar processes. Rees (1961) introduced the notion of primary fabric, which is characterised by a horizontal distribution of the long and intermediate axes, a near-horizontal magnetic foliation plane and the K_{max} well grouped within this plane with the minimum axis orientated vertically and a q value of around 0.4. In some sedimentary deposits two directions of grain orientation have been observed using thin-section petrofabric analysis, one perpendicular to the current flow and the other parallel. The measurement of magnetic susceptibility can only record one single lineation within the magnetic foliation plane. This is because the shape anisotropy is fitted with a resultant tensor which is itself a measure of the orientation of all the magnetic grains in a given sample. AMS will reveal only the stronger orientation and the other will remain undetected unless other methods of microfabric analysis are used, e.g. SEM and/or micromorphology.

RESULTS OF ANISOTROPY OF MAGNETIC SUSCEPTIBILITY (AMS)

Results of AMS, Section 10

Section 10 was located just above the sandstone blocks of one of the piers of a 13th

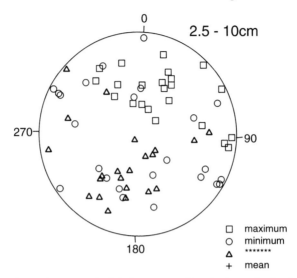

Figure 10.6 Section 10 magnetic fabrics where K_{max} is the principal axis of magnetic susceptibility, K_{int} the intermediate and K_{min} the minimum.

century bridge (Figure 10.2). One monolith sample was taken from section 10 and from this a series of specimens were removed. The specimens were grouped together by depth (Table 10.2). **R** (resultant vector; Butler, 1992) exceeds **R**o (probability level significant point; Irving, 1964) in all the samples, demonstrating that all the calculated mean directions are not due to chance at the 95% confidence level (Table 10.2). The mean directions for the upper half of unit 3 are SE–NW and W–E for the lower portion. Samples derived from unit 3 generally show well-clustered K_{max} vector points and lower mean inclination values than those for unit 2 (Table 10.2). The K_{min} vector points for unit 3 samples are also generally well clustered about the vertical. In these samples foliation is dominant, the degree of this domination being reflected in lower q values (Table 10.3). The K_{max} vector points of samples 1 to 5/6 lie in oblique planes with respect to the horizontal and the K_{min} vector points tend to deviate from the vertical (Figure 10.6). These samples show the minimum-axes preferred orientation of samples taken down the profile, the largest slope angles and the largest q values (Table 10.2 and 10.3).

The samples are characterised by a domination of prolate ellipsoids, low mean intensity of susceptibility values and low anisotropy magnitudes. Lineation is also dominant over foliation (Table 10.3). The mean directions down through the profile are NE–SW, NW–SE and NE–SW. The mean direction of unit 2/3, the location of the boundary between units 2 and 3, is NW–SE. The volume susceptibility values increase down the profile; the increase is particularly marked at the junction of units 2/3 and 3 (54.72 to 233.6 SI $\times 10^{-5}$), the cause being increased pyritisation and particularly the formation of framboids (Ellis and Brown, 1998). The geochemistry of these sediments indicates that sediment deposition occurred under conditions of anoxic organic decomposition in a floodplain aquifer (oxbow lake) with relatively little buffering capacity and rich in sulphur and iron oxides. These conditions which favour the production of pyrite and greigite, and which can also produce extreme acidity if oxidation occurs, are not uncommon in the gypsum-rich Triassic landscapes of north-west Europe.

Table 10.2 Fabric analysis results for section 10 where R is the resultant vector, Ro is the significant point, α_{95} is two standard errors of the mean, κ is the precision parameter (Fisher, 1953) and δ is the angular standard deviation (Butler, 1992)

Sample no. and unit	Depth (cm)	n	Declination	Inclination	R	σ_{95}	κ	Ro	δ
1, unit 2	0–2	11	66	32	8.968	22.9	4.921	5.28	35.39
2–3, unit 2	2.5–10	23	28	42	17.891	16.6	4.306	7.69	38.93
4, unit 2	10.5–12.5	8	344	27	6.766	25.5	5.673	4.48	32.35
5–6, unit 2/3	13–17.5	15	206	46	11.738	21	4.192	6.19	38.51
7–9, unit 2/3	18–25	25	321	12	22.548	9.8	9.787	8.02	25.59
10, unit 3	26–28	9	149	12	8.579	12.1	18.984	4.76	17.59
11–13, unit 3	28–35	27	157	7	25.314	7.3	15.423	8.34	20.35
14, unit 3	37–39	9	270	8	6.998	29.5	3.995	7.76	38.96
15–17, unit 3	39–46	27	67	20	22.557	12.6	5.851	8.34	33.34

Table 10.3 Section 10 magnetic parameters

Sample	K mean	SD	A mean	SD	L mean	SD	F mean	SD	% F	q mean	SD
(1)	4.11	0.27	0.018	0.008	0.086	0.032	0.097	0.062	45.5	0.71	0.26
(2/3)	3.96	0.41	0.019	0.013	0.102	0.100	0.087	0.066	56.5	0.72	0.43
(4)	5.72	1.43	0.009	0.005	0.039	0.019	0.047	0.034	62.5	0.66	0.17
(5/6)	19.91	14.32	0.035	0.017	0.014	0.010	0.021	0.013	73.3	0.55	0.36
(7/8/9)	17.60	7.94	0.033	0.023	0.014	0.014	0.020	0.014	76.0	0.53	0.29
(10	33.56	12.55	0.032	0.009	0.012	0.005	0.020	0.005	100.0	0.44	0.10
(11/12/13)	35.49	27.13	0.038	0.010	0.010	0.005	0.028	0.009	100.0	0.31	0.14
(14)	40.53	14.62	0.022	0.005	0.007	0.004	0.015	0.006	77.8	0.43	0.30
(15/16/17)	44.6	25.39	0.032	0.014	0.012	0.007	0.021	0.011	78.6	0.50	0.30

Discussion of AMS, Section 10

Gravel surrounded the stone blocks of the bridge piers and this was clearly deposited under high velocity conditions (Figure 10.3). The boundary between the gravels and unit 3 was sharp, indicative of a marked change in velocity and hydrological conditions. Archaeomagnetic data (Ellis, 1995) indicate that the bridge, built in AD 1260, was destroyed in c. AD 1375 ± 50 and floods are documented as having severely affected the area in the 1390s and 1401 (Brown, 1998). It is probable that avulsion of the channel was an immediate consequence of one of these floods.

The mean q value increases with height from the base of unit 3 and these values reflect the sedimentary environment; namely initial deposition within a deep pool with a calm sediment/water interface. It is postulated that following cut-off, silt and organic detritus laden water (unit 3) entered the palaeochannel and as indicated by the foliar or primary nature of the fabric, the major force acting upon the particles was gravity; thus deposition was dominated by settling from suspension. Analysis of insect remains has revealed that the water conditions at the base of unit 3 were stagnant, becoming more so with time (Pitts, 1995). The lowermost sample in unit 3 shows a larger inclination value of the K_{max} vector points. This sample was located approximately 2 cm above sandstone blocks which formed the bridge pier, and the fabric is reflecting the undulating nature of these blocks over which the fine sediment was draped.

The dipping and scattered nature of principal susceptibility axes of unit 2 samples, coupled with the domination by lineation and low values of volume susceptibility and anisotropy, are indicative of a weakly developed secondary, or physically disturbed fabric. Therefore settling out of suspension was not the primary mode of deposition (Hamilton and Rees, 1970); rather hydrodynamic forces prevailed in what would have been shallower water. It is probable that further disturbance of the microstructure of unit 2 was caused by post-depositional rootlet penetration and microbiological activity (see discussion of micromorphology). The inferred direction of flow of the palaeochannel at the sampling location was NNW. Three samples from unit 2 and one from unit 3 showed mean directions parallel to the trend of this palaeochannel; the remaining samples show mean directions 90° or less to the west. It is probable that what is being recorded are two flow directions. The change and fluctuation of the mean directions of the samples and the domination of lineation in unit 2 is probably the result of secondary currents caused by occasional overbank flow, or even the result of multidirectional wind-induced currents.

AMS Results, Section 11

Section 11 (Figure 10.2) is located at the east side of the gravel pit; the section was taken through a palaeochannel fill. Sixty-one specimens were cut from a monolith sample (Figure 10.3). All the samples have a significant mean direction at the 95% confidence level, as **R** exceeds **R**o. All the samples show considerable scatter of the K_{max} principal axis of susceptibility (Table 10.4). A decrease in the amount of scatter of the principal axes of susceptibility from the base of unit 4 upwards is apparent (Table 10.3); this is mirrored by an increase in foliar ellipsoids and an increased domination by foliation (Table 10.5). The K_{max} vector points in unit 4 dip much less than those of unit 4/9 (see

Table 10.4 Fabric analysis results for Section 11, where R is the resullant vector, α_{95} is two standard errors of the mean, κ is the precision parameter R_0 is the significant point and δ is the angular standard deviation

Sample no. and unit	Depth (cm)	n	Declination	Inclination	R	σ_{95}	κ	R_0	δ
36–44, unit 4/9	426–428	9	246	61	6.352	35.8	3.021	4.76	45.11
45–52, unit 4/9	428.5–430	9	146	43	6.834	24.7	6.003	4.48	31.32
53–61, unit 4/9	431–434	9	267	59	6.878	30.7	3.771	4.76	40.16
1–9, unit 4	453–455	9	38	11	7.026	29.3	4.052	4.76	38.68
10–18, unit 4	456.5–458.5	9	103	23	7.099	28.6	4.207	4.76	37.93
19–27, unit 4	459–461	9	194	22	6.06	38.8	2.721	4.76	47.68
28–35, unit 4	461.5–463.5	8	151	8	5.76	37.6	3.125	4.48	43.94

Table 10.5 Section 11 AMS parameters

Sample	Mean K	SD	Mean A	SD	Mean L	SD	Mean F	SD	% F	Mean q	SD	Mean v°	SD
36–44	8.04	1.68	0.03	0.006	0.013	0.004	0.012	0.005	55.56	0.72	0.27	44.1	8.77
45–52	8.36	1.58	0.03	0.024	0.017	0.013	0.015	0.012	37.5	0.81	0.35	41.48	10.21
53–61	9.79	1.11	0.03	0.009	0.013	0.007	0.017	0.008	55.56	0.60	0.29	48.24	9.85
1–9	44.00	12.07	0.01	0.003	0.003	0.001	0.007	0.002	88.89	0.32	0.16	59.3	7.24
10–18	15.55	2.72	0.01	0.004	0.004	0.002	0.01	0.003	100	0.34	0.14	58.5	8.08
19–27	11.66	2.78	0.02	0.006	0.008	0.003	0.013	0.004	88.89	0.49	0.12	51.6	4.63
28–35	37.29	18.17	0.02	0.003	0.006	0.002	0.011	0.003	100	0.44	0.18	54.1	8.31

below), reflecting deposition on a horizontal versus deposition on a sloping bed. Of interest is the difference in the mean intensities of susceptibility within unit 4, which is probably a consequence of differing sand contents.

The upper two samples from unit 4/9 exhibited scattered and dipping principal axes of susceptibility and both are dominated by lineation with high values of q. In-field observations noted that unit 10, which lay directly above unit 4/9, dipped 45° south into the section. The lower sample of this unit is dominated by foliation; however, the distribution of the axes of principal susceptibility do not conform to the understood definition of primary or undisturbed fabric. The mean intensity and anisotropy of unit 4/9 is low and this is a consequence of the high sand content coupled with domination by lineation.

The inferred direction of flow for this channel from sedimentary structures is from the SW towards the NE. The samples were taken from a section cut obliquely to the trend of the channel. The mean directions of unit 4/9, from the top to the base, are W–E, SE–NW and W–E. The mean directions of unit 4, from the top to the base, are: NE–SW (however, the K_{max} vector points occur in two distinct clusters, one with an approximate N–S alignment and the other with an E–W alignment), E–W, S–N and SSE–NNW (again with potentially two directions, one SE–NW and the other SW–NE).

Discussion of AMS, Section 11

Gradual abandonment of the palaeochannel can be inferred from the stratigraphic log (Figure 10.3), which shows a general fining-up sequence. After the deposition of the basal unit (11), the channel was abandoned as a major waterway. Water continued to enter into the channel, but this was of a much lower velocity than that under which unit 11 was transported and deposited. It is apparent from the interbedded nature of the sediment and the AMS data (especially unit 4) that the generally low energy environment of deposition within the palaeochannel was occasionally interrupted by sudden and high velocity influxes of sand-laden water (especially unit 4/9). The slower water carried more fine sand, silt, clay and organic material (deposited as unit 4) and the AMS data have demonstrated that the force acting on the grains and detritus was primarily gravitational, acting under non-turbulent water conditions. The abrupt boundary between unit 4 and unit 4/9 represents the scouring and erosion of unit 4 and the subsequent deposition of slightly coarser material under moderate flow conditions and where the affects of hydrodynamic forces dominate the AMS data. The abrupt boundary between unit 9/4 and unit 10 is again indicative of an abrupt change in the flow regime from moderate to low velocity. Unit 9 represents rather regular influxes of sand-laden water into more quiet conditions where silt deposition prevailed.

MICROMORPHOLOGY

The conventional method of determining the fabric of fine-grained sediments and soils is by using micromorphology. This is also the method routinely used in the investigation of sediments on archaeological sites, and has been shown to yield detailed evidence of the mode of deposition and subsequent hydrological conditions (Courty *et al.*, 1989; Miedema and van Oort, 1990; Solo-Benet *et al.*, 1990). Although most archaeological thin-section work on alluvial soils has focused on the identification of cultivation (French, 1988; Limbrey, 1992) and the creation of water-meadow systems (Limbrey, 1992); these studies have also yielded information concerning the hydrological causes of biomineralisation.

Thin-section analysis was undertaken on key strata using a slow alcohol-replacement epoxy resin method. In the case of section 1, unit 2 (Table 10.6) the initial deposit comprised the mineral component and detrital organic matter.

Unit 2 had a weakly striated b-fabric and horizontal bedding as indicated by the alignment of detrital organic matter; however, the laminae were disturbed by later rootlet penetration and the movement of fines into channels and voids. The presence of goethite, a slight striated b-fabric, granostriation around many of the voids and moderate to weak mottling indicated that the sediment had experienced episodic post-depositional wetting and drying. The wetting and drying (oxidising and reducing conditions) had caused localised stress and strain, resulting in compaction, and was also the mechanism for the dissolution and re-precipitation of iron mottles, Fe^{2+} from solution to Fe^{3+}. This process occurred in particular around the passages of rootlets and voids where the passage of pore water and groundwater was less restricted. A worm vermiform postdates the alternating reducing and oxidising cycles.

Table 10.6 Section 1, top monolith, unit 2 (depth 39 cm)

Microstructure	5% voids, circular and planar. Some voids have organic linings. Mostly discrete aggregates of sediment, separated totally or partially by planar voids.
Basic mineral component	Quartz grains, fine sand to silt sized. straight extinction, monocrystalline Muscovite. Black pyrite framboids.
Fine-grained component	Quartz and silt/clay. Probably goethite. Red brown grains, possible colloidal amorphous organics or altered magnetite grains.
Groundmass	Little to no preferred orientation of the clay/silt particles; slight striated b-fabric. Goethite, grey and yellow, makes up majority of groundmass. Slight indication, by organic fragments, of horizontal bedding.
Organic component	Diatoms and phytoliths. Amorphous organic matter. Charcoal.
Fabric pedofeatures	Worm vermiform. Granostriation around voids. Goethite and haematite radiating crystal growth, in the form of lining of voids, associated with Fe^{3+} accumulation.
Amorphous pedofeatures	Red-stained clay coating. Fe^{3+} around root channel, moderate impregnation with sharp boundary. Mottled groundmass, pseudogley. Mottles show moderate to weak impregnation; these occur around voids and within the groundmass.

In the case of unit 4 from sections 1 (Table 10.7) and 4 (Table 10.8) the basic components, quartz, sandstone, feldspars, clay and silt clasts, are probably derived from reworked alluvium; originally derived from Mercia mudstone.

The fabric is clearly of a horizontal nature, indicating that any physical or post-depositional disturbance was minimal. Of interest is the presence of blue vivianite which is a secondary phosphate occurring in oxidised zones containing pyrite and organic fragments. Related to the vivianite by environmental conditions is the occurrence of pyrite framboids which were observed in voids, root channels (with a clear association with organic matter; Figure 10.7) and to a lesser extent within the matrix. The pyrite framboids indicate that the post-depositional environment, prior to the drainage of the section of floodplain for gravel extractions, was an oxygen-poor, reducing, stagnant one, with the source of sulphate being gypsum in the underlying Mercia Mudstones. The presence of chlorite associated with organic matter is also an indication of chemical alteration (in this case of amphiboles or micas); however, detrital chlorite is also sometimes present in alluvial sediments. The occurrence of pyrite framboids within voids and root channels indicates that the framboids formed after initial root penetration and that this must have been before the deposition of the 13th century gravels. It is probable that slight oxidising conditions were created temporarily after the deposition of the sediment by the penetration of root channels and the movement of oxygen-rich pore water through these channels. This would account for the presence of Fe^{3+} concentrations around roots and root channels, and it may also have been at this time that the vivianite was formed. Following root penetration, reducing conditions prevailed, with the formation of the pyrite framboids. The timing of the pyrite framboid formation is difficult to determine, but was probably initiated by the erosion of the top of the channel and the deposition of the 13th century

Table 10.7 Section 1, bottom monolith, unit 4 (depth 100 cm)

Microstructure	Apedal. Channels visible in thin section; some are modern, but others appear to be original features primarily because pyrite framboids occur along and attached to channel walls. Voids, elongate and lenticular, perhaps vughs? 1%, spherical, smooth sphericity. Pyrite framboids occur in some voids forming clusters and partially infilling the individual voids.
Basic mineral components	Quartz grains. Angular to rounded. Simple extinction common, occasional undulose extinction, most are monocrystalline. One quartz grain showing polycrystalline form with straight edge contact indicative of a metamorphic origin. Inclusions in some of the grains. Well-rounded sandstone grains 2 mm, 1 mm and 1.01 mm diameter composed of quartz and small amount of clay. Plagioclase feldspar with sericite alteration. Feldspar with multiple twinning, highly weathered.
Fine fraction	Very fine, less than 10 mm; muscovite, clay PPL, beige. Mostly first-order white/yellow. The clay infill in root channels is a slightly darker brown in XPL. It contains black framboids and therefore must be relatively old. Also contains quartz grains.
Organic component	Charcoal and other unidentifiable fragments. 0.5–5 mm. Angular to well rounded. Brown colour in PPL, same in XPL. Elongate organic matter, reddish brown in PPL and exhibits slight change of colour in XPL, perhaps root fibres. Organic matter, dark reddish brown forming oval shapes. The centre of this organic structure is infilled by clay minerals and other small semi-spherical amorphous organic material or pyrite. Snail shell outline with organics.
Inclusions	Well-rounded, smooth sphericity black framboids, isotropic and opaque. These black framboids occur singly within the groundmass, clustered within the groundmass and most prominently as clusters in pores/voids. Exhibit at 400 × magnification very faint square lighter patches, giving the impression of cubes which make up the whole. Occasionally associated with mottling, as seen in one infilled void. There does not seem to be any preferred orientation of these black framboids.
Groundmass	Silt and clay. Anisotropic. Beige. Preferred orientation shown by quartz grains and clay maximum interference 140/160° and 224/229°. Unistial fabric, basically horizontal relative to section face. Fine fraction infilling in root channels shows a referred orientation to the channel direction.
Pedofeatures	Very slight granostriation around larger organic fragments.

gravels. The depositional and post-depositional sequence suggested above involves the movement of water through the profile, which is attested by the slight striated b-fabric. Within some of the voids and channels, fine silt and clay were observed with a preferred orientation to the channel direction. This suggests the post-depositional movement of fine material through the profile, or perhaps the movement of fine material into voids during compaction.

Table 10.8 Section 4, Unit 4

Microstructure	2% voids. Massive. Matrix supported.
Basic mineral component	Quartz grains, silt and fine sand-sized. Sub to moderately rounded, moderate sphericity; straight extinction. Occasional large polycrystalline quartz grains. Vivianite, granular and blue. Muscovite. Plagioclase feldspars with multiple twinning. Chlorite, associated with organic matter. Black octahedral and spherical mineral grains. Pyrite framboids (spherical in form, occurring in voids and root channels); these are associated with organic matter.
Fine-grained component	Quartz grains and muscovite. Bright green, well-rounded mineral, glauconite.
Compound minerals and rock fragments	Clay clast, possibly calcite-rich. Centre of clasts not effected by Fe^{3+} staining; however, clasts are coated with Fe^{3+}. Clay and silt clasts occur throughout the matrix. Snail shell, lined with organics.
Groundmass	Matrix supported, clay and silt-sized material. Slight striated b-fabric. Amorphous, colloidal organics.
Organic component	Much organic amorphous prolate fragments showing good horizontal orientation. Wood and leaf fragments. Vivianite associated with organics, voids and black framboids. Diatoms and phytoliths. Some pseudomorphs of organic material by Fe^{3+}, forming nodules. Dark red brown faecal pellets, occur in clusters. Quasi-coatings of Fe^{3+} superimposed on matrix.
Fabric pedofeatures	Burrow or channel infilled with fine silt/clay material; cuts the line of dark black grains.
Amorphous pedofeatures	Fe^{3+} impregnation concentration around roots and ancient root channels. Also occurs around voids and possible plucked out grains. Clasts that occur in void spaces of root channels are also affected by Fe^{3+} staining.

SCANNING ELECTRON MICROSCOPY

Despite attempts to develop genetic classifications for clay fabrics (Sergeyev et al., 1980) the applicability of such approaches to real sediments has met with limited success. SEM analysis, although a standard methodology for glacial and aeolian sediments, has been relatively little used on alluvial sediments. One reason for this is that at present it is difficult to apply laboratory-created microstructures to natural sediments, and a second is the practical problem of impregnating clays without destroying the natural fabric. Sergeyev et al. (1980) devised a scheme of clay microstructure classification based on the observation of 300 samples; however, many of these were laboratory produced. Six main types of clay microstructure in soils of sedimentary origin were identified. The aim of the study was to describe and identify clay microstructures and link these with sediment properties and processes (mineral

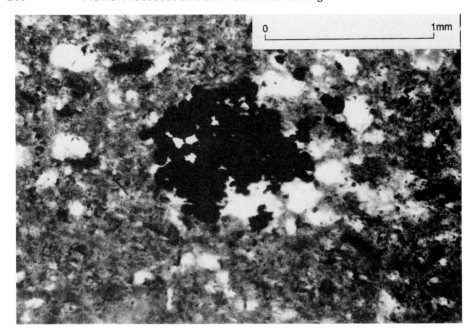

Figure 10.7 Thin-section micrograph of framboids lining a channel from section 10

composition, granulometry, microaggregate composition, physical and mechanical properties, and the mode and environment of deposition). This classification was to form the basis for microfabric description using SEM and micromorphology; however, because of the fine-grained nature of the Hemington sediments and the artificiality of many of those fabrics produced by Sergeyev, this was found to be impossible.

The SEM analysis revealed no discernible preferred orientation of grains in unit 3 (sections 7 and 10) or unit 4 (sections 9 and 11); this was a consequence of the fine-grained nature of the sediment and the domination of a platy, face-to-face, foliar structure and the flocculation or binding of silt, clay and occasional sand grains. A number of diatoms were identified within unit 4; these were *Campylodiscus noricus*, species of *Nitzschia* and *Navicula*, and other small unidentified forms were noted in the sediment.

Framboids (spherical mineral aggregates) were observed from unit 4 (section 4) in thin section and units 3 and 4 (sections 4, 7, 9 and 10; Figure 10.8). The framboids were most common in organic-lined voids and root channels (Figure 10.7). However, some were observed in the matrix and these were regularly associated with vivianite. The framboids were constructed from individual octahedral crystals arranged in spheres of 10–50 μm in diameter. The framboids were elementally analysed using EDXA, which identified the major elements as being iron and sulphur (see Ellis and Brown 1998 for further discussion).

The presence of *Campylodiscus noricus* is indicative of alkaline conditions of pH 7 or greater (Haworth, 1976). They are common in standing and nutrient-rich water. The *Nitzchia* species are a large group, most commonly occurring in eutrophic water. The *Narvicula* species are a varied group of diatoms and are generally found in running

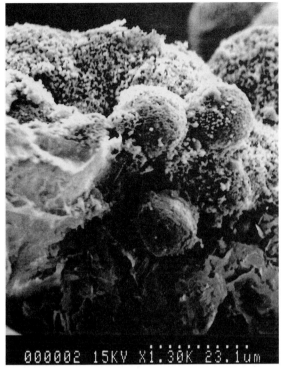

Figure 10.8 Scanning electron micrograph of framboids from section 10

water with a high nutrient content and a wide range of pH. However, some do prefer alkaline or neutral pH water (Haworth, 1976). The diatoms confirm the alkaline nature of the water and sediment. Factors that influence the formation of pyrite and other iron sulphide minerals in sediment are the supply of dissolved sulphate (at Hemington the elemental sulphur was derived from local gypsum layers); the concentration of sulphide in the sediment (where sulphide formation is primarily a biological process requiring dissolved sulphate); soluble iron (derived from iron-rich clay and silt particles of detrital origin); availability of organic matter which can be broken down by sulphate-reducing bacteria; the pH of pore waters; the redox state of the sediment; and the sedimentation rate (see Ellis and Brown in press for further discussion). Conditions suitable for iron sulphide production can be envisaged within the Hemington Palaeochannels, where trapped flood waters and seepage from the main channel in an organic-rich environment resulted in the deposition of organic-rich silt under reducing conditions.

SUMMARY OF THE DEPOSITIONAL ENVIRONMENT

The depositional environment of unit 4 (all palaeochannels) was one of high organic content, both allochthonous and autochthonous in origin. Low velocity flow is indicated by the foliated nature of the magnetic fabric, where stagnant conditions prevailed and where sediment and organic matter settled out in very oxygen-poor to anaerobic water created by the trapping of water in the sinks of the palaeochannels. The occasional lineation of unit 4 is perhaps caused by the rare influx of water into the palaeochannels. Unit 3 was deposited under very similar conditions to that of unit 4. The post-depositional histories of unit 3 and unit 4 are slightly different. Anoxic conditions prevailed within unit 4, although oxygen-rich water flowing around rootlet channels and within some voids allowed pyrite formation. Post-depositional chemical changes, the formation of pyrite framboids and other iron monosulphides, were more marked in unit 3 because of the presence of marginally more oxygen. Unit 2 is more likely to have been deposited under predominantly oxidising conditions. The sediment was subjected to post-depositional wetting and drying episodes caused by changes in the level of the watertable. This produced periods of oxidation and reduction resulting in the reddish brown colour, the presence of mottles and iron nodules. Some of unit 2 was deposited under low velocity conditions, but where weak hydrodynamic forces contributed to the alignment of the particles and grains.

The analysis of the microfabrics of the palaeochannel sediments provides a history of channel abandonment, when combined with dating control, and the development of post-depositional sedimentary properties. Palaeochannels remained important sediment and geochemical sinks and biological habitats for some time after channel avulsion. At Hemington this suggests that the floodplain would have contained several, probably inter-linked, oxbow lakes until most were infilled by post-17th century overbank deposition. The details obtainable from sediment microfabric analysis would probably be even more valuable in situations where the environment of deposition is less easily determined, such as from floodplain sediments in the geological record.

CONCLUSIONS

The results of this analysis of channel fill sediments can be summarised by four main findings:

(1) Analysis of anisotropy of susceptibility can reveal sedimentological information, with foliation indicative of gravitational fallout and also compaction, and lineation indicative of the direction of water flow. Relic primary fabrics can also be detected when dependent on the relationship of the three principal axes of susceptibility. A methodology for the investigation of alluvial magnetic fabrics is presented in Figure 10.9. Further work is needed to relate fabrics to the processes of sediment conveyance and deposition on floodplains.
(2) Remnant features or relics of depositional and post-depositional processes (physical and chemical) can be identified using AMS, micromorphology and the scanning electron microscope.
(3) The formation of pyrite framboids can occur in cut-offs where reducing conditions persist with organic matter and occasional inputs of oxygenated water and available Fe and dissolved S. They are therefore indicative of these conditions of deposition.

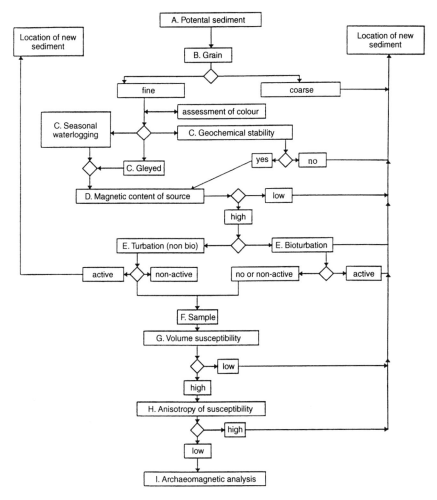

Figure 10.9 A methodology for magnetic fabric studies of alluvial sediments. Recommended in-field procedure:

A. Identification of potential sediment.
B. Preliminary analysis of grain size; coarse sediments are unsuitable for archaeomagnetism.
C. Assessment of the geochemical stability of the sediment; geochemically unstable sediments must be rejected. The identification of gleyed or seasonally waterlogged sediments.
D. Assessment of the magnetic content of sediments; those sediments with a very low magnetic content may be excluded from further analysis.
E. Assessment of turbation and bioturbation; those sediments in which these processes are active must be rejected.
F. Sample for archaeomagnetism.
G. Measurement of volume susceptibility; those samples with very low values may be rejected.
H. Measurement of anisotropy of susceptibility; those samples with anisotropy above 5% may be rejected. However, if they are included in the study the results must be treated with caution.
I. Archaeomagnetic analysis

(4) Microfabric studies can give both fabric and chemical information concerning both depositional and post-depositional processes.

The work at Hemington has illustrated how important geochemical conditions are in the formation of channel abandonment sedimentary characteristics, and how they can yield detailed evidence concerning the conditions of deposition and changes in those conditions even after deposition. The role of groundwater in both the formation and transformation of sedimentary characteristics has yet to be examined in detail.

ACKNOWLEDGEMENTS

The authors must thank the archaeologists involved with Hemington, particularly P. Clay, C. Salisbury, L. Cooper and S. Ripper for all their help and advice, and the owners of the site, Ennemix Co. Ltd., for access. This work was begun whilst one of us (C.E.) was in receipt of a NERC studentship. Finally we must thank K. Shrapel for preparing the thin sections, T. Bacon and H. Jones for their drafting skills and the referees for valuable comments.

REFERENCES

Bagnold, R. A. and Barndorff-Nielson, O. 1980. The pattern of natural size distributions. *Sedimentology*, **27**, 199–207.
Bathal, R. S. 1971. Magnetic anisotropy in rocks. *Earth Science Reviews*, **7**, 227–253.
Brewer, R. 1964. *Fabric and Mineral Analysis of Soils*. J. Wiley, New York
Brown, A. G. 1985. Traditional and multivariate techniques in the interpretation of floodplain sediment grain size variations. *Earth Surface Processes and Landforms*, **10**, 281–291.
Brown, A. G. 1996. Floodplain palaeoenvironments. In Anderson, M., Walling, D. E. and Bates, P. (Eds) *Floodplain Processes*. John Wiley, Chichester, 95–138.
Brown, A. G. 1997. *Alluvial Geoarchaeology: Floodplain Archaeology and Environmental Change*. Cambridge University Press, Cambridge.
Brown, A. G. in press. Geomorphology. In Cooper, L. and Ripper, S. (Eds) *Hemington Bridges Project*. University of Leicester Archaeological Services.
Brown, A. G. and Brookes, S. 1997. Floodplain vegetation and overbank erosion and sedimentation. In Large, A. (Ed.) *Floodplain Rivers: Hydrological Processes and Ecological Significance*. British Hydrological Society Occasional Paper.
Brown, A.G. and Salisbury, C. 1998. Geomorphology. In Cooper, L. and Ripper, S. (Eds) *The Medieval Bridges at Hemington*. Leicester Archaeological Reports, Leicester, 76–98.
Brown, A. G., Keough, M. K. and Rice, R. J. 1994. Floodplain evolution in the East Midlands, United Kingdom: the Lateglacial and Flandrian alluvial record from the Soar and Nene valleys. *Philosophical Transactions of the Royal Society, London, Series A*, **348**, 261–293.
Bullock, P., Federof, N., Jongerius, A., Stoops, G. and Tursina, T. (Eds) 1985. *Handbook of Thin Section Description*. Waine Research.
Burnham, C. P. 1970. The micromorphology of argillaceous sediments particularly calcareous clays and siltstones. *Soil Survey Technical Monograph*, 83–96.
Butler, R. F. 1992. *Palaeomagnetism: Magnetic Domains and Geological Terranes*. Blackwell Scientific, Boston.
Courty, M. A., Goldberg, P. and Macphail, R. 1989. *Soils and Micromorphology in Archaeology*. Cambridge University Press, Cambridge.
Ellis, C. 1995. The archaeomagnetism of fine-grained alluvial sediment. PhD Thesis, University of Leicester.

Ellis, C. and Brown, A. G. 1998. Archaeomagnetic dating and palaeochannel sediments: data from the Medieval channels fills at Hemington, Leicestershire. *Journal of Archaeological Science*, **25**, 149–163.
Ellwood, B. B. 1984. Anisotropy of magnetic susceptibility: empirical evaluation of instrument precision. *Geophysical Research Letters*, **11**, 645–548.
Erskine, W., Melville, M., Page, K. J. and Mowbray, P. D. 1982. Cutoff and oxbow lake. *Australian Geographer*, **15**, 174–180.
Fisher, R. 1953. Dispersion on a sphere. *Proceedings of the Royal Society*, **A217**, 295–305.
French, C. A. I. 1988. Aspects of buried prehistoric soils in the lower Welland valley and the fen margin north of Peterborough, Cambridgeshire. In Groenman-van Waateringe, W. and Robinson, M. (Eds) *Man-Made Soils*. British Archaeological Reports, International Series 410, Oxford, 115–128.
Gillot, J. E. 1969. Study of the fabric of fine-grained sediments with the scanning electron microscope. *Journal of Sedimentary Petrology*, **39**(1), 90–105.
Greene-Kelly, R. and Mackney, D. 1970. Preferred orientation of clay in soils: the effect of drying and wetting. In Osmond, D. A. and Bullock, P. (Eds.) *Micromorphology Techniques and Applications*. Technical Monograph No. 2, Rothamsted Experimental Station, 43–51.
Hamilton, N. 1967. The effect of magnetic and hydrodynamic control on the susceptibility anisotropy of redeposited silts. *Journal of Geology*, **75**, 738–743.
Hamilton, N. and Rees, A. I. 1970. The use of magnetic fabric in palaeocurrent estimation. In Runcorn, S. K. (Ed.) *Palaeogeophysics*. Academic Press, London, New York, 445–464.
Haworth, E. 1976. Two Lateglacial (Late Devensian) diatom assemblage profiles from northern Scotland. *New Phytologist*, **77**, 227–256.
Hrouda, F. 1978. The magnetic fabric in some folds. *Physics Earth Planetary Interiors*, **17**, 89–97.
Hrouda, F. 1982. Magnetic anisotropy of rocks and its application in geology and geophysics. *Geophysical Survey*, **5**, 37–82.
Irving, E. 1964. *Palaeomagnetism and Its Application in Geological and Geophysical Problems*. John Wiley, New York.
Johns, M. and Jackson, M. 1991. Compositional control of anisotropy of remanent and induced magnetisation in synthetic samples. *Geophysical Research Letters* 18, 1293–1296.
Limbrey, S. 1992. Micromorphological studies of buried soils and alluvial deposits in a Wiltshire river valley. In Needham, S. and Macklin, M. (Eds) *Alluvial Archaeology in Britain*. Oxbow Monograph 27, Oxford, 53–64.
Marriott, S. B. 1996. Analysis and modelling of overbank deposits. In Anderson, M. G., Walling, D. E. and Bates, P. D. (Eds) *Floodplain Processes*. John Wiley, Chichester, 63–94.
Miedema, R. and van Oort, F. 1990. Significance of soil microfabric for soil physical characteristics and behaviour of Late Weichselian and Holocene Rhine deposits in the Netherlands. In Douglas, L. A. (Ed.) *Soil Micromorphology: A Basic and Applied Science*. Developments in Soil Science 19, Elsevier, Amsterdam, 97–105.
Murphy, C. P. 1986. *Thin Section Preparation of Soils and Sediments*. AB Academic Publishers.
Pitts, F. 1995. A palaeoecological reconstruction of the River Trent using subfossil insect remains. BSc Dissertation. University of Loughborough.
Rees, A. I. 1961. The effects of water currents on the magnetic remanence and anisotropy of susceptibility of some sediments. *Geophysical Journal of the Royal Astronomical Society*, **5**, 235–251.
Richards, K., Chandra, S. and Friend, P. 1993. Avulsive channel systems: characteristics and examples. In Best, J. L. and Bristow, C. S. (Eds) *Braided Rivers*. Geological Society Special Publication 75, Geological Society, London, 195–203.
Sergeyev, Y. M., Grabwska-Olszewwska, B., Osipov, V. I., Sokilov, V. N. and Kolomenski, Y. N. 1980. The classification of the microstructure of clay soils. *Journal of Microscopy* 120, 237–260.
Solo-Benet, A., Gisbert, J. and Larque, P. 1990. Pedogenic microlaminated clay in Palaeogene sediments. In Douglas, L. A. (Ed.) *Soil Micromorphology: A Basic and Applied Science*. Developments in Soil Science 19, Elsevier, Amsterdam, 689–696.
Theocharopoulos, S. P. and Dalrymple, J. B. 1987. Experimental reconstruction of illuviation

cutans (channel argillans) with differing morphological and optical properties. In Federoff, N., Bresson. L. M. and Courty, M. A. (eds.) *Micromorphologies Des Sols – Soil Micromorphology*, Afes Plaisir, paris, 687–698.

Todd, S. 1996. Process deduction from fluvial sedimentary structures. In Carling, P. A. and Dawson, M. R. (Eds) *Advances in Fluvial Dynamics and Stratigraphy*. John Wiley, Chichester, 299–350.

11 Changing Rates of Overbank Sedimentation on the Floodplains of British Rivers During the Past 100 Years

D. E. WALLING and Q. HE
Department of Geography, University of Exeter, UK

INTRODUCTION

The floodplains of most lowland rivers in Britain are characterised by extensive deposits of fine overbank sediments resulting from the deposition of suspended sediment during flood events that inundate the floodplain. For these rivers, such overbank sediments represent the major component of the floodplain deposits and they play a key role in floodplain development. Viewed in the context of the catchment sediment budget, such deposits also represent an important sink for fine sediment transported through the river system (cf. Walling *et al.*, 1996). Lambert and Walling (1987), for example, have reported that conveyance losses of suspended sediment moving through an 11 km reach of the lower River Culm in Devon could be as high as 28% of the annual load entering the reach, and Walling and Quine (1993) have estimated that overbank floodplain sedimentation on the floodplains bordering the main channel of the River Severn accounted for 23.6% of the suspended sediment load transported by the river during the three-year period 1986–1989. Where it has proved possible to date individual horizons within such overbank deposits, the findings generally indicate that deposition has occurred at relatively low rates over extended periods of time. In the case of two British rivers, Shotton (1978) reports an average rate of deposition of 0.5 cm year^{-1} over the past 3000 years on the floodplain of the River Avon in Warwickshire, and Brown (1987) cites typical sedimentation rates of 0.14 cm year^{-1} over the past 10 000 years for the floodplain of the River Severn. These estimates of accretion rates are in close agreement with those obtained by studies of contemporary floodplain sedimentation using both sedimentation traps and radiocaesium measurements. Thus, for example, Simm (1995) reports contemporary sedimentation rates in the range 0.01–0.81 cm year^{-1} for a series of sedimentation traps installed on the floodplain of the lower River Culm in Devon, and He and Walling (1996a) document

---
Fluvial Processes and Environmental Change. Edited by A. G. Brown and T. A. Quine.
© 1999 John Wiley & Sons Ltd.

Table 11.1 Some potential environmental changes affecting British rivers and their catchments during the past 100 years

(a) Changes in agricultural land use and management
(b) Afforestation
(c) Urbanisation
(d) Encroachment of residential and commercial development onto floodplains
(e) River training and flood control works
(f) Climate change

---

mean annual overbank sedimentation rates of 0.11–0.95 g cm^{-2} year^{-1} for a selection of British lowland rivers over the past c. 30 years based on ^{137}Cs measurements.

For some rivers, detailed analysis of sediment stratigraphy has indicated that rates of floodplain deposition have varied through time and, for example, increased during periods of forest clearance and agricultural expansion within the upstream catchment. Thus, Brown (1990) presents evidence from the River Perry, a tributary of the River Severn, which indicates that increased rates of alluviation occurred c. 1400 years ago in response to the expansion of agriculture. However, little is currently known about more recent trends in rates of overbank deposition and whether accretion rates have changed in reponse to the many important changes that have occurred both within the upstream catchments and on the floodplains themselves (cf. Table 11.1). It is, for example, possible that sedimentation rates may have increased on some floodplains in response to increased erosion and sediment yields associated with the expansion of arable cultivation during the 1939–1945 war and the more recent expansion in the area under autumn-sown cereals. Similarly, increases in flood magnitude associated with the expansion of urban areas, and reductions in the areal extent of inundated areas, due to changing floodplain management and the encroachment of building development, could also be expected to have caused changes in sedimentation rates. Recent advances in the use of the fallout radionuclides ^{137}Cs and unsupported ^{210}Pb to estimate rates of floodplain sedimentation over the past c. 40 years and 100 years respectively (e.g. He and Walling, 1996a, 1996b; Walling et al., 1996; Walling and He, 1997), now afford a valuable opportunity to address this question by comparing the estimates of sedimentation rates derived for these two different periods (cf. Walling and He, 1994). This contribution reports the results of applying this approach to the floodplains of a representative selection of British rivers.

USING FALLOUT RADIONUCLIDES TO ESTIMATE RATES OF FLOODPLAIN SEDIMENTATION

The basis for using measurements of the ^{137}Cs content of floodplain sediments to derive estimates of mean annual sedimentation rates over the past 30–40 years is described in detail in He and Walling and He (1992, 1993, 1997) Walling et al. (1992) and Walling (1996a). In outline, the approach makes use of the fallout of the artificial radionuclide ^{137}Cs, resulting from the atmospheric testing of nuclear weapons during the middle years of the 20th century. Most of the fallout occurred during the decade

1955–1965, with significant deposition commencing in c. 1954 and maximum deposition occurring in 1963, the year of the Nuclear Test Ban Treaty. In some areas of the world, an additional short-term fallout input occurred in 1986 as a result of the Chernobyl accident, which released radiocaesium into the atmosphere, but in this study, attention will focus on bomb-test fallout. Caesium-137 (^{137}Cs) has a half-life of 30.17 years and although more than 30 years have elapsed since the main period of fallout, radiocaesium derived from bomb fallout is still readily detectable in most regions of the world. In most environments, the fallout input was rapidly and strongly adsorbed by clay particles in the surface horizons of the soil, and floodplain surfaces will have received inputs of radiocaesium both directly from atmospheric fallout, and in association with deposition of sediment eroded from the upstream drainage basin. Estimates of the rate of deposition can be obtained by comparing either the depth distribution or the total inventory of ^{137}Cs in the floodplain sediment with that of an adjacent undisturbed site above the level of inundating floodwater. The depth at which peak concentrations are found can be used to establish the level of the floodplain surface in 1963, the year of peak fallout. Alternatively, the "excess" ^{137}Cs inventory associated with the floodplain site can be used to determine the mass of sediment deposited, and therefore the mean sedimentation rate, during the period since the commencement of ^{137}Cs fallout in the mid-1950s. In the latter case, only a single measurement of the total ^{137}Cs inventory of a floodplain sediment core is required, as distinct from measurements on the individual slices of a sectioned core. However, it is necessary to take account of the influence of grain-size composition of the deposited sediment when deriving estimates of the mass of sediment deposited from the "excess" inventory, because of the preferential association of ^{137}Cs with fine-grained sediment particles (cf. Livens and Baxter, 1988; He and Walling, 1996c; Walling and He, 1997).

Unsupported or fallout ^{210}Pb differs from ^{137}Cs in that it is a natural radionuclide. It similarly reaches the land surface via atmospheric fallout, but the annual fallout can be viewed as being essentially constant through time. Since it is again rapidly and strongly adsorbed by the surface soil or sediment, it affords a means of estimating deposition rates over somewhat longer time periods (c. 100 years, i.e. about five times the half-life of ^{210}Pb). ^{210}Pb is a product of the ^{238}U decay series, with a half-life of 22.26 years. It is derived from the decay of gaseous ^{222}Rn, the daughter of ^{226}Ra. ^{226}Ra occurs naturally in soils and rocks. Diffusion of a small proportion of the ^{222}Rn from the soil introduces ^{210}Pb into the atmosphere and its subsequent fallout provides an input of this radionuclide to surface soils and sediments which is not in equilibrium with its parent ^{226}Ra. This component is designated unsupported ^{210}Pb, since it cannot be accounted for (or supported) by decay of the *in situ* parent. The amount of unsupported or atmospherically derived ^{210}Pb in a sediment sample can be calculated by measuring both ^{210}Pb and ^{226}Ra and subtracting the supported or *in situ* component. As in the case of ^{137}Cs, floodplain surfaces will receive inputs of unsupported ^{210}Pb from both direct fallout and in association with the deposition of sediment mobilised from the upstream catchment area.

Although ^{210}Pb measurements have been extensively used for establishing lake sediment chronologies extending over the past 50–150 years (cf. Robbins, 1978; Appleby and Oldfield, 1983; Oldfield and Appleby, 1984a, 1984b), their application to floodplain sediments introduces some further complexities. Neither the constant flux

(CF) (also referred to as the constant rate of supply, CRS) model, nor the constant activity (CA) (also referred to as the constant initial concentration, CIC) model, which are commonly used to derive age–depth relationships for lake sediments, are directly applicable to floodplain situations. The CA or CIC model assumes that accumulation of unsupported ^{210}Pb in the sediment profile results from the deposition of sediment with a constant ^{210}Pb activity or concentration, but in the case of floodplain sediments, a substantial proportion of the ^{210}Pb will be deposited directly onto the surface as atmospheric fallout. The final concentration in the surface sediment will reflect the interaction of the two components and the rate of sedimentation and may not be constant through time. The CRS or CF model departs less from physical reality. However, although it is possible to treat the atmospheric flux component of unsupported ^{210}Pb input as essentially constant, the annual flux of catchment-derived unsupported ^{210}Pb associated with deposited sediment is likely to vary from year to year in response to variations in rates of sediment deposition. In addition, both the above models assume that post-depositional mixing is limited and this again may not hold true for some floodplain sediment deposits.

To overcome some of the limitations of existing models and also to avoid the need for detailed information on the down-profile variation of unsupported ^{210}Pb activity in a sediment core (which is both costly and time-consuming to obtain), the authors have favoured the application of a constant initial concentration–constant sedimentation rate (CICCS) model to derive the age–depth relationship for floodplain sediment cores. This approach has been documented in detail in He and Walling (1996b) and Walling et al. (1996). In this case, the total unsupported ^{210}Pb inventory is divided into two components, representing that associated with atmospheric fallout and that associated with deposition of sediment-associated ^{210}Pb eroded from the upstream catchment. Values of unsupported ^{210}Pb inventory obtained from neighbouring undisturbed pasture sites above the level of flood inundation and where neither erosion nor deposition will have occurred, can be used to estimate the atmospheric fallout component. The "excess" inventory associated with sediment deposition can be calculated as the difference between the measured total unsupported ^{210}Pb inventory associated with a floodplain sediment core and the estimated local ^{210}Pb reference inventory. The value of "excess" unsupported ^{210}Pb can in turn be used to estimate the sedimentation rate, if the concentration of unsupported ^{210}Pb in the deposited sediment at the time of deposition, otherwise referred to as the "initial concentration", is known. This initial concentration can be estimated from measurements of the unsupported ^{210}Pb activity of either samples of freshly deposited sediment collected using sedimentation traps, or suspended sediment. In the latter case, it is necessary to take account of potential contrasts in the grain-size composition of suspended sediment and the deposited sediment because of the strong affinity of fallout ^{210}Pb for fine sediment particles (cf. He and Walling, 1996c). Since the CICCS model assumes a constant sedimentation rate, the model estimates the average rate of deposition over the period involved. The model also has the important advantage that an estimate of the average sedimentation rate can be derived from a single measurement of the total unsupported ^{210}Pb inventory of a floodplain core and that it is therefore possible to study the spatial patterns of sedimentation rate involved by collecting and analysing a substantial number of cores. Use of the CICCS model is, however, dependent upon the availability of an estimate of the initial unsupported ^{210}Pb content of deposited sediment derived from the analysis of suspended sediment or sediment retrieved from sedimentation traps. Where such information is unavailable, the CIC model can be applied if the sedimentation rate is relatively high and both the direct atmospheric ^{210}Pb fallout and the sediment-associated ^{210}Pb may be assumed to be uniformly distributed within the deposited sediment.

By undertaking both ^{137}Cs and unsupported ^{210}Pb measurements on the same core, or on immediately adjacent floodplain cores, it is possible to estimate average rates of sedimentation relating to both the past 30–40 years (^{137}Cs) and the past 100 years (^{210}Pb) (cf. He and Walling, 1996a, 1996b; Walling and He, 1997). Comparison of the two estimates affords a means of establishing whether significant changes in rates of deposition have occurred during this period. The results of applying this approach to a selection of sites on the floodplains of British rivers are reported in this contribution. In four cases, comparison of the average sedimentation rates for the two periods was based on a large number (70–274) of cores collected from small areas of the floodplains. This provided a means of assessing changes in both rates of sedimentation and in the spatial patterns involved. In order to extend the investigation to other rivers, where it was not possible to collect multiple cores, single cores were also collected from representative points at 21 further floodplain sites in order to compare the estimates of average sedimentation rate for the two periods.

THE STUDY SITES

For convenience, the study sites have been divided into two groups. The first comprises the four sites where detailed investigations involving multiple cores were undertaken. The second group includes those sites where single representative cores were collected. Further details of the sites and their locations are provided in Table 11.2 and Figure 11.1. Taken together, the two groups of sites afford a representative sample of the floodplains of the middle and lower reaches of British rivers, in terms of upstream basin characteristics, hydrological regime and the nature of the floodplains themselves. At those sites where only a single core was collected, the core was obtained from a point selected as representative of areas of active overbank deposition away from the immediate vicinity of the channel.

SAMPLING AND MEASUREMENT PROCEDURES

All cores were collected using a purpose-built motorised percussion corer. A 12 cm diameter core tube was used to collect cores for sectioning (i.e. all cores collected from sites with single cores, and primary cores collected from the detailed study sites), and a 6.9 cm diameter core tube was used to collect bulk cores within the detailed study sites. In the case of the latter sites, the bulk cores were collected at the intersections of a grid, with the spacing ranging from 12 m × 12 m at the site on the River Culm to 25 m × 25 m at the site on the River Severn. Coring depths were $c.$ 70 cm, and in all instances the cores included the total depth of ^{137}Cs bearing sediment and reached the

Table 11.2 The study sites

Site number	River/location	Number of sediment cores
I Detailed study sites		
1.	River Culm at Silverton Mill	274
2.	River Severn at Buildwas	124
3.	Warwickshire Avon at Eckington Bridge	80
4.	River Rother at Shopham Bridge	70
II Single-core sites		
1.	River Ouse near York (SE597458)	1
2.	River Vyrnwy near Llanymynech (SJ268206)	1
3.	River Severn near Atcham (SJ538095)	1
4.	River Wye near Preston on Wye (SO367432)	1
5.	River Severn near Tewkesbury (SO887325)	1
6.	Warwickshire Avon near Pershore (SO956466)	1
7.	River Usk near Usk (SO383985)	1
8.	Bristol Avon near Langley Burrell (ST947757)	1
9.	River Thames near Dorchester (SU582934)	1
10.	River Torridge near Great Torrington (SS475217)	1
11.	River Taw near Barnstaple (SS569283)	1
12.	River Tone near Bradford on Tone (ST163227)	1
13.	River Exe near Stoke Canon (SX923963)	1
14.	River Culm near Silverton (SS984012)	1
15.	River Axe near Colyton (SY261936)	1
16.	Dorset Stour near Spetisbury (ST033028)	1
17.	River Rother near Fittleworth (SU998183)	1
18.	River Arun near Billingshurst (TQ069261)	1
19.	River Adur near Partridge Green (TQ179187)	1
20.	River Medway near Penshurst (TQ528436)	1
21.	River Start near Slapton (SX813447)	1

base of detectable levels of unsupported ^{210}Pb. Additional cores were collected from undisturbed reference locations above the level of flood inundation, adjacent to all sites, in order to provide estimates of the local fallout inventories. The sectioned cores were sectioned into 2-cm increments prior to analysis, but the bulk cores were analysed as bulk samples in order to determine their total ^{137}Cs and unsupported ^{210}Pb inventories. In addition, for the detailed study sites and several of the single-core sites, suspended sediment samples representative of major storm events and/or samples of contemporary overbank floodplain deposits were also collected. All samples were air-dried and disaggregated prior to gamma assay using high purity, low background, low energy N-type germanium detectors linked to a multi-channel analyser. Count times were typically c. 40 000 s. ^{137}Cs activities were measured directly at 662 keV and unsupported ^{210}Pb concentrations were calculated from measurements of total ^{210}Pb, ^{226}Ra and ^{214}Pb activity, according to standard procedures (cf. Joshi, 1987).

Some typical analytical results are presented in Figure 11.2 which presents ^{137}Cs and unsupported ^{210}Pb depth distributions for the sites on the River Tone (site 12, Table 11.2) and the River Arun (site 18, Table 11.2) and where single sectioned cores

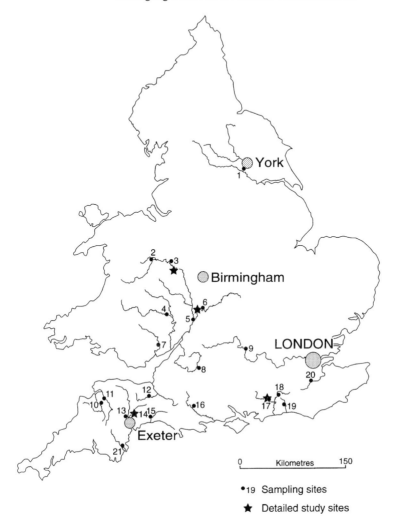

Figure 11.1 The location of the study sites

were collected. In both cases, the ^{137}Cs depth distribution evidences a clearly defined subsurface peak, about 18 cm and 13 cm below the surface for the two cores respectively, which reflects the floodplain surface in 1963 at the time of peak bomb fallout. Estimates of the precise depth of the floodplain surface in 1963, and therefore of the average sedimentation rate over the period extending to the time of collecting the cores, were derived using the model proposed by Walling and He (1997) which takes account of post-fallout redistribution of ^{137}Cs within the sediment profile. The mean annual sedimentation rates over the past 33 years for these two sites, estimated from the ^{137}Cs data are 0.56 and 0.39 g cm^{-2} year^{-1} for the Rivers Tone and Arun respectively. The unsupported ^{210}Pb depth distributions illustrated in Figure 11.2 are characterised by progressive decline in concentration with depth reflecting sediment

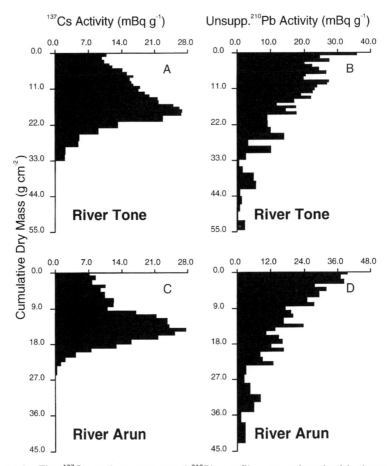

Figure 11.2 The ^{137}Cs and unsupported ^{210}Pb profiles associated with the sediment cores collected from the floodplains of the River Tone near Bradford on Tone, Somerset, and the River Arun at Billingshurst

accumulation and the radioactive decay of ^{210}Pb. Since no samples of suspended sediment or contemporary overbank floodplain deposits were available for these sites, information on the initial unsupported ^{210}Pb concentration of deposited sediment is not available and the CIC model was therefore used to establish the sedimentation rate. The resulting estimates of mean annual sedimentation rate for the past 100 years are 0.43 g cm^{-2} year^{-1} for the River Tone and 0.48 g cm^{-2} year^{-1} for the River Arun. The age–depth relationships for these two sites, based on the estimates of the average sedimentation rates for the past 33 years and 100 years provided by the ^{137}Cs and unsupported ^{210}Pb data respectively, are presented in Figure 11.3. Figure 11.3 indicates that average sedimentation rates have increased towards the present in the case of the site on the floodplain of the River Tone, whereas the core from the site on the River Arun floodplain shows evidence of a decreasing sedimentation rate towards the present.

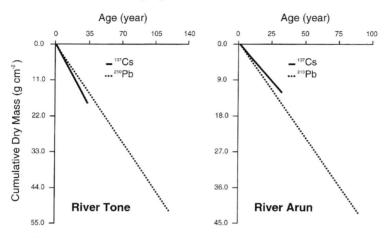

Figure 11.3 The age–depth relationships established for the sediment cores depicted in Figure 11.2 using the estimates of average sedimentation rate provided by the ^{137}Cs and unsupported ^{210}Pb measurements

In the case of the four detailed study sites, where multiple bulk cores were collected, estimates of the average sedimentation rates over the period since the commencement of ^{137}Cs fallout and the past 100 years were derived from the whole core values of "excess" ^{137}Cs and unsupported ^{210}Pb inventory respectively, using the procedure described by Walling and He (1997) for ^{137}Cs and the CICCS model for unsupported ^{210}Pb (cf. He and Walling, 1996b). The spatial patterns associated with the grid-based data were mapped using the UNIRAS computer-based interpolation algorithm.

RESULTS

The Four Detailed Sampling Sites

The spatial patterns associated with the mean annual sedimentation rates within the four detailed study sites over the past c. 40 and 100 years derived using the ^{137}Cs and unsupported ^{210}Pb measurements are presented in Figures 11.4 and 11.5. The rates of overbank sedimentation documented for these four floodplain areas are in close agreement with the relatively low rates of accretion found by other studies undertaken on British rivers discussed above, and in most cases fall within the range 0.05–0.5 g cm^{-2} year^{-1}. There is little difference in the sedimentation rates documented for the rivers Culm, Severn and Rother, but those reported for the Warwickshire Avon are considerably lower. This contrast is probably largely a reflection of the choice of a study site characterised by relatively low sedimentation rates, but it may also reflect the relatively low suspended sediment concentrations encountered on this river, the low gradient of the river channel and its associated floodplain, and the construction of weirs to control water levels for navigation. For each site, there is clear evidence of a reduction in sedimentation rates with increasing distance from the river channel, but the detailed patterns shown by the sedimentation rates also reflect the floodplain

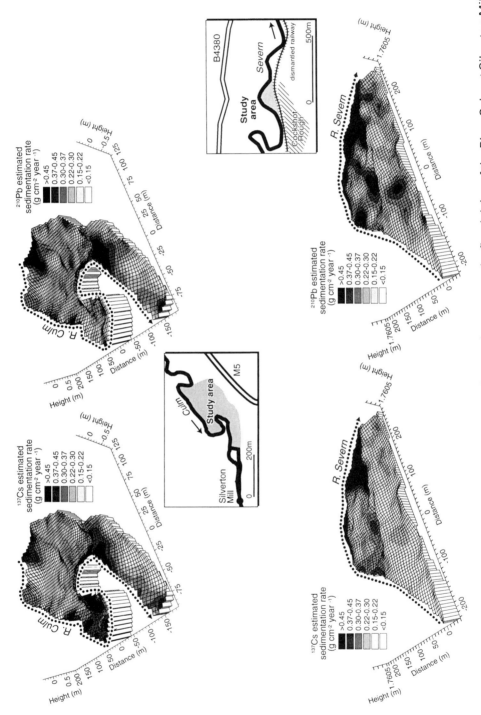

Figure 11.4 The spatial pattern of mean annual overbank sedimentation rates on the floodplains of the River Culm at Silverton Mill and the River Severn at Buildwas, established using ^{137}Cs and unsupported ^{210}Pb measurements

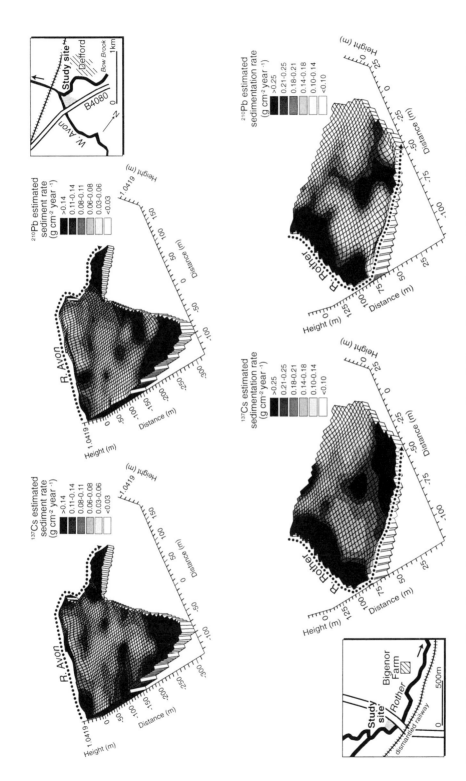

Figure 11.5 The spatial pattern of mean annual overbank sedimentation rates on the floodplains of the Warwickshire Avon at Eckington Bridge and the River Rother at Shopham Bridge, established using ^{137}Cs and unsupported ^{210}Pb measurements

Table 11.3 The average mean annual sedimentation rates estimated for the detailed study sites from the ^{137}Cs and unsupported ^{210}Pb measurements

	Mean annual sedimentation rate ($g\,cm^{-2}\,year^{-1}$)	
River/location	Past 40 years	Past 100 years
1. River Culm at Silverton Mill	0.29	0.27
2. River Severn at Buildwas	0.28	0.33
3. Warwickshire Avon at Eckington Bridge	0.09	0.08
4. River Rother at Shopham Bridge	0.22	0.20

microtopography and associated depths of inundation, the channel pattern and the influence of tributary inputs (cf. Walling *et al.*, 1996). Further discussion of these patterns and their associated controls lies beyond the scope of this contribution. Here, emphasis is placed on a comparison of the sedimentation rates for the past 40 years with those for the past 100 years. In all four cases, the patterns evidence a high degree of similarity between the two periods, indicating that little change has occurred in the spatial distribution of sedimentation rates. This in turn emphasises the stability of the floodplain microtopography and the channel pattern over the past 100 years and demonstrates that there is little feedback between the spatial variation of sedimentation rates and associated changes in microtopography and the pattern of sedimentation, over a 100-year timescale. To assist further comparison of the sedimentation rates for the two time periods, mean sedimentation rates for the two periods have been calculated for each site using the data for the complete sets of cores collected from each site. These values are presented in Table 11.3. In each case, there is relatively little difference between the mean sedimentation rates for the two periods. The maximum contrast is found in the values for the site on the River Severn at Buildwas, where the mean annual sedimentation rate for the past 40 years averages $0.28\,g\,cm^{-2}\,year^{-1}$ across the site, whereas the equivalent value for the past 100 years averages $0.33\,g\,cm^{-2}\,year^{-1}$. When comparing the sedimentation rate estimates for the two periods, it is important to take account of the precision of the procedures used to derive those estimates. Taking the longer-term 100-year sedimentation rate as the reference, a difference of $> \pm 10\%$ has been used as the threshold to indicate a significant change in sedimentation rate in recent years. Using this criterion, the results presented in Table 11.3 indicate that there has been no significant change in sedimentation rates at the sites on the Rivers Culm and Rother, whereas a very small increase is evident for the site on the River Avon and a small decrease is shown by the site on the River Severn. These changes are, however, very small and only slightly in excess of the 10% threshold introduced above ($+ 12.5\%$ and $- 15\%$ respectively). The results obtained from the four detailed study sites therefore provide no clear evidence of changing rates of overbank sedimentation at these sites over the past 100 years. Figures 11.4 and 11.5 show that the patterns of sedimentation at these sites have remained essentially similar over the past 100 years, and comparison of the magnitude of the sedimentation rate estimates for the two periods suggests that little or no change has occurred.

The Single-Core Sites

The results obtained from the 21 additional floodplain sites, where only single cores were collected, are clearly less reliable than those provided by the detailed study sites, since they are based on estimates of sedimentation rate for a single point. Nevertheless, because these points were carefully selected to be representative of areas of active sedimentation on the individual floodplains, and because emphasis is placed on comparing estimates of sedimentation rate for two different periods, rather than the absolute values of sedimentation rate involved, the data are judged to provide a valid means of extending the results obtained from the detailed study sites to a wider sample of British floodplains. The estimates of overbank sedimentation rates for these sites, for the past 33 years and 100 years, are listed in Table 11.4. The sedimentation rates listed cover the ranges 0.04–1.22 g cm^{-2} year^{-1} and 0.04–1.42 g cm^{-2} year^{-1} for the ^{137}Cs and unsupported ^{210}Pb measurements respectively. They are therefore again consistent with existing evidence concerning the magnitude of sedimentation rates on British floodplains. The site-specific nature of the individual estimates of sedimentation rate precludes detailed consideration of the factors influencing the magnitude of the values listed in Table 11.4, but there is some evidence that maximum values are associated with rivers draining catchments with headwaters in upland areas (e.g. Severn, Usk, Yorkshire Ouse, Taw and Torridge), whilst rivers draining lowland catchments underlain by more permeable rocks are characterised by lower values (e.g. Dorset Stour,

Table 11.4 The mean annual sedimentation rates estimated for the single-core sites from the ^{137}Cs and unsupported ^{210}Pb measurements

River/location	Sedimentation rate (g cm^{-2} year^{-1})		Trend
	Past 33 years	Past 100 years	
1. River Ouse near York	0.95	1.04	stable
2. River Vyrnwy near Llanymynech	0.21	0.46	decrease
3. River Severn near Atcham	1.22	1.42	decrease
4. River Wye near Preston on Wye	0.15	0.28	decrease
5. River Severn near Tewkesbury	0.86	0.95	stable
6. Warwickshire Avon near Pershore	0.46	0.66	decrease
7. River Usk near Usk	0.88	1.01	decrease
8. Bristol Avon near Langley Burrell	0.39	0.33	increase
9. River Thames near Dorchester	0.51	0.64	decrease
10. River Torridge near Great Torrington	0.70	0.93	decrease
11. River Taw near Barnstaple	0.60	0.65	stable
12. River Tone near Bradford on Tone	0.56	0.43	increase
13. River Exe near Stoke Canon	0.45	0.42	stable
14. River Culm near Silverton	0.35	0.32	stable
15. River Axe near Colyton	0.51	0.40	increase
16. Dorset Stour near Spetisbury	0.04	0.04	stable
17. River Rother near Fittleworth	0.11	0.14	decrease
18. River Arun near Billingshurst	0.39	0.48	decrease
19. River Adur near Partridge Green	0.51	0.71	decrease
20. River Medway near Penshurst	0.15	0.23	decrease
21. River Start near Slapton	0.51	0.45	increase

Rother, Arun and Medway). Again, however, emphasis is placed on comparison of the sedimentation rate estimates for the two different periods, rather than on consideration of their absolute magnitude. If the $> \pm 10\%$ threshold introduced above is used as a criterion for identifying significant change in sedimentation rates over the past 100 years, decreasing sedimentation rates are evidenced by 11 floodplain sites, increasing sedimentation rates by four sites and stable sedimentation rates by six sites. In all cases, however, both the increases and decreases in sedimentation rate are of limited magnitude and the data presented in Table 11.4 again suggest that rates of sedimentation on the floodplains of British rivers have remained relatively constant over the past 100 years. Where evidence of change does exist, small *decreases* in sedimentation rate predominate. Such decreases could be accounted for by the progressive increase in the height of the floodplain surface, consequent upon sediment accretion, which would cause a reduction in the frequency and depth of overbank inundation if it was not accompanied by equivalent aggradation of the channel bed. However, they could equally reflect a reduced magnitude and/or frequency of overbank flooding, reduced suspended sediment loads and concentrations, and other factors. Interestingly, all four of the floodplain sites showing evidence of a small but significant increase in sedimentation rate towards the present (Axe, Tone, Start and Bristol Avon) are located in south-west England. However, other rivers in this region (Taw, Torridge, Exe and Culm) are characterised by stable or reduced sedimentation rates and the increases are therefore more likely to reflect changes in land use or other local conditions, rather than climate change which could be expected to exert a more region-wide influence. The 11 floodplain sites characterised by reduced sedimentation rates towards the present, include both rivers draining upland areas (Vyrnwy, Severn, Wye, Usk and Torridge) as well as lowland areas (Warwickshire Avon, Thames, Rother, Medway and Arun) and no clear regional differentiation is apparent.

PERSPECTIVE

Although the study reported above demonstrates that the fallout radionuclides ^{137}Cs and unsupported ^{210}Pb offer a valuable means of quantifying medium-terms rates and patterns of floodplain sedimentation, for which more traditional techniques offer little by way of an alternative, the limitations of the approach and its assumptions must be recognised. More particularly, the resulting estimates of sedimentation rates represent *average* values for the period under consideration (i.e. c. 33 or 40 years and 100 years), and changes could have occurred within these periods. Furthermore, it should be appreciated that the average sedimentation rate for a period will reflect the magnitude and frequency characteristics of floodplain inundation and that essentially random differences in the incidence of flood events between the two periods could either mask or generate contrasts in the average rate of sedimentation.

Accepting these limitations, the results presented in Figures 11.4 and 11.5 and in Tables 11.3 and 11.4 suggest that rates of overbank sedimentation on the floodplains of British rivers have remained essentially constant over the past 100 years, despite the potential for change in sedimentation rates associated with the various aspects of environmental changes identified in Table 11.1. This conclusion must clearly be qualified by the fact that for many sites the estimates of sedimentation rates were based on a single core and that only a limited number of floodplain sites were investigated. Nevertheless, the consistency of the findings for the range of sites investigated adds credence to the results presented. Care should, however, be exercised in attempting to extrapolate the above conclusions to provide a more general assessment of the erosional response of UK river basins to the various facets of environmental change identified in Table 11.1. Land-use change could, for example, cause appreciable increases in on-site erosion rates, but such increases may not result in significant increases in downstream sediment loads and associated floodplain sedimentation, due to the complexity of the sediment delivery system (cf. Trimble, 1983; Walling, 1983). Relatively little is currently known about the suspended sediment yields of UK river basins and their variation through time (cf. Walling and Webb, 1987), but the evidence obtained from this study of floodplain sedimentation rates over the past 100 years suggests that no major changes have occurred.

ACKNOWLEDGEMENTS

The work reported in this contribution was undertaken as part of a NERC-funded investigation of contemporary sedimentation rates on river floodplains (GR3/8633). The authors gratefully acknowledge this financial support, the co-operation of landowners in allowing access to sampling sites, the valuable contribution of Mr Jim Grapes to the gamma spectrometry analysis and the assistance of Mr Terry Bacon in producing the diagrams.

REFERENCES

Appleby, P. G. and Oldfield, F. 1983. The assessment of ^{210}Pb data from sites with varying sedimentation rates. *Hydrobiologia*, **103**, 29–35.

Brown, A. G. 1987. Holocene floodplain sedimentation and channel response of the lower River Severn, UK. *Zeitschrift für Geomorphologie*, **31**, 293–310.

Brown, A. G. 1990. Holocene floodplain diachronism and inherited downstream variations in fluvial processes: a study of the River Perry, Shropshire, England. *Journal of Quaternary Science*, **5**, 39–51.

He, Q. and Walling, D. E. 1996a. Rates of overbank sedimentation on the floodplains of British lowland rivers documented using fallout ^{137}Cs. *Geografiska Annaler*, **78A**, 223–234.

He, Q. and Walling, D. E. 1996b. Use of fallout Pb-210 measurements to investigate longer-term rates and patterns of overbank sediment deposition on the floodplains of lowland rivers. *Earth Surface Processes and Landforms*, **21**, 141–154.

He, Q. and Walling, D. E. 1996c. Interpreting particle size effect in the adsorption of ^{137}Cs and unsupported ^{210}Pb by mineral soils and sediments. *Journal Environmental Radioactivity*, **30**, 117–137.

Joshi, S. R. 1987. Nondestructive determination of Lead-210 and Radium-226 in sediments by direct photon analysis. *Journal of Radioanalysis and Nuclear Chemistry, Articles*, **116**, 169–182.

Lambert, C. P. and Walling, D. E. 1987. Floodplain sedimentation: a preliminary investigation of contemporary deposition within the lower reaches of the River Culm, Devon, UK. *Geografiska Annaler*, **69A**, 47–59.

Livens, F. R. and Baxter, M. S. 1988. Particle size and radionuclide levels in some west Cumbrian soils. *Science of the Total Environment*, **70**, 1–17.

Oldfield, F. and Appleby, P. G. 1984a. Empirical testing of ^{210}Pb dating models for lake sediments. In Haworth, E. Y. and Lund, J. W. (Eds) *Lake Sediments and Environmental History*. Leicester University Press, 93–114.

Oldfield, F. and Appleby, P. G. 1984b. A combined radiometric and mineral magnetic approach to recent geochronology in lakes affected by basin disturbance and sediment redistribution. *Chemical Geology*, **44**, 67–83.

Robbins, J. A. 1978. Geochemical and geophysical applications of radioactive lead. In Nriagu, J. O. (Ed.) *Biogeochemistry of Lead in the Environment*. Elsevier, Amsterdam, 285–293.

Shotton, F. W. 1978. Archaeological inferences from the study of alluvium in the lower Severn–Avon valleys. In Limbrey, S. and Evans, J. G. (Eds) *Man's Effect on the Landscape in the Lowland Zone*. Council for British Archaeology Research Report No. 21, 27–32.

Simm, D. J. 1995. The rates and patterns of overbank deposition on a lowland floodplain. In Foster, I. D. L., Gurnell, A. M. and Webb, B. W. (Eds) *Sediment and Water Quality in River Catchments*. John Wiley, Chichester, 247–264.

Trimble, S. W. 1983. A sediment budget for Coon Creek basin in the Driftless Area, Wisconsin, 1853–1977. *American Journal of Science*, **283**, 454–474.

Walling, D. E. 1983. The sediment delivery problem. *Journal of Hydrology*, **65**, 209–237.

Walling, D. E and He, Q. 1992. Interpretation of caesium-137 profiles in lacustrine and other sediments: the role of catchment-derived inputs. *Hydrobiologia*, **235/236**, 219–230.

Walling, D. E and He, Q. 1993. Use of caesium-137 as a tracer in the study of rates and patterns of floodplain sedimentation. In *Tracers in Hydrology*. IAHS Publication No. 215, 319–328.

Walling, D. E. and He, Q. 1994. Rates of overbank sedimentation on the flood plains of several British rivers during the past 100 years. In *Variability in Stream Erosion and Sediment Transport*. IAHS Publication No. 224, 203–210.

Walling, D. E. and He, Q. 1997. Use of fallout ^{137}Cs in investigations of overbank sediment deposition on river floodplains. *Catena*, **29**, 263–282.

Walling, D. E. and Quine, T. A. 1993. Using Chernobyl-derived fallout radionuclides to investigate the role of downstream conveyance losses in suspended sediment budget of the River Severn, United Kingdom. *Physical Geography*, **14**, 239–253.

Walling, D. E. and Webb, B. W. 1987. Suspended load in gravel-bed rivers: UK experience. In Thorne, C. R., Bathurst, J. C. and Hey, R. D. (Eds) *Sediment Transport in Gravel-bed Rivers*. John Wiley, Chichester, 691–732.

Walling, D. E., Quine, T. A. and He, Q. 1992. Investigating contemporary rates of floodplain sedimentation. In Carling, P. A. and Petts, G. E. (Eds) *Lowland Floodplain Rivers: Geomorphological Perspectives*. John Wiley, Chichester, 166–184.

Walling, D. E., He, Q. and Nicholas, A. P. 1996. Floodplains as suspended sediment sinks. In Anderson, M. G., Walling, D. E. and Bates, P. D. (Eds) *Floodplain Processes*. John Wiley, Chichester, 399–440.

12 Floodplain Evolution and Sediment Provenance Reconstructed from Channel Fill Sequences: The Upper Clyde Basin, Scotland

J. S. ROWAN,[1] **S. BLACK**[2] **and C. SCHELL**[3]
[1] *Department of Geography, University of Dundee, UK*
[2] *PRIS, University of Reading, UK*
[3] *School of Geography, University of Leeds, UK*

INTRODUCTION

Floodplains play a vital role as sediment sinks within the fluvial system. With appropriate understanding of alluvial histories, floodplain sequences offer considerable potential as archives of changing sediment yields or supply sources within drainage basins (e.g. Macklin *et al.*, 1994; Passmore and Macklin, 1994; Collins *et al.*, 1997a). Such information gives a basis on which to reconstruct the effects of historic climate and land-use changes on sediment delivery, and provides the foundation upon which to model the effects of future catchment management strategies. Metalliferous mining and associated processing activities typically result in elevated metal concentrations in downstream floodplain sediments. Metals have been widely used as tracers to elucidate floodplain dynamics because of their strong affinity for particulate-phase transport (e.g. Lewin *et al.*, 1983). As a corollary, an appreciation of physical transport processes is vital to predicting and managing the fate of pollutants (Miller, 1997). For example, significant temporal discontinuities in the delivery of sediment-associated pollutants can result from the remobilisation of historically contaminated floodplain sediment stores by laterally active channels (e.g. Marron, 1992; Rowan *et al.*, 1995).

This study offers a preliminary assessment of the recent alluvial history of the upper Clyde valley. This is one of the most important water supply basins in Scotland, which in its lower reaches flows through the populous Clydeside conurbation. The headwaters of the Clyde have experienced a long, but intermittent, history of mining spanning over 700 years. However, the nature and scale of geochemical and geomorphological impacts, both within the mining district and downstream in the main channel–floodplain system, remain poorly understood.

The present work sought to characterise the distribution of metals within the floodplain and to explain the patterns found in terms of governing geomorphological processes. A single site on the main river, situated 40 km downstream of the mining field, was selected for this pilot study. Sediment cores were collected from a series of well-defined palaeochannels and analysed for heavy metal and trace element compositions. The value of metals to provide an accretion chronology was tested against map-based morphological evidence and ^{210}Pb dates. A quantitative multivariate fingerprinting approach was then used to investigate temporal variations in fine-grained sediment provenance and in particular to determine the relative importance of the mining legacy in relation to floodplain alluviation.

FLOODPLAIN SITE AND STUDY CATCHMENT

This study focused on the extensive floodplain developed on the upper reaches of the River Clyde, in southern Scotland (Figure 12.1(a)). The Clyde–Medwin meander complex is designated as a site of special scientific interest (SSSI) under the Geological

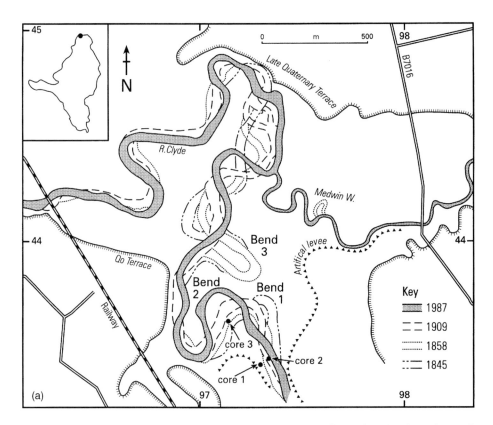

Figure 12.1 (a) Floodplain study site and coring location from dated palaeochannels; (b) topography and solid geology of the upper Clyde basin

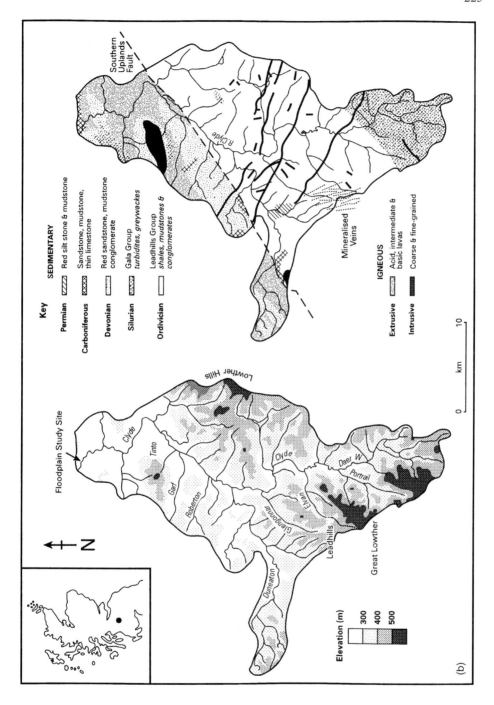

Conservation Review because it is a rare example in the British Isles of a large lowland river system actively migrating and reworking its floodplain (Brazier et al., 1993). The floodplain here comprises a wide valley floor, locally exceeding 1 km and containing a sinuous and highly active channel corridor c. 400 m wide. Bounded by late Quaternary terraces, the main Holocene floodplain is relatively flat with relative relief rarely exceeding 3 m. The meandering character of the channel is expressed in the form of numerous oxbows, palaeochannels and scroll bars. Eroding cut-banks and extensive unvegetated point bars are testimony to continued lateral instability. Changes in channel pattern were readily reconstructed from historical maps, most notably since the appearance of Ordnance Survey maps (i.e. 1852, 1858, 1896, 1909, 1957 and 1977), and aerial photographs (e.g. 1946, 1954, 1968 and 1988).

The 650 km^2 catchment area above the main floodplain site can be divided into southern and northern elements along the line of the Southern Uplands Fault (Figure 12.1(b)). The southern headwaters, mainly lying above 300 m, comprise a series of fault-bounded blocks of Lower Palaeozoic age. The Ordovician and Silurian country rocks (Leadhills and Gala Groups) are dominated by highly folded and faulted mudstones, shales, greywackes and conglomerates. These are traversed by a series of NW–SE trending dykes, mainly basic and fine-grained in composition. Mean annual precipitation totals exceed 2000 mm. The major soil types are ferric stagnopodsols and cambic humic gleys in the valley bottoms, with brown earths on the well-drained valley sides and extensive areas of blanket peat on the broad interfluves and undulating plateaux of the highest ground. Land use is dominated by sheep and cattle grazing along with heather management for grouse.

The lithology of the northern part of the catchment is dominated by Upper Palaeozoic sequences. Devonian sandstones, mudstones and conglomerates are abundant, whereas Carboniferous and Permian rocks (limestones and breccias respectively) are much rarer and outcrop only locally. The Devonian was also a period of major volcanic activity causing andesetic and basaltic lavas and tuffs to be extruded widely. Intrusive volcanics are also represented by the Tinto Hills. These northern lithologies have produced a more subdued and rolling topography with elevations typically below 200 m. Mean annual precipitation similarly declines to values of c. 900 mm. The volcanic series in particular are characterised by well-drained loamy soils. Arable cropping, especially for beet crops, is common in the northern and north-eastern areas of the basin.

MINERALISATION AND THE LEADHILLS MINING LEGACY

The primary zone of mineralisation occurs in the headwaters of the Elvan and Glengonnar Waters which join the River Clyde approximately 25 km upstream of the main floodplain study site (Figure 12.1(b)). The Leadhills mining district along with the adjacent Wanlockhead District has dominated Scottish base metal production. Together they accounted for over 90% of all lead, silver and zinc ever produced in Scotland and as much as 10% of the national UK output of lead (Smoult, 1967).

More than 70 mainly N–S trending vein sets have been recorded (see Figure 12.1(b)). Minerals were emplaced into "geofractures" during the Caledonian and again during the Hercynian phases (430/390 Ma). Deposits are typically filled fissures between 0.05 and 5 m wide. The first generation of sulphides (galena, sphalerite, chalcopyrite and pyrite) were associated with dolomite, calcite and baryte. Subsequent replacement of the carbonates and sulphates by quartz was accompanied by a second generation of sulphides (BGS, 1993). Most of the secondary minerals contain lead, zinc and copper in their structures, and several of these minerals are rare, or particular to the district, i.e. Leadhillite, Susannite and Scotlandite. Gold deposits are more disseminated and most abundant in the pyritic black shales associated with high Zr content vein systems in the southern half of the mining field.

Mining at Leadhills was first documented in the 13th century, but is generally accepted to predate the Romans. Medieval lead mining was small scale and intermittent. Until the early 17th century, mining was mainly focused on gold extraction from gravel fills within the valley floors of the Glengonnar and Elvan Waters (Gillanders, 1981). Relict landforms associated with the widespread use of hushing techniques, i.e. flood releases from check dams to scour valley fills, are still obvious today and testify to major sediment mobilisation during this period (cf. Hunter, 1884). Commercial gold working largely ceased after 1620 when attention turned to lead mining within the vein systems.

Large-scale lead mining operations commenced in the second half of the 18th century, the development being enhanced by advances in mining and processing technologies, and peaked during the Napoleonic Wars. The introduction of Free Trade in the 1830s brought about a decline in the district's lead mining as ore prices fell due to greater imports. Modernisation and capital investments led to a brief revival in the 1860s which continued until the early part of this century. Mining ceased in 1928 when lead prices went into further decline (Smoult, 1967). A brief revival occurred in the 1950s when lead and zinc were obtained by reworking of old spoil heaps (Gillanders, 1981).

A reconstruction of the lead production history of the Leadhills mining field is shown in Figure 12.2(a). The output pattern reflects variations in demand and technology such that before the 19th century ore-processing was restricted to relatively inefficient hand-sorting, crushing and smelting processes and much of the ore was wasted. Extraction efficiency improved especially with the introduction of water and steam power for washing and smelting plants and later electricity (Gillanders, 1981). Different processing techniques produced waste of varied characteristics such that early mining wastes were coarse-grained and ore-rich, whereas later more efficient crushing and extraction methods generated finer-grained, less metalliferous waste. According to Moffat (1991), about 0.5 Mt of metalliferous waste remains in over 1700 ha of contaminated land in the Leadhills region. Rowan et al. (1995) reported Pb concentrations in excess of 57 500 mg kg^{-1} in floodplain sediments of the Glengonnar water, demonstrating that the mining era continues to have a persistent pollution legacy.

FIELD METHODS AND ANALYTICAL TECHNIQUES

The floodplain study concentrated on the inside of a large meander 1 km upstream of the Clyde–Medwin confluence. Numerous palaeochannel remnants were identified in

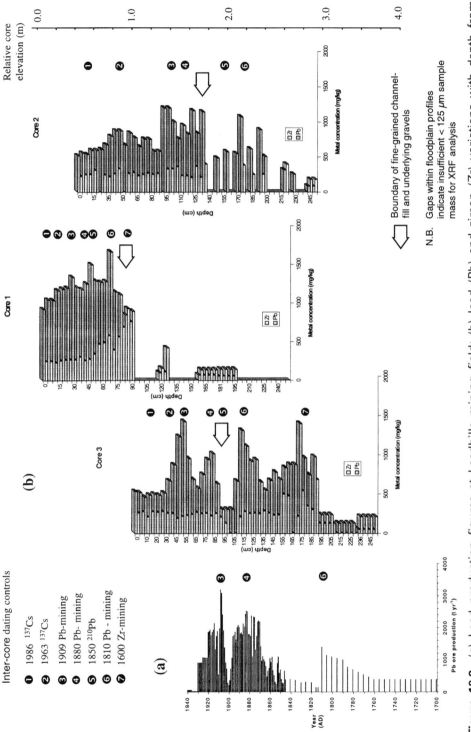

Figure 12.2 (a) Lead production figures at Leadhills mining field; (b) lead (Pb) and zircon (Zr) variations with depth from palaeochannel cores

the field and provisionally dated from historical maps (cf. Brazier *et al.*, 1993). The period 1850–1910 is particularly relevant because it spanned the era of reliable mining records (Burt *et al.*, 1981) and represents the period of reliable map coverage. Three coring locations were selected from palaeochannel remnants (Figure 12.1(a)) which were well defined in the field and identifed on historical maps. Sediment cores were collected to a depth of 3 m using a stainless steel auger driven by a motorised percussion hammer. The cores were stratigraphically logged and representative samples obtained at 5 cm increments, or smaller depending on the recognition of facies boundaries. Samples were taken from the centre of the core to avoid any potential contamination with the auger lining.

An extensive survey of potential catchment sediment source groups was initiated to investigate possible changes in the provenance of the fine-grained constituent of the floodplain sequence. The catchment was divided into four lithological source groups, i.e. (a) Ordovician–Silurian, (b) Devonian, (c) volcanics and (d) a range of sediment sources including spoil and metalliferous floodplain sediments within the mining district. Within each group the dominant sediment sources were sampled ($n = 49$), including eroding surface soils, channel banks and locally important erosion features such as channel bluffs cutting into Pleistocene till and outwash terraces.

All sediment samples were air-dried, disaggregated and mixed thoroughly before screening through a 2 mm sieve. Water content, loss on ignition, pH and cation exchange capacity were determined using standard methods (Avery and Bascomb, 1974). Because of the limited amount of fine-grained matrix present in some of the floodplain gravels, geochemical analyses were conducted on the $< 125\ \mu m$ fraction (cf. Horowitz, 1991). Trace element composition of the mineral fraction was determined by X-ray fluorescence (Phillips PW1400) generating data on the following elements: Cr, La, Rb, Zr, Pb, Ni, Ba, Sr, V, W, Zn, Cu and Y, along with numerous associated ratios. Calibration on representative matrix material was ensured using USGS standard river sediment with an accuracy of $\pm < 2\%$.

In order to establish independent dating control for floodplain aggradation within the channel fills, radiometric dating was undertaken using a combined γ- and α-spectrometry approach for ^{210}Pb. ^{210}Pb is a naturally occurring radionuclide produced from the ^{238}U decay series with a half-life ($t_{1/2}$) of 22.2 years. Its presence in the atmosphere results from the decay of radon gas (^{222}Rn, $t_{1/2} = 3.8$ days) which diffuses from the lithosphere (^{222}Rn is itself the decay daughter of ^{226}Ra, $t_{1/2} = 1602$ years). Atmospheric ^{210}Pb returns to the land surface mainly by wet deposition associated with precipitation. Accordingly the activity levels of ^{210}Pb within a sediment sequence can be resolved into two components. "Unsupported" ^{210}Pb is the fallout equivalent and represents the difference between total ^{210}Pb and that supplied by ^{226}Ra (supported). Because annual fluxes of ^{210}Pb from the atmosphere are assumed constant over time, the levels of ^{210}Pb$_{unsupp}$ within an aggrading sediment profile should vary as a function of age (cf. Oldfield and Appleby, 1984).

Gamma-spectrometry was used to measure supported ^{210}Pb$_{supp}$ (via ^{214}Pb at 352 MeV) after sealing sediment samples for 30 days to achieve equilibrium with ^{226}Ra. Gamma-spectrometry was carried out using low background EG&G Ortec HPGe co-axial detectors giving analytical errors of 2–3% (1 σ). Count times were typically in the order of two days. Alpha-spectrometry was used to measure

$^{210}Pb_{unsupp}$ via ^{210}Po using EG&G silicon surface barrier detectors. This was required at depths where the decay of $^{210}Pb_{unsupp}$ reduced activity concentrations to levels below the sensitivity of conventional γ-spectrometry. Using this combined γ- and α-spectrometry approach, the dating envelope was extended to its maximum with reproducible results and very low analytical uncertainties of 2–5% (1 σ).

FLOODPLAIN CHANNEL PATTERN CHANGES AND FACIES ANALYSIS

The historical adjustments of the main channel within the Clyde–Medwin meander complex is shown in Figure 12.1(a). Since 1845 the sinuosity of the total river length shown has varied between 1.76 and 2.04, which is well above the threshold of 1.5 that distinguishes sinuous from truly meandering rivers (R. A. Davies, 1983). Some reaches, particularly those furthest downstream, have been relatively stable whilst others have been highly active. Bend 1 has been the most consistently dynamic (with a mean migration rate of 0.75 m year^{-1}) and its development evidences elements of extension, translation and rotation (cf. Hooke, 1977). Channel abandonment has resulted from both lateral migration, channel avulsion and neck cut-offs. The oxbow lake at bend 3 was formed between 1858 and 1896. The incipient oxbow at bend 2 will almost certainly be completed within the next 10 years.

All three channel remnants cored showed common stratigraphies (Table 12.1) comprising a fine-grained unit (0.8–1.3 m thick) overlying gravels of undetermined depth (> 3 m). The upper units were composed of weakly laminated sandy-silts representing floodwater incursion and overbank deposits. Occasional lenses of sandy gravel were noted, probably representing discrete flood events. In all units the contact between the channel fill unit and underlying gravels was sharp. The nature of the auguring technique prevented detailed logging of gravel architecture. However, cross-bedding was occasionally observed and fabrics varied between well-sorted open-framework gravels to poorly-sorted matrix-supported units. Maximum clast size in the gravel units varied from 10 to 70 mm (b-axis), consistent with the gravels of the present channel bed.

Table 12.1 Summary of palaeochannel features from the Clyde–Medwin study site

Site properties	Core 1	Core 2	Core 3
Relative elevation (m)	3	2.6	2
Channel-fill depth (m)	0.82	1.32	0.92
Metalliferous fill depth (m)	1.28	3	2.5
Inferred map-derived cut-off date	> 1848	1909	1850
Absolute date (^{210}Pb and metals)	> 1800	1895	1856
Mean fill accretion rate (cm year^{-1})	0.45	1.33	0.66
Peak accretion rate (cm year^{-1})	1.33	2.56	10

GEOCHEMISTRY OF CHANNEL FILLS

The behaviour of heavy metals in fluvial sediments is controlled by a range of physico-chemical properties including grain size, pH, organic matter and cation ion exchange capacity. A detailed account of the core geochemistry is inappropriate here but a summary of the XRF analysis is given in Table 12.2. Geochemical gradients exist within the profiles, i.e. pH typically increased with depth (4.5–6.0), but organic matter content was generally low (0.6–9.1%), as was the CEC (4.6–11.5 meq/100 g dry mass), reflecting the predominantly silty texture of the sediment analysed. Some metals such as Zn and Cu evidenced considerable translocation and therefore offer limited potential for chronology purposes (cf. Farmer, 1991). By contrast, Pb and particularly Zr (present as zircon) are effectively immobile in this sedimentary environment and down-profile variations in concentration reflect temporal variations in supply.

Lead levels peaked above 1500 mg kg^{-1} but were highly variable, as shown by the minimum value of 53 mg kg^{-1}. Similarly, variable levels of Ba and Zr contrasted with the consistency of rare earth elements such as Rb; the latter confirms that restricting the analysis to the < 125 μm fraction minimised possible grain-size dependency in the results. Metal concentrations in the upper 1–2 m of each core evidenced significant enhancement relative to the background values, i.e. 42 mg kg^{-1} for Pb (cf. B. E. Davies, 1983). However, towards the base of each core, metal levels declined to background values suggesting "uncontaminated" pre-mining phase sediments were reached (cf. Macklin and Klimek, 1992).

Down-profile variations in metals within the cores are exemplified using Pb and Zr (Figure 12.2(b)). It is important to note that elevated metal levels extended below the fine-grained channel fills representing a total metalliferous fill depth approaching 4 m. The metalliferous sequence is shallower in core 1 and down-profile variability is less marked when compared to the other cores (though this is in part a function of the 5 cm sampling interval). Core 3 evidences the most distinctive variation in Pb levels with four distinct Pb peaks containing over 1000 mg kg^{-1} (labelled 3, 4, 6, 7 in Figure

Table 12.2 Summary of trace element composition within palaeochannel sequences

Site		Cr	La	Rb	Zr	Pb	Ni	Ba	Sr	V	W	Zn	Cu	Y
Core 1	Mean	219	34	104	382	984	55	646	2	149	2	108	52	6
1.9 m	Min	130	22	86	135	55	48	468	< dl[a]	79	< dl[a]	72	36	2
	Max	431	43	118	927	1662	65	766	11	177	6	135	63	11
	C.V.%	30	17	9	58	47	9	13	170	14	69	17	16	42
Core 2	Mean	204	32	107	273	768	58	681	9	160	3	136	55	8
2.5 m	Min	152	20	92	181	84	44	462	< dl[a]	127	< dl[a]	86	40	1
	Max	309	42	117	402	1197	68	904	28	176	7	168	63	13
	C.V.%	16	17	5	21	31	10	17	82	8	44	13	8	39
Core 3	Mean	213	31	110	277	749	62	825	7	161	3	129	57	8
3 m	Min	130	20	92	128	53	43	560	< dl[a]	138	1	80	43	2
	Max	297	46	134	686	1421	81	1050	30	176	7	188	78	17
	C.V.%	13	20	11	38	42	15	13	135	6	42	19	15	42

[a] < dl = Below detection limits.

12.2(b)). All three profiles evidenced significant Zr peaks at the base of the metalliferous sequence, most obviously in core 1.

CHANNEL FILL CHRONOLOGY

Several workers have used heavy metals to formulate alluvial chronologies by matching sediment profiles to historical mining production data (e.g. Macklin, 1985). However, in complex multi-peaked profiles potential ambiguities in "curve-matching" requires the use of independent dating procedures (Dearing, 1992). The starting point of the present investigation was to core channel remnants identified from historical maps, i.e. it was inferred that channel 3 was abandoned in the period 1858–1890 thus giving a minimum age for the base of the fine-grained channel fill. In core 3 the upper two Pb peaks (labelled 3 and 4) can thus be tentatively assigned to 1908 and 1880 (Figure 12.2). However, the lower two Pb peaks (labelled 6 and 7) occur within the underlying gravels and indicate an earlier episode of accretion involving metalliferous gravels. These peaks could not be dated unequivocally because of the lack of detail in the output history of the Leadhills mining field and thus require an independent dating control such as provided by ^{210}Pb dating.

The ^{210}Pb$_{unsupp}$ profile for core 3 is shown in Figure 12.3. Two distinct segments are shown reflecting significantly different sedimentation rates. This distinct non-linearity necessitated the use of a ^{210}Pb sedimentation model that was sensitive to changing sedimentation rates. The CRS (constant rate of supply) model was used because variable ^{210}Pb accumulation rates must be modelled precisely to construct reliable chronologies (Ivanovich and Herman, 1992). The utility of the CRS model was independently validated using ^{137}Cs peaks corresponding to bomb-test and Cher-

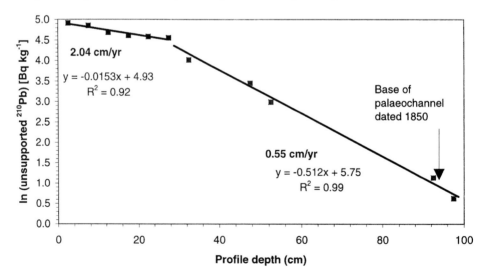

Figure 12.3 Unsupported ^{210}Pb profile for core 3

nobyl fallout spikes in 1963 and 1986 respectively (Rowan, 1995). Using the ^{210}Pb CRS model for core 3, the contact between the fine member and underlying gravels was dated at 145 ± 7 years (between AD 1842 and 1856) which was in excellent agreement with the inferred age obtained from the map data (1858–1890). In core 2 this interface occurred in the 1890s (also broadly agreeing with the map data), but channel 1 was found to be considerably older than originally thought because the interface of channel fines and the underlying gravel lay beyond the temporal range of the ^{210}Pb technique, i.e. earlier than AD 1800 (Table 12.2).

The ^{210}Pb dating procedure confirmed that the upper two Pb peaks in core 3 were associated with peaks in mining output around 1909 and 1880. The ^{210}Pb$_{unsupp}$ measurements confirmed that the third Pb peak within core 3 (labelled 6 in Figure 12.2(b)) corresponds to the Napoleonic period. The fourth Pb peak (labelled 7) was beyond the time range of the ^{210}Pb technique and has no obvious source in the established mining output data. The basal Pb peak was, however, associated with a significant increase in the Zr levels and so may tentatively be linked to gold mining during the Medieval period. The accessory mineral zircon is closely associated with the gold-bearing shales within the Leadhills mineralised zone. The earliest phase of commercial mining concentrated on alluvial gold deposits which may have preferentially flushed out zircon (specific gravity 4.7), which is considerably lighter than native gold (specific gravity 19.3). This interpretation is consistent with the fact that the Zr anomalies are found within the gravel layers thus predating fine-grained channel fills. The utility of this anomaly is limited for dating because the specific timing of the Zr maximum is uncertain. However, it is generally agreed that the main phase of gold mining took place between 1580 and 1620 (cf. Hunter, 1884), allowing the Zr peak to be tentatively assigned the intermediate date of AD 1600.

The extension of the metalliferous sequence into the underlying gravel units suggests they were laid down under high energy conditions, i.e. as extensive point- and mid-channel bars within a channel system frequently switching position (cf. Lewin et al., 1983; Bradley and Cox, 1987). The depth of this fill is easily accounted for by the sinuous nature of the channel which means that channel bed height differences between the upstream and downstream limbs of kilometre-scale meander loops can readily exceed 3 m.

Using a combination of ^{210}Pb dating and heavy metal variations, a composite accretion history for core 3 was produced (Figure 12.4). The composite chronology resolved to 12 discrete intervals ranging from 2 to 150 years. It is notable that the metallo-chronology appears to remain viable within the gravel sequences where the sedimentation rate averages $c.$ 2 mm year^{-1}. Deposition within the fine-grained palaeochannel fill accelerated from 0.55 to 2 cm year^{-1} towards the top of the profile. Several distinctive flood horizons were also recognised, e.g. those assumed to be associated with the major catchment-wide floods in 1910, 1955 and 1983 (McEwan, 1990; Acreman, 1991). The latter event was responsible for a 10 cm thick sediment unit and highlights the importance of individual floods in floodplain construction.

The preservation of individual horizons and their significance to local deposition rates depends on both allogenic influences such as climate change, changing flood frequency or variable sediment yields, and autogenic controls, i.e. the changing relationship between a channel cut-off and inundation from a mobile main channel. Thus

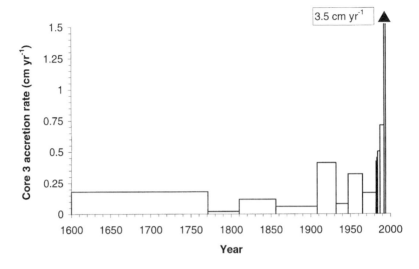

Figure 12.4 Composite metal and ^{210}Pb accretion chronology for core 3

the accelerated deposition rate observed at core 3 is believed to reflect the progressive migration of the active channel towards the cut-off site causing more frequent and prolonged overbank flows and enhanced suspended sediment deposition. Cores 2 and 3 display comparable geochemical profiles, but substantially different accretion histories reflecting the complexity of spatio-temporal deposition patterns even within lateral distances of 50 m. Cut-off 1 was significantly older than the other channel remnants examined and incised into uncontaminated Holocene gravels. Thus the relatively thin unit of metalliferous gravel indicates that the channel was active in the

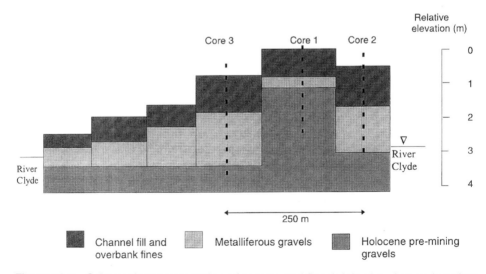

Figure 12.5 Schematic representation of segmented floodplain showing coring sites

Medieval period (evidenced by the distinctive Zr peak) but that the rate of infilling was slow, reflecting its relative height and associated reduced inundation frequency.

These preliminary results indicate that the deposition of mining sediments within the Clyde–Medwin meander complex has been controlled by both lateral instability of the channel system and sediment supply dynamics. Meandering has developed by neck cut-offs, channel avulsion and lateral migration, and has produced a relatively narrow, highly segmented floodplain corridor, containing metalliferous sediments to depths approaching 4 m adjacent to the main channel (Figure 12.5). By contrast, the bulk of the floodplain surface has remained relatively stable, receiving only a thin veneer of overbank sedimentation from high magnitude, low frequency events (cf. Taylor and Lewin, 1996).

FINE-GRAINED SEDIMENT PROVENANCE

The final element of this work examined variations in the provenance of the fine-grained component of the Clyde floodplain using the temporal framework provided by the Pb–Zr and the ^{210}Pb chronology. The ability to identify and quantify the supply of sediment from its respective sources is a key element in effective catchment management. In the absence of direct measurements, indirect estimation techniques such as sediment "fingerprinting" offer the greatest potential to pursue this goal (cf. Peart and Walling, 1986). In principle the fingerprinting procedure is relatively simple in that the properties of the suspended sediment incorporated into the floodplain are compared with the equivalent values of potential sources. A useful tracer should be able to differentiate between potential sources and exhibit conservative behaviour during erosion and fluvial transport. Because no single tracer can reliably distinguish all potential sources, composite signatures offer the most powerful discrimination (Dearing, 1992; Walling et al., 1993; Collins et al., 1997b).

Fingerprinting techniques have been applied over a range of timescales, from the event level (Slattery et al., 1995) to extended reconstructions involving sediment sinks such as reservoirs (Foster and Walling, 1994), estuaries (Yu and Oldfield, 1989) and floodplains (Passmore and Macklin, 1994; Collins et al., 1997a). The current generation of fingerprinting models use multivariate mixing models based on constrained linear equations (cf. Yu and Oldfield, 1993). Such models are over-determined if $m \geq n$ (where m is the number of tracer properties, and n is the number of distinct source groups) and thus require optimisation procedures to determine the relative contributions made by each source group. In this preliminary study a spreadsheet optimisation tool was used to optimise model performance, i.e. a likelihood function which in this case was explained variance (cf. Nash and Sutcliffe, 1970):

$$\text{Efficiency} = 1 - \frac{\sum_{i=1}^{m} (\hat{x}_i - x_i)^2}{\sum_{i=1}^{m} (x_i - \bar{x}_i)^2}$$

where m is the total number of sediment properties, $\bar{x}_i$ is the mean of the source group properties, and x_i and $\hat{x}_i$ are the measured and predicted values for property i (where $i = 1, 2, \ldots, m$). The linear mixing model used to calculate $\hat{x}_i$ is of the form:

$$\hat{x}_i = \sum_{j=1}^{n} (a_{ij} b_j) \qquad \text{subject to the constraints} \qquad \sum_{j=1}^{n} b_j = 1$$

$$0 \leq b_j \leq 1$$

where n is the number of source groups, a is the mean value of property i ($i = 1, 2, \ldots, m$) of source group j ($j = 1, 2, \ldots, n$) and b is the contributory coefficient of source group j.

To minimise over-parametrisation it is therefore necessary to define the least number of tracers capable of unequivocally discriminating between the pre-selected source groups. Multivariate discriminant analysis (MDA), specifically the stepwise Mahalanbois procedure, was used to identify the optimal composite signature. This stepwise procedure operates through selection of sediment properties in order of their relative discriminating power until all the variables have either been included in the discriminant function, or excluded because they are incapable of contributing any further significant information to the model. The best fit of the data is associated with low within-group variability relative to high between-group variability (cf. Hair et al., 1987).

Four lithologically based source groups were identified in the catchment sampling programme. XRF analysis produced geochemical profiles for each sample comprising 13 trace elements and associated ratios. These data were entered into the MDA stepwise selection procedure within SPSS (v. 6.1) which identified six variables capable of discriminating between the pre-defined source groups, i.e. offering greater between-group variability than within-group variability as measured by the Wilks' Lamda Coefficient. Pb was the best predictor variable in the discriminant function, followed by Rb, V, W, N and Cr/V. Together these variables accounted for 100% of the cumulative variance in the sample set. Subsequent analysis using the multivariate mixing model was restricted to these statistically significant variables.

The mixing model outlined above was applied directly to selected dated horizons within core 3. Mean values for source groups were used without correction for grain size or organic matter enrichment effects (cf. Collins et al., 1997a). Figure 12.6 illustrates that all the main lithologies are represented in the fine fraction ($< 125 \mu$m) of the floodplain. However, it is clear that Devonian and volcanic sources are the most prevalent, particularly the volcanics, which on average contribute over 45% in each sample. This most likely reflects the more intensive agricultural practices and the higher erodibility of local soils in the northern parts of the basin. Southern sediments from the Ordovician–Silurian uplands typically accounted for around 30% of the floodplain sediments. Despite their much larger areal coverage and generally steeper slopes, these upland areas are characterised by low intensity farming practices, and sediment supply is primarily from channel bank erosion.

One of the most significant results from this preliminary study was to demonstrate the relatively minor contribution made by the mining field. Prior to the large-scale mining period (pre-Medieval?), sediment was derived from throughout the catchment area. As the scale and intensity of mining operations grew, the proportion of mining-

Floodplain Evolution and Sediment Provenance 237

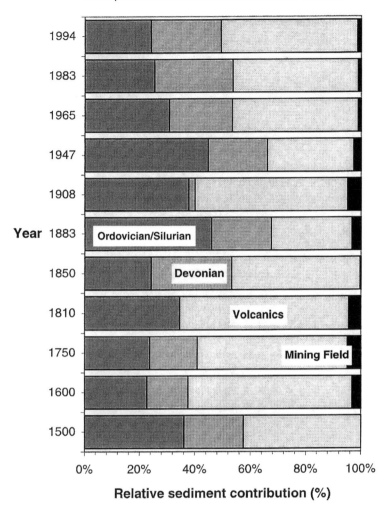

Figure 12.6 Relative sediment contribution from major catchment lithology groups

derived sediments rose and peaked at 5.1% around the same time as the hiatus of the upstream mining operations. These findings confirm that metals were incorporated into the sequence in a manner analogous to the "passive transport" mechanism described by Lewin and Macklin (1987).

CONCLUSIONS

This preliminary study sought to investigate the alluvial history of the Clyde–Medwin meander complex. Morphological development was observed through analysis of historical maps and air photographs, but greater insight was gained from a consideration of the metalliferous palaeochannel infills. The distribution of metals revealed a high degree of spatial and temporal variability reflecting the complexity of the floodplain environment. More than 3 m of gravels and channel fines have been incorporated into the floodplain by a suite of meander processes. In the case of the Clyde floodplain, heavy metals ultimately offer dating opportunities, but the need for an independent control such as ^{210}Pb was demonstrated to avoid ambiguity in multi-peaked sequences.

The application of a quantitative sediment fingerprinting approach was used to explore variations in sediment supply from the catchment and showed unequivocally that mining sediments comprise a small fraction, by mass, of the fine sediments within the floodplain. Future work will seek to employ further fingerprinting variables such as mineral magnetics capable of resolving source *type* as well as spatial origin (cf. Collins *et al*., 1997b). This has particular application in the investigation of sediment sources for discrete flood horizons within the high-resolution palaeochannel sequences. Clearly there is scope to recognise these events and to build up the event stratigraphy and investigate longer-term flood frequency within these important water supply basins. It follows that the changing sedimentation rates within the palaeochannels may owe more to climatic and autogenic influences than variable sediment supply rates from the mining belt. During the remainder of the 20th century the proportion of mining-derived sediments has fallen to around 1%. The bulk of the mining waste continues to be stored in the tributary systems. These are only slowly released by the reworking of severely contaminated floodplain deposits, especially within the Glengonnar Water (cf. Rowan *et al*., 1995).

ACKNOWLEDGEMENTS

The authors wish to express their thanks to the many local landowners for granting access, and to Robert Rowe, Vicky Burnett and Chris Murdoch for their assistance in sample collection, XRF and ^{210}Pb analysis respectively.

REFERENCES

Acreman, M. 1991. The flood of July 25th 1983 on the Hermitage Water, Roxburghshire. *Scottish Geographical Magazine*, **107**, 170–178.

Avery, B. W. and Bascomb, C. L. (Eds) 1974. *Soil Survey Laboratory Methods*. Rothamsted Experimental Station, Harpenden.

Bradley, S. B. and Cox, J. J. 1987. Heavy metals in the Hamps and Manifold valleys, North Staffordshire, UK: partitioning of metals in floodplain soils. *Science of the Total Environment*, **65**, 135–153.

Brazier, V., Kirkbride, M. and Werritty, A. 1993. Scottish landform examples: the Clyde–Medwin meanders. *Scottish Geographical Magazine*, **109**, 45–49.

BGS (British Geological Survey) 1993. *Regional Geochemistry of Southern Scotland and Part of Northern England*. BGS, Keyworth.

Burt, R., Waite, P. and Atkinson, M. 1981. Scottish metalliferous mining 1845–1913: detailed returns from the mineral statistics. *Industrial Archaeology*, **16**, 9–11, 141–157.

Collins, A. L., Walling, D. E. and Leeks, G. J. L. 1997a. Use of the geochemical record preserved in floodplain deposits to reconstruct recent changes in river basin sediment sources. *Geomorphology*, **19**, 151–167.

Collins, A. L., Walling, D. E. and Leeks, G. J. L. 1997b. Source type ascription for fluvial suspended sediment based on a quantitative composite fingerprinting technique. *Catena*, **29**, 1–27.

Davies, B. E. 1983. A graphical estimation of the normal lead content of some British soils. *Geoderma*, **29**, 67–75.

Davies, R. A. 1983. *Depositional Systems. A Genetic Approach to Sedimentary Geology*. Prentice-Hall, London.

Dearing, J. A. 1992. Sediment yields and sources in a Welsh upland lake-catchment during the past 800 years. *Earth Surface Processes and Landforms*, **17**, 1–22.

Farmer, J. G. 1991. The perturbation of historical pollution records in aquatic sediments. *Environmental Geochemistry and Health*, **13**, 76–83.

Foster, I. D. L. and Walling, D. E. 1994. Using reservoir deposits to reconstruct changing sediment yields and sources in the catchment of the Old Mill Reservoir, South Devon, UK, over the past 50 years. *Hydrological Sciences Journal*, **39**, 347–368.

Gillanders, R. J. 1981. Famous mineral localities: the Leadhills–Wanlockhead district, Scotland. *The Mineralogical Record*, July–August, 235–250.

Hair, J. F., Anderson, R. E. and Tatham, R. L. 1987. *Multivariate Data Analysis*. Macmillan, New York.

Hooke, J. M. 1977. The distribution and nature of changes in river channel patterns. The example of Devon. In Gregory, K. J. (Ed.) *River Channel Changes*. John Wiley, Chichester, 265–280.

Horowitz, A. J. 1991. *Sediment-Tracer Element Chemistry*, 2nd edition. Lewis, Michigan.

Hunter, J. R. S. 1884. The Silurian Districts of Leadhills and Wanlockhead, and their early and recent mining history. *Transactions Geological Society of Glasgow*, **7**, 373–392.

Ivanovich, M. and Herman, R. S. (Eds) 1992. *Uranium-series Disequilibrium: Applications to Earth, Marine and Environmental Sciences*, 2nd edition. Clarendon Press, Oxford.

Lewin, J. and Macklin, M. G. 1987. Metal mining and floodplain sedimentation in Britain. In Gardiner, V. (Ed.) *International Geomorphology 1986, Part 1*. John Wiley, Chichester, 1009–1028.

Lewin, J., Bradley, S. B. and Macklin, M. G. 1983. Historical valley alluviation in mid-Wales. *Geological Journal*, **18**, 331–350.

Macklin, M. G. 1985. Floodplain sedimentation in the upper Axe Valley, Mendip, England. *Transactions of the Institute of British Geographers*, N.S., **10**, 235–244.

Macklin, M. G. and Klimek, K. 1992. Dispersal, storage and transformation of metal-contaminated alluvium in the upper Vistula basin, southwest Poland. *Applied Geography*, **12**, 7–30.

Macklin, M. G., Ridgway, J., Passmore, D. G. and Rumsby, B. T. 1994. The use of overbank sediments for geochemical mapping and contamination assessment: results from selected English and Welsh floodplains. *Applied Geochemistry*, **9**, 689–700.

Marron, D. C. 1992. Floodplain storage of mine tailings in the Belle Fourche river system: a sediment budget approach. *Earth Surface Processes and Landforms*, **17**, 675–685.

McEwen, J. J. 1990. The establishment of a historical flood chronology for the River Tweed catchment, Berwickshire, Scotland. *Scottish Geographical Magazine*, **106**, 37–48.

Miller, J. R. 1997. The role of fluvial geomorphic processes in the dispersal of heavy metals from mine sites. *Journal of Geochemical Exploration*, **58**, 101–118.

Moffat, W. E. 1991. Restoration of metalliferous mine waste at Leadhills: 1985–1990. Report to the Scottish Development Agency, Glasgow.

Nash, J. E. and Sutcliffe, J. V. 1970. River flow forecasting through conceptual models 1. A discussion of principles. *Journal of Hydrology*, **10**, 282–290.

Oldfield, F. and Appleby, P. G. 1984. Empirical testing of ^{210}Pb dating models for lake sediments. In Haworth, E. Y. and Lund, J. W. G. (Eds) *Lake Sediments and Environmental History*. Leicester University Press, 93–114.

Passmore, D. G. and Macklin, M. G. 1994. Provenance of fine-grained alluvium and late Holocene land-use change in the Tyne basin, northern England. *Geomorphology*, **9**, 127–142.

Peart, M. R. and Walling, D. E. 1986. Fingerprinting sediment sources: the example of a small drainage basin in Devon, UK. In Hadley, R. F. (Ed.) *Drainage Basin Sediment Delivery*. IAHS Publication No. 159, 41–55.

Rowan, J. S. 1995. The erosional behaviour of radiocaesium in catchment systems: case-study of the Exe basin, Devon, UK. In Foster, I. D. L., Gurnell, A. M. and Webb, B. W. (Eds) *Sediment and Water Quality in River Catchments*. John Wiley, Chichester, 331–354.

Rowan, J. S., Barnes, S. J. A., Hetherington, S. L., Lambers, B. and Parsons, F. 1995. Geomorphology and pollution: the environmental impacts of lead mining, Leadhills, Scotland. *Journal of Geochemical Exploration*, **52**, 57–65.

Slattery, M. C., Burt, T. P. and Walden, J. 1995. The application of mineral magnetic measurements to quantify within-storm variations in suspended sediment sources. In *Tracer Technologies for Hydrological Systems*. IAHS Publication No. 229, 143–151.

Smoult, T. C. 1967. Lead-mining in Scotland, 1650–1850. In Payne, P. L. (Ed.) *Studies in Scottish Business History*. Frank Cass, Edinburgh, 103–135.

Taylor, M. P. and Lewin, J. 1996. River behaviour and Holocene alluviation: the River Severn at Welshpool, Mid-Wales, UK. *Earth Surface Processes and Landforms*, **21**, 77–82.

Walling, D. E., Woodward, J. C. and Nicholas, A. P. 1993. A multi-parameter approach to fingerprinting suspended sediment sources. In Bogen, J., Walling, D. E. and Day, T. (Eds) *Erosion and Sediment Transport Monitoring Programmes*. IAHS Publication No. 215, 329–338.

Yu, L. and Oldfield, F. 1989. A multivariate mixing model for identifying sediment sources from magnetic measurements. *Quaternary Research*, **32**, 168–181.

Yu, L. and Oldfield, F. 1993. Quantitative sediment source ascription using magnetic measurements in a reservoir-catchment system near Nijar, S.E. Spain. *Earth Surface Processes and Landforms*, **18**, 441–454.

13 Siberian-type Quaternary Floodplain Sedimentation: The Example of the Yenisei River

A. F. YAMSKIKH,[1] A. A. YAMSKIKH[2] and A. G. BROWN[3]
[1] Laboratory of Paleogeography, Krasnoyarsk State Pedagogical University, Russia
[2] Department of Ecology, Krasnoyarsk State University, Russia
[3] Department of Geography, University of Exeter, UK

INTRODUCTION

It is necessary to define the term "floodplain" in order to understand the nature of polycyclic alluvial sedimentation. The floodplain is a geomorphological element of a river valley that is covered by high water stages and formed by the erosional and depositional activity of rivers. In this study the modern floodplain is considered to have formed under the influence of seasonal river level fluctuations during the last 400–500 years. This definition differs from the standard definition usually used because here only those landforms that have been formed during this period of time are regarded as floodplain. Using this definition, an alluvial basement could be formed earlier than the contemporary floodplain. This way of defining the floodplain allows us to exclude alluvial deposits accumulated by high floods on the surfaces of the low terraces. This deposition occurred during different periods of the Holocene. Using the standard definition of the floodplain the 8–20 m terraces would be included; however, this would be incorrect because, as this chapter shows, such deposits were not formed under the present hydrological regime.

The cyclic theory of terrace formation (Yamskikh, 1992a, 1996) was the working paradigm during the geological and geomorphological investigations of the Siberian river valleys. Dating was achieved using floodplain deposits (mainly palaeontological remains) from different horizons of the terraces with a wide age range (Figure 13.1). Figure 13.1 gives ages for the Yenisei terraces estimated by different researchers, and (on the right) age estimates based on the polycyclic theory of the terrace formation (Yamskikh 1992a, 1996). Polycyclic terrace formation occurred as a result of simultaneous alluvial accumulation on terraces of different heights. As a result of this process, the stratigraphy of the polycyclic terraces includes alluvial deposits accumulated during different glacial–interglacial climatic cycles. Sedimentation occurred during

Fluvial Processes and Environmental Change. Edited by A. G. Brown and T. A. Quine.
© 1999 John Wiley & Sons Ltd.

242   Fluvial Processes and Environmental Change

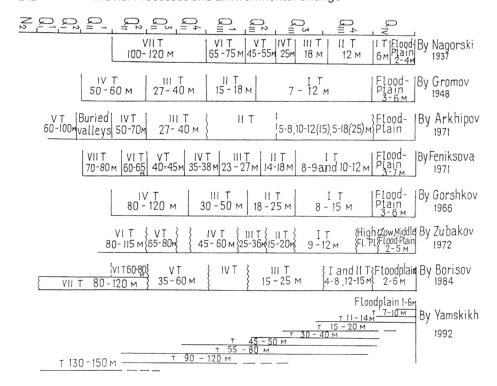

Figure 13.1 Correlation of the terrace ages in the middle part of the Yenisei River Valley according to various authors

high floods and/or as a result of high dams caused by river ice and ice sheets with consequent outflow of the dammed reservoirs. Holocene flood levels in the low part of the Yenisei River Basin exceeded the mean levels of 10–20 m during floods and 35–45 m during dam formation. In the middle part of the Yenisei River Basin, Holocene flood levels were up to 20–25 m above the channel. The highest flood levels occurred during periods of cool climate. New data, especially the results of absolute dating, contradict the traditional theories of terrace formation. One of the most interesting problems is the explanation of the origin of the so-called "problem horizons". These horizons, which are a major part of the Pleistocene terraces, are formed by loess deposits. The age of the lower part of the loess strata differs greatly from that of the upper part. These sediments have alluvial and lacustrine–alluvial origins but they include intercalated layers of the flood and aeolian deposits (Yamskikh, 1992b). For example, the lowest part of the loess strata of the 35–45 m high terrace of the Yenisei river was formed about 70 000 years BP (Frechen and Yamskikh, 1998), whereas the upper part of this stratum was formed about 13 000 years BP (Yamskikh, 1992b). Such composition of the terrace sections is typical for intracontinental Siberia. Diagenetic transformation by subaerial processes associated with loess deposition of the original alluvial and lacustrine–alluvial deposits has caused the disappearance of alluvial laminations. Detailed investigations have revealed the secondary role of the colluvial and aeolian deposits in the terrace stratigraphy. The polycyclic theory of terrace formation explains contradictions in the terrace horizon dating by the simultaneous alluvial accumulation on different geomorphological surfaces during different glacial–interglacial periods.

The polycyclic origin of the majority of terraces in the river valleys of southern Siberia have been confirmed for not only the Pleistocene but also for the Holocene (Yamskikh, 1993). The simultaneous deposition of overbank sediments onto terraces has also been noted in other environments, a classic example being Brackenridge's (1984) terrace-overlap sedimentation on the Duck River, Tennessee. It can also be caused by relative sea-level or base-level rise, such as is the case for Pleistocene sub-alluvial terraces in the lower zones of many western European rivers, by excessive sediment deposition causing channel–floodplain aggradation (Brown, 1997) and by slackwater deposits associated with mega-floods (Baker et al., 1983).

METHODS OF INVESTIGATION

The reconstruction of alluvial sedimentary records has been performed using a combination of geological, geomorphological and palaeogeographical methods and absolute dating techniques. Lithological, palaeopedological and palynological investigations have been applied to floodplain sediments intercalated in the different terraces. Also taken into consideration were neotectonic fluctuations. Investigations of the alluvial soil sequences prompted detailed investigations of the late Pleistocene and Holocene hydrological river regime and periods of soil formation. Results of archaeological investigations have helped to reveal some peculiarities of the polycyclic dynamics of sedimentation. Several concentrations of multilayer Palaeolithic, Neolithic, Bronze and Iron Age archaeological sites, located on the different geomorphological surfaces, have been studied in the Yenisei River Valley (A. F. Yamskikh, 1993; A. A. Yamskikh, 1995).

TERRACE COMPLEXES IN THE SIBERIAN RIVER VALLEYS

All the large Siberian rivers flow into the northern seas of the Arctic Ocean. The upper parts of their valleys are located in the mountains of southern Siberia, and the middle and lower parts in the depressions and uplifted plains. The upper mountainous reaches have also experienced Pleistocene glaciations, whilst the lower reaches were affected by Pleistocene ice sheets (Figure 13.2). The longest stretch of the Yenisei River Valley (2000–2800 km) is located within the boundaries of the Pleistocene periglacial belt. This part of the river valley has the largest number of terraces. Different researchers have identified between 5 and 12 terraces. Because the floodplain and terraces are multi-level, the terraces have gradually transitional boundaries and it is difficult to separate one terrace from another. The Yenisei is considered to be a typical Siberian river in terms of geomorphology. Generally, the following terraces have been recognised: 8–10 m, 11–14 m, 15–20 m, 25–30 m, 30–40 m, 45–50 m, 55–80 m, 90–120 m, 130–150 m, 170–190 m and 200–240 m (in mountains, 350–400 m).

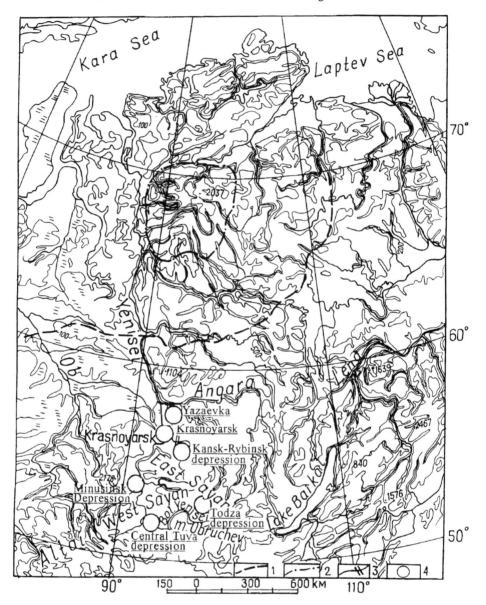

Figure 13.2 Hydrographic map of central and southern Siberia

The major periods of incision correspond to the Neogene–Quaternary tectonic phases. They occurred at the Oligocene–Pliocene boundary, the Pliocene–Pleistocene boundary and during the middle and late Pleistocene (Seliverstov, 1966; Zyatkova and Rakovec, 1969; Zubakov, 1972; Yamskikh, 1972). Strata of the basal channel alluvium were formed during these incision phases. Basal channel gravel in the mountains is located on the high geomorphological surfaces. Usually there are no floodplain deposits on the surfaces of these gravel deposits. The basal alluvium is the basement for the terrace complexes in the intermontane depressions and plains of southern and central Siberia. The lower boundary of the basal alluvium of the low terrace complex (from 8–10 m to 15–20 m) is 5–15 m below the present-day river level (in neotectonical depressions, 80–120 m below the present-day river level). The middle terrace complex (30–120 m) is 10–80 m higher than the modern mean river level. The thick strata of the floodplain loess alluvium is typical for these terrace complexes (except the 8–10 m terrace).

Low terraces (8–10 m up to 18 m) were formed during the Holocene. They are located in the depressions and plain parts of the river valleys. Holocene terraces and floodplains were formed by the "normal" alluvium (basal sediments) covered by the multilayer overbank strata of sands, loamy sands and loams and horizons of buried soils.

CAUSES OF THE POLYCYCLIC SEDIMENTATION OF FLOODPLAIN ALLUVIUM

Floodplain sedimentation and the formation of landforms in river valleys in intracontinental Siberia occurred during periods of tectonic activity and (rhythmic) climate changes. Middle Pliocene, Pliocene–middle Pleistocene and middle–late Pleistocene phases of tectonic uplift caused erosion, incision and the accumulation of channel gravels.

Hydroclimatic factors affect floodplain formation through changes in the palaeohydrological regime. Interactions between snow and rain and cold seasons favourable for ice-dam formation (due to the northerly direction of river flow) determine floodplain response. High floods in the upper parts of the river valleys were caused by the melting of enormous volumes of snow which accumulated in the mountains during cold periods.

Interaction between both groups of factors has produced peculiarities of floodplain form and stratigraphy in Siberia and channel processes determine the dynamics of alluvial accumulation. Nevertheless, the ratio between channel and floodplain deposits is stable. Channel sediments deposited on the convex banks with subsequent vertical floodplain sediment typically occurs in meandering rivers. Channel sediment deposition on the upper parts of the islands and partial erosion of their lower parts typically occurs in braided rivers. Accumulation of the floodplain deposits occurs on the surface of the channel alluvium and previously existing floodplain sediments of the islands.

Floodplain form is also affected by landscape features. The influence of the ponding in the river valley bottoms is increased by tree dams and debris dams in the taiga regions. Such dams usually form in the upper parts of the stream network. They determine alluvium accumulation upstream. Well-sorted alluvium with a low content of silt and clay grains is another feature of the river valley deposits in the taiga zone. Sediments of oxbow lakes and horizons of wood peat with tree stumps are also typical for this zone.

THE PECULIARITIES OF THE HYDROLOGICAL RIVER REGIME

Climatic and landscape structure determines the intensity, value and levels of the floodplain alluvium accumulation. The seasonal distribution of river runoff is irregular in intracontinental Siberia (Figure 13.3). About 50–65% of the annual river runoff occurs during the spring and early summer, and is related to snow melting (May–early June). Snow accumulates during the five to six coldest months (late October–early April). In the Sayan Mountains and the mountains of the Baikal region there are on average eight cold months (late September–early May). Winter runoff is about 5–13% of the annual total; summer–autumn runoff is 25–35% of the annual total.

Flood levels from the end of April to the beginning of May are caused by snow melting in the depressions and lower mountains in southern Siberia, resulting in river-ice movement, ice-dam formation and erosion and subsequent alluvial accumulation on the low and middle floodplain. The highest stage of the spring flood is caused by snow melting in the middle and high mountains. All the relief forms (channel, floodplain and some low terraces) of the river valley bottom are reworked. The largest discharges in the middle part of the Yenisei River Valley are $30\,000$–$57\,000\,\mathrm{m}^3\,\mathrm{s}^{-1}$. Ice break-up begins here 20–25 days earlier than the atmospheric temperature rise. This is because the Yenisei River has a northerly direction of flow. Atmospheric warming and consequent ice break-up in the upper part of the valley occurs 15–25 days earlier than in the middle and lower parts of the river valley, and so broken ice moves from the upper part of the Yenisei River Valley to the middle and lower reaches. Thick unbroken ice in the middle and low (northern) parts of the Yenisei River Valley prevents ice moving from the upper part of the river and causes river ice-dam formation. It has been calculated that the power of a flood with a discharge $100\,000$–$120\,000\,\mathrm{m}^3\,\mathrm{s}^{-1}$ is not great enough to break the ice cover of Siberian rivers. The biggest

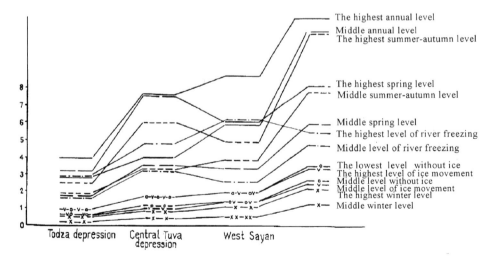

Figure 13.3 Modern seasonal river level fluctuations in the Upper Yenisei valley montane and inter-montane locations

river-ice dams formed in the Pleistocene periglacial zone where rivers were fully frozen and permafrost was widespread. Tectonic activity and ice-dams are both important for the Yenisei River in the Obruchev Ridge, West and East Sayan Mountains and Yenisei Ridge (Yazaevka location) (Figure 13.2). Similar ice-dams are common on tributaries of the Yenisei River, and also the Ob River and rivers of the Baikal region. Modern ice-dams on the Yenisei River and its tributaries are from 10–15 to 45 m high. Water level rises, caused by the dams, influence the processes of valley formation upstream, and in the depressions causing additional alluvial accumulation, floodplain formation and widening of the valley bottom. Terrace extent increases in the foothills.

MODERN FLOODPLAIN FORMATION

Floodplains in Siberian river valleys exist at several altitudinal levels. Usually there are low and high levels of the floodplain corresponding to typical flood levels: low, high and several transitional stages (Figure 13.4). Modern floodplain formation is the result of simultaneous alluvial accumulation on geomorphological surfaces of different heights over the last 400–500 years. Low terrace levels are affected by the intensive erosion–accumulation activity of the spring and autumn floods. Shrubs and meadow vegetation (*Cyperaceae*, *Salix*) grow on its surface. High floodplain is usually flooded only during the second (high) stage of the spring flood at the beginning of June. Since the building of the Krasnoyarsk hydroelectric power station, Yenisei river-level fluctuations have been suppressed and the high floodplain is very rarely flooded. Trees, shrubs and meadow vegetation grow on its surface (including *Populus*, *Picea obovata* and *Abies sibirica* in the taiga zone). The velocity of the shallow water during floods is reduced by the vegetation, creating favourable conditions for the deposition of silt and clay.

Both low and high levels of the modern floodplain were formed during the last several hundred years, but floodplain deposits have also accumulated over a longer period of time. Basal horizons of the channel alluvium were deposited during the middle Holocene (6000–5000 years BP). Floodplain alluvium accumulated on the floodplain and on the low terraces.

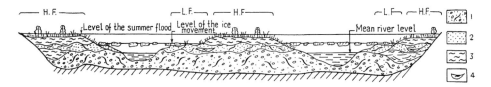

Figure 13.4 A generalised stratigraphy of Siberian rivers showing relations between morphology and hydrological regime. HF = high flood; LF = low flood. 1 = channel sandy ground; 2 = sand; 3 = overbank and alluvial loam and loamy sand, 4 = silty sand, loamy-sand and loam

QUATERNARY SEDIMENTATION OF THE FLOODPLAIN ALLUVIUM

Periodic climate cooling characterises the Quaternary history of Siberia. The severity of the climate tended to increase from the early to late Pleistocene, with maximum cooling during the Last Glacial (22 000–10 500 years BP). The Siberian hydrological river regime, caused by the climatic conditions, determined floodplain formation during the whole of the Quaternary. Increasing climate severity caused increased flood levels.

Climatic conditions during the interstadials were not stable. The late Pleistocene Mega-interstadial (Middle Wurm Interstadial) (50 000–22 000 years BP) was periodically cold with cryogenic processes repeated rhythmically in the course of the 2000-year-long cycles. Alluvial sedimentation on the surfaces 30–50 m high occurred during this period. The amplitude of the river-level fluctuations decreased during the stadials.

Pleistocene hydrological and climatic conditions caused a trend of floodplain alluvial accumulation on the high geomorphological surfaces in the river valleys. Floodplain alluvial accumulation occurred periodically on the low and middle height terrace surfaces and valley slopes. Buried soils in the floodplain alluvium mark interruptions of the alluvial accumulation on the high geomorphological surfaces.

The tendency to alteration of high floods by changes in the mean runoff was also typical for Holocene. River stages of 15–18 m (maximum 25 m) were typical during this period, whereas the modern floods reach only 5–7 m due to dam formation.

Formation of the 9–11 m high terrace in the river valleys of central and southern Siberia began in the late Pleistocene to early Holocene. Local incision of the channels, accumulation of the channel alluvium, and floodplain alluvium sedimentation occurred during this period. Terrace stratigraphy contains cultural horizons of the Upper Palaeolithic, Mesolithic, Bronze and Iron Ages. Simultaneous floodplain alluvium accumulation on the floodplain and terrace surfaces (during high floods) began 6300–5000 years BP. Interruptions in alluvial accumulation occurred simultaneously on the first terrace and high floodplain. Morphological changes to the first terrace were caused mainly by the floodplain alluvial accumulation on its surface during high floods. The surface of the 14–17 m high terrace had been influenced by these processes only during rare extreme floods. Holocene alluvial accumulation continued during the Little Ice Age. Up to 5 m of alluvial deposits accumulated in some places. These sediments buried meadow-chernozem soil dated by ^{14}C at 1960 ± 80 years BP. Upper horizons of this alluvium are reworked by the wind and one can see on their surface aeolian landforms. A decrease in flood levels was caused by the rise in temperature after AD 480 ± 60 and since the end of the Little Ice Age.

During the Holocene there were periods of low flood levels (2–6 m) and periods of high floods and dammed levels (8–12 m, and up to 25 m). A fall in the height of the Holocene flood levels is marked by the buried soil horizons intercalated in the strata of the low terraces and floodplain (Figure 13.5). Synchronous interruptions in sedimentation in the different parts of the Yenisei River Basin have been established and a trend has been revealed with a lag in soil formation from the south to north. In the intermontane depressions of the west and east Sayan Mountains the oldest Holocene soil horizons formed 10 500–10 200 years BP. On the Krasnoyarsk Plain and

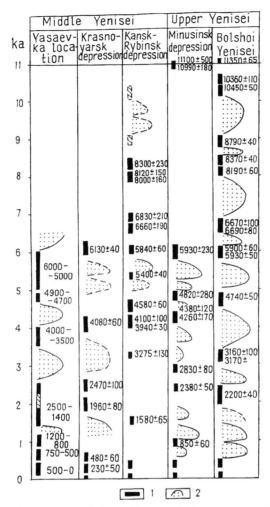

Figure 13.5 Geochronology of the Middle and Upper Yenisei showing cyclicity of floodplain accumulation over the last 11 000 years. 1=Buried soils in the terrace and floodplain strata; 2=periods of high flood and dam formation

Yazaevka location (1000–1300 km to the north of the Sayan Mountains) soil formation on the surface of the 9 m terrace began only at the end of the Atlantic period. Frequent interruptions of soil formation during the late Holocene are typical for the upper part of the Yenisei River Valley. Protracted periods of soil formation were typical for the middle part of the Yenisei River Valley in the late Holocene. Nevertheless, periods of Holocene soil formation generally coincide in the different parts of the Yenisei River Valley. Interruptions in active sedimentation on the floodplain and low terrace surfaces occurred during epochs, marked by soil formation about 10 500–10 200, 8900–8700, 8400–8000 (8800–8000 in the upper part of Yenisei River), 6800–6600, 6100–5800, 4800–4000, 3300–2800, 2500–2000, 1200–800, 500–0 years BP (A. F. Yamskikh, 1983, 1993; A. A. Yamskikh, 1995).

CONCLUSIONS

The dynamics of the floodplain formation discussed here differ significantly from the traditional models of floodplain formation. Pleistocene accumulation of floodplain alluvium occurred simultaneously on terraces of different height. The detailed investigations of the low Holocene terrace and floodplain stratigraphy have revealed a lag in the beginning of soil formation from south to north. Nevertheless, periods of the Holocene soil formation generally coincide in the different parts of the Yenisei River Valley.

A combination of climatic, hydrological and tectonic factors explain the polycyclic dynamics of floodplain accumulation. In regions where rivers cross the tectonic structures of intensive neotectonic uplift, antecedent type valleys are formed on solid basement with erosionally modified floodplains. As has been discussed earlier, landslide-dams usually form upstream of these reaches of the river valleys. The formation of multi-level floodplain and polycyclic terraces also occurs in the plains and intermontane depressions, where neotectonic movements are not so significant. Here the main factors causing polycyclic terrace formation are the periods of increased flood heights caused by climate change. These were interrupted by low flood phases of river runoff. For intracontinental Siberian regions there is a metachronous dependence on temperature and precipitation changes. Generally, an increase in precipitation corresponds to climatic cooling with the short periods of high precipitation at the end of the warmer periods continuing into periods of climatic cooling. This is why periods of high floods corresponded to cold periods with increased water storage in snow and ice-masses and widespread permafrost. In the regions with more marine and temperate continental climates, such as in the north-western part of the Russian Plain, increases in precipitation usually corresponded to the warm Holocene climatic stages (Klimanov, 1994). In the Karelia region (eastern Finland), decreases in precipitation have coincided with climatic cooling both at the end of the late Pleistocene and during the Holocene. Similar patterns have been revealed for Byelorussia during the late Pleistocene, and early and late Holocene; whereas the reverse (increasing precipitation corresponding to climatic cooling) was typical for the mid Holocene interval (5000–3500 years BP; Klimanov, 1994). In Central Europe increasing precipitation corresponded to climatic cooling during the Holocene (Starkel, 1994). Also in northern Siberia, increased precipitation coincided with warm climatic stages during the early Holocene. In the middle and late Holocene increasing precipitation has been revealed not only for warm climatic stages but for some cold ones as well (Nikolskaya *et al.*, 1989). The formation of river ice-dams during these periods is another factor raising flood levels, especially in narrow parts of the river valleys which provide favourable conditions for ice-dam formation.

A combination of these factors determines a specific Siberian-type of river valley morphogenesis. In the polycyclic model, in contrast to the cyclic model, the accumulation of floodplain deposits periodically occurred simultaneously on the floodplain surface and on terraces of different altitudes.

This model of polycyclic sedimentation is supported by the results of absolute dating, investigations of multilayer archaeological sites and the results of complex palaeogeographical investigations. Remaining questions about the age and genesis of the terraces can only be solved on the basis of the theory of the polycyclic terrace formation. This model is presumed applicable to stratigraphical subdivision of the Quaternary deposits in other regions of the periglacial belt.

REFERENCES

Arkhipov, S. A. 1971. *Chetvertichnyi Period v Zapadnoi Sibibri*. Novosibirsk, Nauka (in Russian).

Baker, V. R., Kochel, R. C., Patton, P. C. and Pickup, H. A. 1983. Palaeohydrologic analysis of Holocene slack-water sediments. In Collinson, J. D. and Lewin, J. (Eds) *Modern and Ancient Fluvial Systems*. International Association of Sedimentologists, Special Publication 6, Blackwell, London, 229–239.

Borisov, B. A. 1984. Altae-Sayanskaya gornaya oblast. In Nalivkin, D. V. and Sokolov, B. S. (Eds) *Stratigraphiya SSSR. Chetvertichnaya Systema*, Vol. 2. Moskva. Nedra, 331–351 (in Russian).

Brackenridge, G. R. 1984. Alluvial stratigraphy and radiocarbon dating along the Duck river, Tennessee. Implications regarding floodplain origin. *Bulletin of the American Geological Society*, **95**, 9–25.

Brown, A. G. 1997. *Alluvial Geoarchaeology: Floodplain Archaeology and Environmental Change*. Cambridge University Press, Cambridge.

Feniksova, V. V. 1971. *Verkhnii Pleistocen Ugo-vostoka Zapadno-Sibirskoi Nizmennosti*. Moskva Izd-vo MGU (in Russian).

Frechen, M. and Yamskikh, A. F. 1998. Loess stratigraphy of Yenisei, Siberia. *Journal of the Geological Society*, In prep.

Gorshkov, S. P. 1966. O stratigraphii antropogenovykh otlozenii vnelednikovoi zony Prieniseiskoi Sibiri. In Saks, V. N. (Ed.) *Chetvertichnyi Period Sibiri*. Moskva, Nauka, 71–82 (in Russian).

Gromov, V. I. 1948. Palynologicheskoe i Arkheologicheskoe Obosnovanie Stratigraphii Chenvertichnogo Perioda na Territorii SSSR. *Trudy Geologicheskogo Instituta*, **64**, Seriya geol., No. 17 (in Russian).

Klimanov, V. A. 1994. Izmenenie kliata na territorii Vostochnoi Evropy v Golocene. In Velichko, A. A. and Starkel, L. (Eds) *Paleogeographcheskaya Osnova Sovremennykh Landshftov*. Moskva, Nauka, 150–152 (in Russian).

Nagorski, M. P. 1937. Materialy po Geologii Chetvertichnykh Otlozeni Centralnoi Chasti Krasnoyarskogo Kraya. *Vestnik ZSGU*, **5**, 17–34 (in Russian).

Nikolskaya, M. V., Borisova, Z. K., Kaplanskaya, F. A., Klimanov, V. A., Stefanovich, E. K., Tornogradsky, V. D., Cherkalova, M. N. and Shofman, I. L., 1989. Klimaticheskie izmeneniya v nekotorykh raionakh Severnoi Asii. In Velichko, A. A. (Ed.) *Palaeoklimaty Pozdnelednikoviya i golocena*. Moskva, Nuaka, 141–145 (In Russian).

Seliverstov, Y. P. 1966. Neogen-Chetvertichnye obrazovaniya i nekotorye voprosy paleogeografii voprosy paleogeographii gor i vpadin uga Sibiri (Altay, Sayan, Tuva). In Saks, V. N. (Ed.) *Chetvertichnyi Perion Sibiri*. Moskva, Nauka, 117–127 (in Russian).

Starkel, L. 1994. Izmenenie klimata i landshaftov Polshi v Golocene. In Velichko, A. A. and Starkel, L. (Eds) *Paleogeographcheskaya Osnova Sovremennykh Landshftov*. Moskva, Nauka, 147–150 (in Russian).

Yamskikh, A. A. 1995. Nekotorye osobennosti golocenovykh iskopaemykh pochv v doline r. Yenisei (na Bobrovskom uchastke). In *Paleogeographiya Srednei Sibiri*, Vol. 2. Krasnoyarsk, 49–66 (in Russian).

Yamskikh, A. F. 1972. Stratigraphiya Kainozoiskikh otlozenii v basseine Verkhnego Yeniseya. In Golovin, V. F. (Ed.) *Materialy po Geologii i Geographii Srednei Sibiri*. Krasnoyarsk, 51–82.

Yamskikh, A. F. 1983. Paleogeographicheskie usloviya Todzinskoi kotoloviny v Golocene. In

Yamskikh, A. F. (Ed.) *Prirodnye Usloviya i Resursy Uga Srednei Sibiri*. Krasnoyarsk, 3–19.

Yamskikh, A. F. 1992a. *Polycyclovoe Terrasoobrazovanie i Stratigraphicheskoe Raschlenenie Chetvertichnykh Otlozenii Rechnykh Dolin*. Krasnoyarsk (in Russian).

Yamskikh, A. F. 1992b. *Lessovye Porody v Rechnykh Dolinakh Prieniseiskoi Sibiri*. Krasnoyarsk (in Russian).

Yamskikh, A. F. 1993. *Osadkonakopleneie i Terrasoobrazovanie v Rechnykh Dolinakh Uznoi Sibiri*. Krasnoyarsk (in Russian).

Yamskikh, A. F. 1996. Late Quaternary intracontinental river paleohydrology and polycyclic terrace formation: the example of South Siberian river valleys. In Branson, J., Brown, A. G. and Gregory, K. J. (Eds) *Global Continental Changes: the Context of Paleohydrology*. Geological Society Special Publication, Geological Society, London, No. 115, 181–190.

Zubakov, V. A. 1972. *Noveishie Otlozeniya Zapadno-Evropeiskoi Nizmennosti*. Leningrad, Nedra (in Russian).

Zyatkova, L. K. and Rakovec, O. A. 1969. Minusinskaya vpadina. In Strelkovand, S. A. and Vdovin, V. V. (Eds) *Altae-Sayanskaya Gornaya Oblast*. Moskva, Nauka, 240–275 (in Russian).

Section 4

FLOODPLAIN RESPONSE

14 Long-Term Episodic Changes in Magnitudes and Frequencies of Floods in the Upper Mississippi River Valley

JAMES C. KNOX
University of Wisconsin, Madison, Wisconsin, USA

INTRODUCTION

Traditional methods of flood frequency analysis are based heavily on statistical probability theory. A common assumption underlying this statistical methodology is that environmental conditions remain relatively constant over time. A relatively uniform set of environmental conditions in turn promotes a stationary statistical mean and variance in hydrological series. Although most in the scientific community acknowledge that the stationarity assumption for the mean and variance is violated on geological timescales, there are few quantitative data that document such examples. This chapter examines how magnitudes and frequencies of floods responded to late glacial, late glacial transition to post-glacial, post-glacial, and historical environmental conditions of the Upper Mississippi Valley (UMV) of Midwest North America. The geomorphological record expressed in sediments and associated landforms provides a longer-term perspective of what has happened in the past and of what might be possible in the future, and this record is the principal source of information for this chapter. As expected, results confirm that the modern instrumental observations of the last century poorly represent rates and magnitudes of fluvial activity operating on geological timescales as well as even the last millennium. Results also show that the human impact on floods and soil erosion is so great that non-stationary behaviour of the mean and variance in hydrological series has also characterised the brief historical period of the last two centuries. The results imply a need for considerable caution in applying modern rates of fluvial activity to past environments. The results also imply that equal caution is necessary when using past environmental fluvial activity as analogues of potential future fluvial activity expected under various climate change scenarios.

Examples are drawn from north-eastern Iowa, south-eastern Minnesota, north-western Illinois, and especially south-western Wisconsin (Figure 14.1). For the purposes of this chapter, four major environmental episodes are recognised:

Fluvial Processes and Environmental Change. Edited by A. G. Brown and T. A. Quine.
© 1999 John Wiley & Sons Ltd.

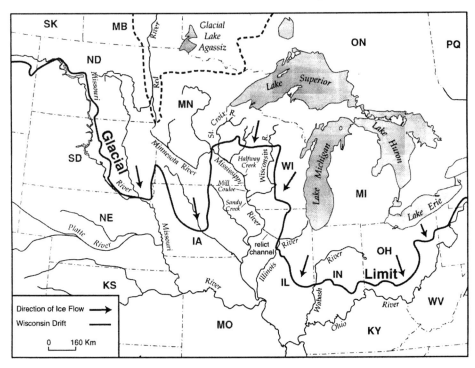

Figure 14.1 The upper Mississippi River drainage system and its relationship to late Wisconsin glaciation. The late Wisconsin glacial limit is generalised from Dyke and Prest (1987) and the extent of Glacial Lake Agassiz deposits of various ages is generalised from Teller (1987)

(1) a late Wisconsin tundra and periglacial regime occurring within the period between about 25 000 and 14 000 ^{14}C years BP, and especially between about 18 000 and 14 000 ^{14}C years BP;
(2) a period of major transition from glacial to post-glacial climate/vegetation regimes between about 14 000 and 9000 ^{14}C years BP;
(3) the Holocene (post-glacial) period of natural vegetation dominance from 9000 ^{14}C years BP to about 200 years ago;
(4) a regime of extreme human disturbance of geomorphological processes following the widespread introduction of agriculture in the early 19th century.

THE UPPER MISSISSIPPI VALLEY: 25 000–14 000 ^{14}C YEARS BP

The Late Wisconsin glaciation in North America readvanced southward into the drainage of the upper Mississippi River by 25 000 ^{14}C years BP (Hallberg and Kemmis, 1986; Johnson, 1986; Leigh and Knox, 1994). This glacial episode, which lasted until about 14 000 ^{14}C years BP (Teller, 1987), was associated with massive aggradation of the UMV and its tributaries that directly drained the ice sheet (Wright, 1987; Knox, 1996). The landform representing the maximum elevation of this aggrada-

Episodic Changes in Magnitudes and Frequencies of Floods 257

Figure 14.2 Fluvially streamlined erosion and sedimentation on the Bagley Terrace, south-western Wisconsin. The landform represents a cut terrace of multiple surfaces that were eroded into the Savanna Terrace, the highest terrace of late Wisconsin age. The Bagley Terrace was probably shaped by a catastrophic flood or floods of unknown magnitude. If the flood occurred before deep incision of the Bagley Terrace, then the discharge(s) probably did not exceed 30 000 m^3 s^{-1}. If the flood occurred after deep incision of the Bagley Terrace, then the discharge(s) could have ranged as high as 60 000–125 000 m^3 s^{-1}. Photo courtesy of the US Army Corps of Engineers

tional sequence is known as the Savanna Terrace in the UMV (Flock, 1983). This high terrace is mainly found in the mouths of tributaries where it was protected from erosion by floods that cut the Bagley Terrace when outlet failures of proglacial lakes occurred during the late glacial to Holocene transition (Figure 14.2). Proglacial outwash sediment extended downstream through the lower Mississippi River valley to the Gulf of Mexico (Saucier, 1994). The total depth of late Wisconsin alluviation in the UMV is unknown but terrace outcrops associated with post-glacial downcutting in Minnesota have exposed at least 45 m of sediment (Wright, 1987). Drill borings across the Mississippi River floodplain at sites along western Wisconsin also indicate that the downcutting was followed by renewed aggradation during the Holocene (Brown Survey, 1931). Many of the borings show a minimum of 15–20 m of Holocene sedimentation, implying that the depth of late Wisconsin aggradation was at least as much as 60–65 m in some reaches of the UMV.

Mississippi River Floods: 25 000–14 000 ^{14}C Years BP

A precise understanding of flood magnitudes that accompanied the late Wisconsin aggradational phase is lacking because it has not been possible to accurately reconstruct former channel cross-section dimensions. Furthermore, the fine-grained composition of terrace sediment is unsuitable for application of competency methods to estimate former flood depths. Some have suggested that between about 18 000 and 13 000 ^{14}C years BP large glacial lake outburst floods occurred frequently, but quantitative estimates of these flood magnitudes and recurrence frequencies have not been demonstrated (Bettis *et al.*, 1992). However, examination of sedimentary structures indicates some relatively large floods but shows most were not of the catastrophic type that characterised the environmental transition period after 14 000 ^{14}C years BP (Knox, 1996).

Slackwater sedimentation from these late glacial floods is represented in Savanna Terrace deposits at tributary mouths and in lower valley reaches of tributaries. These slackwater deposits are usually dominated by sandy facies at the tributary mouths, whereas silt and clay sedimentation becomes dominant farther away in the lower reaches of tributaries. An example of tributary mouth deposition is shown in Figure 14.3. The foreset bedding shown in the bottom third of the 8 m section of Figure 14.3 is noteworthy because the beds are dipping upstream into the tributary mouth documenting that a relatively major passing flood spilled large quantities of water and sediment into the tributary. Using the principle that water depth commonly varies between five and seven times the height of channel bottom foreset beds (Allen, 1984, p. 333), the late glacial flood that deposited these beds probably ranged between 4 and 6 m deep. Since the base of the foreset bedding also represents the approximate level of the Mississippi River floodplain at the time of their deposition, modelling of water surface profiles provides an approximation of the flood discharge. The US Army Corps of Engineers HEC-2 computer program calculates water surface profiles associated with energy losses between adjacent cross-sections under conditions of gradually varied flow (Hoggan, 1989). HEC-2 modelling indicates that a flood discharge probably ranging between 10 000 and 15 000 m^3 s^{-1} was responsible for the foreset bedding in Figure 14.3 (Knox, 1996). The largest historically observed flood since 1828 on the Mississippi River in this locality was only about 7800 m^3 s^{-1} (Anderson and Burmeister, 1970). Therefore, in the context of modern flood recurrence frequencies, the late glacial flood responsible for the foreset beds of Figure 14.3 would be judged as exceptionally rare. Furthermore, since no other comparable sedimentary structures were observed in the Figure 14.3 section, it is apparent that a flood of 10 000–15 000 m^3 s^{-1} was probably also a relatively infrequent event during an extended part of the late glacial period preceding 14 000 ^{14}C years BP. Floods of considerably smaller magnitude seem to have been more typical. Such floods were responsible for deposition of the planar bedded silty clay and clayey silt units ranging from less than 1 cm to 3–4 cm thick shown at the base of the section on Figure 14.3.

THE UPPER MISSISSIPPI VALLEY DURING THE LATE WISCONSIN–HOLOCENE TRANSITION: 14 000–9000 ^{14}C YEARS BP

After about 14 000 ^{14}C years BP the southern margin of the continental glacier generally retreated northward with occasional interruptions by minor readvances (Mickelson *et al.*, 1983; Teller, 1987). Lakes formed between the retreating ice margin and moraines of former ice front positions. The lakes trapped much of the sediment

Figure 14.3 Late Wisconsin Savanna Terrace at the mouth of Mill Coulee, southwestern Wisconsin. The sedimentary sequence represents late Wisconsin flood slackwater sediment that was mostly derived from an aggrading Mississippi River which was then draining the continental glacier. The view is south with the Mississippi River on the immediate right and the tributary headwaters to the left. The 90 cm high foreset beds in the lower section dip upstream into Mill Coulee tributary and record the passing of a large flood of 10 000–15 000 $m^3 s^{-1}$. Photo by the author

being discharged with glacial meltwater and led to channel entrenchment in the alluvial fill of the Mississippi River and its tributaries. At Halfway Creek in the UMV along western Wisconsin, a relict channel with a basal radiocarbon age of 13 545 ± 85 (AA 23384–WG 470) is incised about 20–25 m below the Savanna Terrace (Figure 14.1). The basal age for this channel indicates that major incision occurred quickly following glacial retreat. It is unclear whether this incision of 20–25 m occurred over an extended period of several hundred years in response to discharge of low sediment concentration, or whether it occurred nearly instantaneously in response to single or multiple catastrophic floods. Catastrophic floods are known to have occurred in several sub-basins of the upper Mississippi River drainage in response to the sudden release of water when outlets failed on proglacial lakes (Matsch and Wright, 1967; Willman and Frye, 1969; Clayton, 1983; Matsch, 1983; Teller, 1987; Wright, 1987; Clayton and Attig, 1989; Hajic, 1991; Kehew, 1993; Porter and Guccione, 1994; Becker, 1995; Carney, 1996; Knox, 1996).

Fluvially streamlined landforms, which appear to have resulted from an extreme flood, are indicated by the surface morphology on a cut terrace in the UMV at Bagley, Wisconsin (Figure 14.2). The average elevation of the incised Bagley Terrace is about 15 m below the Savanna Terrace, but where the UMV is in a narrow bedrock gorge of 2.5–3.0 km width the vertical difference is reduced to about 9 m. At many locations, as at Bagley, this cut terrace consists of a series of stepped surfaces indicating progressive downcutting during its formation. Drill borings of the Brown Survey (1931) as well as bridge site borings (Church, 1984), indicate that more than 20 m of additional incision occurred below the main level of the Bagley Terrace. While all of this incision occurred within the period 14 000–9000 ^{14}C years BP, it has not been possible to establish its age relationship with the main Bagley Terrace surface. At Bagley, the relief of the fluvially sculptured surface shown in Figure 14.2 is about 7–10 m, and the tops of the highest surfaces nearly coincide with the surface of the late Wisconsin (Savanna) aggradational terrace preserved within tributaries.

Estimating the flood(s) magnitude that sculptured the Bagley Terrace can be only crudely accomplished because of uncertain cross-section boundaries. However, the lack of sandy flood-deposited sediment on top of the Savanna Terrace suggests that the flood probably did not reach that level. Therefore, a first approximation for the upper limit of the flood is represented by the discharge confined between the general level of the Bagley Terrace and the surface of the Savanna Terrace. Knox (1996), using HEC-2 modelling, estimated that a discharge of about 30 000 m^3 s^{-1} would fill this channel at the Bagley site. However, it is possible that incision may have extended below the general level of the Bagley Terrace at the time of the flood because the Brown Survey (1931) drill borings indicate at least 15–20 m of incision occurred below the presently exposed Bagley Terrace. Morphological discontinuities between the incised buried channel and the main Bagley Terrace surface suggest that most of the deep incision occurred after formation of the main Bagley surface. Nevertheless, it is a possibility that the flood(s) which sculptured the main Bagley Terrace surface might also have scoured some or all of the incised channel. If the flood channel is expanded to accommodate an additional 20 m incision, then the largest flood(s) that eroded the Savanna Terrace could have ranged between 60 000 and 125 000 m^3 s^{-1} (Becker, 1995; Knox, 1996).

Proglacial Lakes

Potential sources of large floods that might have been responsible for much of the scour and removal of sediment underlying the Savanna Terrace are proglacial lakes that formed between the retreating ice front and moraines of earlier ice advances. One of the biggest UMV flood sources for the period between 14 000 and 9000 ^{14}C years BP was Glacial Lake Agassiz (Figure 14.1). Lake Agassiz formed after 14 000 ^{14}C years BP between the northern watershed divide of the Mississippi River and the retreating ice front of the continental glacier. For at least a brief period Lake Agassiz drained southward into the Mississippi River via the Minnesota River valley (Teller, 1987) (Figure 14.1). Matsch (1983) estimated the dimensions of relict outlet channels near the outlet at the head of the Minnesota River and concluded that bankfull discharge averaged about $40\,000\,m^3\,s^{-1}$ during times of stable channel geometry. Matsch (1983) suggested that magnitudes of $1\,000\,000\,m^3\,s^{-1}$ might have been reached during peak floods. Wiele and Moores (1989) used estimates of water depths competent to transport relict boulders that remain in the southern outlet channel and reached a conclusion that the discharge through the southern outlet was about $50\,000\,m^3\,s^{-1}$. Becker (1995) used HEC-2 modelling to re-examine the relict southern outlet channels of Glacial Lake Agassiz and he concluded that discharge estimates of Wiele and Moores (1989) and of Matsch (1983) have been overestimated. Teller (1990) estimated precipitation and meltwater budgets for sectors of the late glacial continental ice sheet and found that average annual runoff to the UMV exclusive of the Lake Superior drainage ranged from a maximum of about $36\,000\,m^3\,s^{-1}$ between 14 000 and 13 500 ^{14}C years BP to a minimum of about $16\,000\,m^3\,s^{-1}$ between 10 000 and 9500 ^{14}C years BP.

Large magnitude floods also entered the UMV from the Lake Superior region via the St Croix River (Figure 14.1). Glacial Lake Duluth occupied the western basin of modern Lake Superior after about 12 000 ^{14}C years BP (Ojakangas and Matsch, 1982). Carney (1996) compared HEC-2 results applied to Lake Duluth outlet channels with estimates of peak summer meltwater production derived from energy balance determinations and he concluded that discharges from Glacial Lake Duluth ranged from 30 000 to $45\,000\,m^3\,s^{-1}$. These various estimates of floods resulting from outlet failures and meltwater runoff during the transition from late glacial to Holocene climate conditions are in broad agreement and suggest that discharges on the upper Mississippi River between 14 000 and 9000 ^{14}C years BP probably exceeded 25 000–$30\,000\,m^3\,s^{-1}$ on several occasions.

Glacial Lakes Agassiz and Duluth are illustrative of several large lakes that formed along the southern margin of the continental ice sheet and produced catastrophic floods in their downstream drainages. The large floods from outlet failures on proglacial lakes appear to have been of significantly greater magnitudes than floods of the glacial climate episode between 25 000 and 14 000 ^{14}C years BP. They have no counterparts in the Holocene record.

HOLOCENE FLOODS: 9000 ^{14}C YEARS BP TO THE TIME OF EURO-AMERICAN AGRICULTURE

Variations in magnitude and recurrence frequencies of Holocene floods have largely been a result of climate and vegetation changes because no glacial meltwaters entered the watershed during this time. Although the northern tributaries to the upper Mississippi River probably have experienced modest uplift in response to isostatic rebound since deglaciation, tributaries in south-western Wisconsin and adjacent areas to the south and west do not seem to have been affected. Human factors capable of influencing flood hydrology begin only about 175–200 years ago with the introduction of

European and American settlement of the region. Perhaps the most noteworthy Holocene geomorphological adjustment on the mainstem upper Mississippi River has been major aggradation associated with remobilisation of late glacial sediment.

Holocene Remobilisation of Sediments

Large amounts of the colluvial and alluvial sediment stored in tributaries during the glacial period have been remobilised and transported downstream to the Mississippi River during the Holocene. Consequently, most of the Holocene record of floods on the upper Mississippi River north of the southern border of Wisconsin is buried by alluvial sediments. The shallow drill borings of the Brown Survey (1931) show that the depth of Holocene alluviation exceeds at least 15–20 m in many reaches of the river. Other investigations indicate that Holocene sedimentation probably exceeds 50 m where large post-glacial alluvial fans have built into the Mississippi River valley at the mouths of rivers with high magnitude sandy bedload (Wright, 1987). The massive Holocene alluviation results in the exposed sections of most islands and channel bars being of relatively young ages, and this situation prevents easy reconstruction of Holocene flood characteristics. Nevertheless, an excellent record of Holocene floods is available from Mississippi River tributaries in south-western Wisconsin and adjacent areas. This record shows a strong association with Holocene climate changes (Knox, 1985, 1993, 1996).

Holocene Climate and Vegetation Changes

A major ecotone extending from north-west Minnesota to north-eastern Illinois, separating forest to the north-east from prairie to the south-west, has traversed the UMV throughout the Holocene (Wright, 1992). This ecotone has been shown to be associated with relatively steep seasonal air mass boundaries (Bryson, 1966; Bryson and Wendland, 1967). The ecotone has shifted slightly in position during the Holocene in response to climate change, but large-scale vegetation change has not occurred during the Holocene in the UMV. Nevertheless, adjustments of the ecotone apparently represent responses to at least hemispheric-scale environmental change because the environmental episodes along the UMV ecotone are broadly similar in timing to environmental changes known to have occurred along the forest–tundra ecotone of north-central Canada. Following the final wastage of the continental ice sheet in Keewatin, Northwest Territories, Canada, around 6500 ^{14}C years BP (Dyke and Prest, 1987), the northern treeline moved northward to a position about 300 km north of its modern position in south-western Keewatin (Sorenson, 1977). However, between 3500 and 2900 ^{14}C years BP the northern treeline was displaced about 350 km southward, apparently in response to global cooling (Sorenson, 1977). Since then shifts in the northern treeline have been more conservative. However, relative to its modern position in south-western Keewatin, the treeline was 100 km further north during a warm period around 1100–1200 ^{14}C years BP and was about 50 km further south during a cool phase beginning around 800 ^{14}C years BP (Sorenson, 1977). These environmental discontinuities on the northern treeline are broadly similar to environmental discontinuities along the UMV prairie/forest ecotone. The similarity suggests that at least hemispheric-scale climate change is responsible.

The relationship of the UMV to the ecotone and seasonal air mass boundaries explains the high sensitivity of the region to even relatively modest climate changes. Dorale *et al.* (1992) and Baker *et al.* (1996) used speleothem calcite from a cave in north-eastern Iowa near the present ecotone position to examine variations in ^{18}O and ^{13}C isotopes. Their analyses suggested three major climate-based stages during the Holocene, represented by major shifts around 5100 ^{14}C years BP and around 3000 ^{14}C years BP. Dorale *et al.* (1992) and Baker *et al.* (1996) concluded that relative to the early Holocene UMV, mean annual temperatures in north-eastern Iowa experienced an abrupt warming around 5100 ^{14}C years BP. By around 3000 ^{14}C years BP, the isotopic data suggest that mean annual temperature had increased about 1.5 °C above early Holocene levels, but the isotopes suggest that after around 3000 ^{14}C years BP mean annual temperature cooled abruptly in association with savanna vegetation replacing prairie in north-eastern Iowa (Baker *et al.*, 1996). Climatic reconstructions from fossil pollen at sites along the ecotone from south-eastern Minnesota to southern Wisconsin suggest that, during the 5100–3000 ^{14}C years BP stage 2, the mean annual precipitation was about 15% less than it is today and the July temperature about 0.5 °C warmer than before or after stage 2 (Bartlein *et al.*, 1984; Winkler *et al.*, 1986). Winkler *et al.*'s (1986) examination of fossil pollen from lakes cores in south-central Wisconsin indicated that accelerated warming started there around 6500 ^{14}C years BP. A study of Holocene aeolian activity and vegetation change near the ecotone in east-central Minnesota showed three distinct episodes of dune activity with temporal boundaries very similar to those described here (Keen and Shane, 1990). These authors described episode 1 from 9100 to 6500 ^{14}C years BP as a time when "prairie development proceeded steadily as precipitation declined and temperature (especially winter temperature) increased." Keen and Shane (1990) noted that during episode 2, 6500–5100 ^{14}C years BP, the climate appeared to experience greater short-term variability than before or after, especially in the January mean temperature. Although they wondered if sampling strategy might have contributed to the apparent increased variability then, the stratigraphic evidence supported the idea of high variability in aeolian flux. Episode 3 of dune activity extended from 5100 to 4000 ^{14}C years BP. Keen and Shane concluded that precipitation decline during episode 3 was smaller than in the prior dune episodes, and they attributed increased aeolian activity to high temperatures producing high evapotranspiration and drought.

Comparison of proxy climate indicators in the UMV with those of the northern forest/tundra ecotone of Canada shows broadly synchronous behaviour of major climate episodes. The accelerated warming and higher variability of climate in the UMV about 6500 ^{14}C years BP corresponds with the final disappearance of the continental ice sheet in Keewatin. The relatively warmer and effectively drier episode in the UMV between about 5000 and 3000 ^{14}C years BP corresponds with the Holocene maximum northerly position of the boreal forest in north-central Canada, and the abrupt southward displacement of the forest/tundra ecotone around 3000 ^{14}C years BP corresponds to an equally abrupt shift to relatively cooler and wetter conditions in the UMV.

Holocene Floods

Estimates of magnitude variations of Holocene floods of both high and low recurrence probability have been determined from the alluvial deposits in river valleys of southwestern Wisconsin and north-western Illinois. Graphical illustrations of the temporal variability of the reconstructed flood records have been presented elsewhere and these sources also include a more detailed explanation of databases and methodology (Knox, 1985, 1993, 1996). Summary statistics describing flood magnitudes for the climate episodes defined in the preceding section are presented here in Table 14.1. Data are presented for high frequency floods of 1–2 years recurrence frequency and for large overbank floods of low recurrence frequency. The 1–2 years recurrence frequency floods mainly result from late winter to spring snowmelt or snowmelt associated with spring rains. The large overbank floods, on the other hand, mainly result from excessive summer rainfalls that are typically the result of stalled or slow moving frontal systems involving the juxtaposition of air masses from polar and tropical source regions.

Reconstructed magnitudes of floods of 1–2 years recurrence probability were determined from the dimensions of relict channels of cut-off meanders (Figure 14.4). Floods of this recurrence frequency in the UMV tributaries tend to fill channels to their bankfull stage, which is defined here as the top of point bars and the upper limit of sand deposition by lateral accretion. The morphology of the relict channels was identified by closely spaced borings perpendicular to the channel and by differentiating channel fill sediment from sediment of channel bed, point bar, and overbank origins. Channel beds are typically represented by a chert-rich cobble and boulder gravel (Figure 14.5). Survey transects across valley floors show that channel bed elevations of the relict channels are relatively uniform across a valley, indicating that movements of Holocene channels have been mainly lateral rather than vertical cutting and filling (Figure 14.6).

Depth magnitudes of relatively low frequency overbank floods were determined by computing the minimum depth of water that would have been competent to transport the largest clasts of flood-transported sediment found in Holocene overbank alluvium (Figures 14.7 and 14.8). The equation and methodology for this procedure have been presented elsewhere (Knox, 1993). The stable channel bed elevations throughout the Holocene and the minimal differences in elevation of Holocene terrace surfaces of different ages (Figure 14.5) imply that stage–discharge relationships for large floods have been relatively constant throughout the Holocene. Therefore, the computed competent depth of a relict large overbank flood can be measured relative to the stable channel bed surface and then compared to the modern 1–2 year flood depth indicated by point bar height above the channel bed at the same site. The comparison provides a dimensionless ratio which is a proxy indicator of the flood recurrence interval (Dunne and Leopold, 1978, p. 648; Knox, 1993). For floods in the small UMV tributaries described here, a flood that is about twice the bankfull depth of 1–2 years return probability would be expected about once in 50 years whereas a flood that is about three times the bankfull depth would be expected only about once in 500 years.

Partitioning the Holocene into episodes with temporal boundaries defined by the major discontinuities in the proxy climate record presented above shows that both high frequency bankfull floods and low frequency overbank floods have been respon-

Table 14.1 Climate-flood episodes in the upper Mississippi River Valley, south-western Wisconsin and north-western Illinois

Climate-flood episode ^{14}C years BP	% Departure of palaeo 1–2 year flood from modern 1–2 year flood			Statistical significance of difference from preceding climate/fluvial episode	T-test statistical probability (p)	Average ratio of depths of palaeo overbank floods to depths of modern 1–2 year floods			Statistical significance of difference from preceding climate/fluvial episode	T-test statistical probability (p)
	n	Mean	SD			n	Mean	SD		
9000–6500	9	−20.1	10.1	—	—	2	3.90	0.30	—	—
6500–5000	7	+14.6	10.7	yes	0.0001	12	1.72	0.51	yes	0.0001
5000–3000	8	−15.4	14.8	yes	0.0007	31	2.25	0.62	yes	0.011
3000–1200	10	−3.6	6.7	yes	0.038	10	2.30	0.50	no	0.96
1200–800	3	+18.1	7.1	yes	0.0005	13	2.67	0.65	no	0.15
800–200	—	—	—	—	—					

Sources of data: Knox (1985, 1993, 1996).

266　Fluvial Processes and Environmental Change

Figure 14.4 View upstream over Pecatonica River valley near Mineral Point, south-western Wisconsin. Many Holocene channels in south-western Wisconsin have migrated laterally on an armoured boulder gravel bed, resulting in a juxtaposed pattern of relict channels and Holocene sediments, but all at nearly the same elevation. The dimensions of the relict channels provide a proxy measure of former floods of approximately 1–2 years recurrence probability. The horseshoe-shaped cutoff channel in front of the relict channel dated 4150 ± 50 is approximately the same age as the cutoff channel dated 2030 ± 70. Photo by the author

sive to quite modest Holocene climate changes (Table 14.1). Note, for example, that between 9000 and 6500 ^{14}C years BP the bankfull floods averaged about 20% smaller magnitude than their modern counterpart. No large overbank floods for this period were observed in the stratigraphic record. The absence of large overbank floods might be due to the destruction of evidence by subsequent fluvial activity, and the absence is possibly due to an increased probability of their evidence being covered by younger sediment that minimises their discovery. However, the most likely explanation is that they were extremely infrequent. Meteorological conditions capable of generating large overbank floods on these UMV tributaries were not favoured in the early Holocene. The continued existence of a relatively large continental ice mass over north-central Canada then (Dyke and Prest, 1987) would have favoured persistent north-westerly atmospheric circulation over the UMV (Knox, 1983). North-westerly wind flow over the UMV brings cool and dry air masses to the region because source regions for the dominant air masses are the Canadian prairies. Major floods of the UMV are associated with warmer and more moist air masses that originate in the Gulf of Mexico and other southerly source regions. Persistence of north-westerly air flow would have displaced the main storm tracks south of the UMV study area in all seasons, reducing both winter snowfall and summer rains.

Figure 14.5 Schermmerhorn Site, Brush Creek, near Ontario, south-western Wisconsin. The base elevations of Holocene point bars and other Holocene alluvial sediments vary little between lateral sections of widely different ages because a very coarse boulder and cobble gravel has prevented Holocene stream channel incision. Holocene deposits of nearly the same vertical thickness, but often of widely different ages, are found juxtaposed in a complex mosaic pattern. The thick colluvial deposits on lower hillslopes mainly accumulated under periglacial climatic conditions of late Wisconsin time. The armouring of channel beds that underlie the Holocene alluvium resulted from winnowing fines from cobble and boulder rich gravel that accumulated on the valley floors also during late Wisconsin time. Photo by the author

The period between 6500 and 5000 ^{14}C years BP experienced a dramatic increase in the magnitude of bankfull stage floods suggesting that snowmelt runoff was increased as climatic warming accelerated after about 6500 ^{14}C years BP. Although there is weak evidence that large overbank floods may also have increased then, the sample size is too small to draw a firm conclusion (Table 14.1). Between about 5000 and 3000 ^{14}C years BP, when proxy climate indicators suggest that mean annual temperature had increased about 1.5 °C above early Holocene levels and prairie grassland was expanding further north-eastward into the forest of north-eastern Iowa, both bankfull stage floods and large overbank floods decreased in magnitude. The bankfull stage floods, which had averaged about 15% larger than those of today between 6500 and 5000 ^{14}C years BP, abruptly decreased to an average of about 15% smaller than those of today (Knox, 1985). The large overbank floods also became smaller then (Table 14.1) (Knox, 1993). The average of the ratios of overbank flood depths to bankfull flood depths for the period 5000 to 3000 ^{14}C years BP was only 1.72. This ratio equates with present-day floods in the study area which have recurrence probabilities between once in 25 years and once in 50 years. These data indicate that the large overbank floods of this

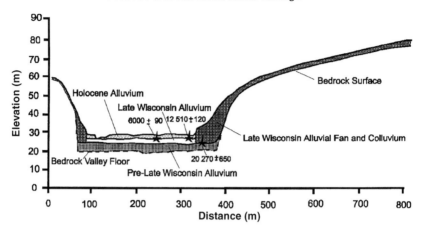

Figure 14.6 A series of drill borings across the floodplain of the Platte River valley headwaters south-west of Montfort, south-western Wisconsin showed that lower valley sides accumulated massive colluvial deposits during the cold periglacial climates of the late Wisconsin while valleys were aggraded with coarse gravel. Holocene streams have laterally migrated across their floodplains without incision because winnowing of fine sediment has resulted in an armoured lag concentration of boulders and cobbles. (After Mason and Knox, 1997)

approximately 2000 year duration warm period probably were in fact very small in comparison to present-day large overbank floods. A relatively well-developed soil buried by about 2 m of overbank sediment is found in island levees of the upper Mississippi River along south-western Wisconsin. The soil formed over an extended period when overbank flooding and levee alluviation was minimal. Charcoal mixed in the near-surface horizon of this soil has an age of 2470 ± 70 ^{14}C years BP (BETA-92063). Since the bulk of the soil development predates the charcoal, it is likely that its evolution occurred during the climate episode 5000–3000 ^{14}C years BP when floods from the contributing tributaries were relatively small and Mississippi River floodplain aggradation was very slow. Therefore, late Holocene magnitude and frequency variations of floods on the upper Mississippi River seem to have been similar to those occurring on the small tributaries.

The decrease in magnitude for both bankfull and overbank floods during the period 5000–3000 ^{14}C years BP is consistent with the type of large-scale atmospheric circulation implied from the proxy climate indictors. Increasing the mean annual temperature 1.5 °C and reducing mean annual precipitation 15% less than modern (Bartlein et al., 1984; Winkler et al., 1986; Baker et al., 1996) implies that a stronger zonal westerly component of the large-scale atmospheric circulation was present then. A strengthened zonal westerly flow over the UMV increases the frequency of warm dry air masses from the western Great Plains and acts as a wedge between polar and tropical derived air masses that would otherwise be more frequent over the UMV (Borchert, 1950; Knox, 1983). Minimising the collision of polar and tropical air masses greatly reduces occurrences of high magnitude snowfalls and rainfalls over the UMV.

Around 3000 ^{14}C years BP an abrupt increase in magnitudes of bankfull and large overbank floods occurred. This change was related to an abrupt climatic cooling

Figure 14.7 Pebbles, cobbles and boulders that can be traced laterally as interruptions in fine-textured overbank alluvium (vertical accretion) can be used to estimate the depth of flood water necessary for their transportation down-valley on the floodplain. Here, the star indicates the top of a gravelly point bar which is buried by 1.8 m of overbank sediment along the Galena River near Benton, south-western Wisconsin. Four pebble and cobble units (white dots) occur in the silty overbank sediment and they represent deposition from large floods (denoted by survey pins). Photo by the author

indicated by savanna vegetation replacing prairie along the ecotone in north-eastern Iowa (Baker *et al.*, 1996), and by the abrupt southward shift of the forest/tundra ecotone in north-central Canada (Sorenson, 1977). The mean bankfull discharge between 3000 and 1200 ^{14}C years BP increased to become only about 4% smaller than present-day bankfull values (Table 14.1). At the same time, the mean ratio of large overbank to bankfull flood depths increased to 2.25, a value that implies floods occurred with discharges well in excess of the modern 100-year probability flood. Flood depth to bankfull depth ratios for several floods approached 3.0 after 3000 ^{14}C

Figure 14.8 Downstream view of the Galena River near Benton, south-western Wisconsin. The bank section of Figure 14.7 is located at the arrow position. Photo by the author

years BP and is evidence that some extremely large floods began occurring then (Table 14.1) (Knox, 1993).

Radiocarbon ages of about 1200 and 800 ^{14}C years BP are frequently inferred as representing climatic discontinuities in mid-continent North America (Knox, 1983), but it is not yet possible to demonstrate this activity had a statistically significant influence on floods in the UMV tributaries discussed here (Table 14.1). Time series plots of the individual flood indices, however, are suggestive of a weak climatic influence. An adequate test of causal associations is hampered by lack of data for the key periods in question (Table 14.1). The nature of climate shifts during this period is illustrated by Laird et al. (1996), who found that just west of the prairie/forest ecotone in eastern North Dakota the age of 800 ^{14}C years BP (c. AD 1200) marks the approximate beginning of a wetter climate. Three extreme droughts, each lasting from 150 to 200 years, occurred in this region of the northern Great Plains during the previous millennium. Laird et al. (1996) suggested that the last of the droughts, which happened around AD 1000–1200, was part of the medieval warm period and that the shift to cooler and more moist conditions after about AD 1200 represented the onset of the Little Ice Age climate in North America. The idea that changes of climate along the prairie/forest ecotone in the northern Great Plains are related to hemispheric activity is consistent with synchronous changes on the forest/tundra ecotone in north-central Canada. During the millennium containing the three extended droughts, the forest/tundra border was significantly further north than its present-day position. Around 800 ^{14}C years BP the forest/tundra ecotone abruptly shifted south of its modern location in response to climatic cooling (Sorenson, 1977).

It is interesting that the Table 14.1 data for the climate episode 1200–800 ^{14}C years BP suggest a trend toward larger floods in the UMV even though the relationship is based on only three observations in the case of bankfull floods and is not statistically significant in the case of large overbank floods. Warm episodes earlier in the Holocene were associated with decreases in flood magnitudes. It is unclear whether this anomaly is due to poor dating control or the effects of small sample size in a period characterised by high variability climate. Large overbank floods, on the other hand, do show a tendency toward further increases in magnitude after 800 ^{14}C years BP as might be expected with a shift toward climatic cooling.

The relationships between Holocene floods and climate changes in the UMV have shown that climate changes need not be large to significantly influence magnitudes and recurrence frequencies of floods. This sensitivity is strong evidence against the idea that major climate change is necessary to invalidate the assumption of statistical stationarity that underlies most methodologies currently employed to estimate recurrence probabilities of floods. The transitions between the major climate episodes and resulting flood regimes occurred abruptly rather than gradually in the UMV. This abruptness might be due to the association of the UMV with a region of steep seasonal climatic gradients along the prairie/forest ecotone where hemispheric and/or global climatic changes become magnified. It is important to acknowledge that changes in *mean* temperature and *seasonal* precipitation were not necessarily the direct causes of the changes in the flood magnitudes between episodes because floods tend to be singular events. Changes in mean climate conditions are symbolic of changes in air mass boundaries and trajectories of storm tracks that either enhance or suppress the probability of floods. Finally, it is noteworthy that bankfull high frequency floods and low frequency large overbank floods have generally responded in similar ways to Holocene climate changes (Table 14.1). The similarity is interesting because the 1–2 year probability bankfull floods are dominated by late winter or spring snowmelt runoff, whereas large overbank floods in the region are mainly a product of extreme summer rainfalls. Therefore, for the long-term average timescale, shifts in seasonal climates responsible for the floods must have been similar. The similarity in time series for frequent and infrequent floods on the tributaries leads to the obvious scale question of whether the history of flooding in the tributaries is also representative of the mainstem upper Mississippi River. The buried soil that apparently formed on Mississippi River levees in the warm dry climate episode predating 3000 ^{14}C years BP, and renewed vertical accretion since then, suggests that at least the late Holocene Mississippi River experienced variations in flood magnitudes that were similar to those occurring in the tributaries.

FLOODS SINCE THE INTRODUCTION OF EURO-AMERICAN AGRICULTURE IN THE EARLY 19TH CENTURY

The replacement of the natural mosaic of forest and prairie with agricultural crops and pasture radically altered the hydrology of watersheds in the UMV. This transformation began during the first decades of the 19th century in the southern part of the UMV and was completed in the late 19th century in the northern extremities of the UMV. Commercial logging followed by agriculture occurred in the more northerly regions of the UMV. The agricultural influence on floods and soil erosion is characterised by three major divisions. The first period extends to the late 19th century and is represented by the time of land clearance, settlement, and agricultural expansion. The second period extends from the late 19th century into the 1940s and represents the period of greatest human impact on accelerating floods and soil erosion. Better land conservation practices and changes in the ways that crops are planted and grown have resulted in major reductions in flooding and soil erosion during the third period from the 1940s to the present, but magnitudes and rates are still much above their natural Holocene counterparts.

The cultivation of a landscape exposes the soil to direct raindrop impact, generally reduces the hydraulic roughness at the ground surface, reduces the level of organic matter in the soil solum, and leads to deterioration of natural soil structure. These changes all favour acceleration of surface runoff, soil erosion and flooding. An additional cause of accelerated flooding in the UMV is stream channelisation. Those parts of the UMV that were covered by late Wisconsin glaciation had a presettlement landscape that was poorly drained and represented by extensive wetlands. Extensive channelisation and artificial drainage was undertaken to render the landscape more suitable for agriculture. Estimated wetland losses for Wisconsin, Minnesota and Michigan range from 32 to 71%, but in Ohio, Iowa, Indiana and Illinois losses are estimated to be 90% or more (McCorvie and Lant, 1993). The development of a human-made channel network greatly increased the efficiency of runoff and thereby significantly increased magnitudes and frequencies of floods that were formerly of small to moderate magnitudes. Human influences on extreme floods seem to have been relatively small by comparison (Knox, 1977).

Because south-western Wisconsin has been the focus of a number of studies that investigated hydrological responses to agricultural land use, it is used here to characterise agricultural influences on flooding, erosion and sedimentation in the middle UMV. South-western Wisconsin is a hilly unglaciated area where moderate to steep slopes separate narrow and gently sloping uplands from narrow valley floors, and local relief averages between 50 and 120 m km^2 (Figure 14.2). The underlying nearly flat-lying carbonate and sandstone bedrock is covered with loess that is 1.5–4 m thick on upland divides. Loess and residuum are usually less than 1 m thick on steep slopes. Small watersheds experienced magnitudes of flood peaks from common summer rainstorms during the late 1800s and early 1900s that were five to six times greater than magnitudes of floods representative of the pre-agriculture natural late Holocene (Knox, 1977). The accelerated surface runoff produced severe soil erosion. Studies of soil profiles and sediment budgets indicate that south-western Wisconsin watersheds have lost 9–16 cm of topsoil during the past 150–200 years of agriculture (Trimble, 1983; Benedetti, 1993; Beach, 1994).

The sedimentary record in alluvial fills documents the three episodes of historical agricultural influence on watershed hydrology noted above. An example is provided by Doyle Site 10 on the Shullsburg Branch, south-western Wisconsin (Figure 14.9). This is a headwater tributary with a contributing drainage area of 25 km^2. The Doyle Site is typical of small headwater tributaries which generally have experienced from about 50 cm to 1.5 m sedimentation on their floodplains in response to increased flooding caused by agriculture, and, as is the case for most tributaries of this size, overbank flooding onto the main valley floor has now become a rare event. Bank erosion from the increased magnitudes and frequencies of floods since agricultural settlement caused rapid expansion of the historical meander belt which has progressively captured and conveyed floods that once spilled across the entire valley floor (Figures 14.9 and 14.10).

Figure 14.9 shows how the Shullsburg Branch has moved laterally on an armoured bed of cobble and boulder gravel since about 1820, in the same fashion that natural Holocene lateral migration occurred. The pre-agriculture banktop was identified by a very well developed mollisol, now buried by about 1.6 m of overbank sediment (Figure 14.9). Initially the overbank flooding and associated sedimentation increased slowly in response to progressive conversion of the prairie and forest to agricultural land. By 1850 all of the land of the Shullsburg Branch watershed had been claimed for settlement, which included both agriculture and mining (Schafer, 1932). Surface runoff and soil erosion greatly increased with agricultural development because of extensive plantings of wheat and corn. The peak of wheat planting occurred in the 1870s, but corn continuously expanded from the 1850s onward into the 20th century (Blanchard, 1924, p. 75–82). By 1890, hydrological response to cultural disturbance was readily apparent as nearly 60 cm of overbank sediment had accumulated on the floodplain. The increasing heights of banks led to deeper and faster waters during floods causing a significant increase in the shear stress at the bank base (Figure 14.10). During the next 35 years, between 1890 and 1925, an additional 90 cm of overbank sedimentation occurred at the Doyle Site on the Shullsburg Branch, and the thalweg channel bottom shear stress increased to 80 N m^{-2}, from its original pre-agriculture 36 N m^{-2}. The added energy accelerated erosion of the right bank. Between 1820 and 1890 the right bank was eroding at about 6 cm year^{-1}, but between 1890 and 1925 the rate increased to 53 cm year^{-1}. The cumulative effects of poor land management further accelerated flooding and soil erosion into the first half of the 20th century when the right bank recession rate increased to 70 cm year^{-1} between 1925 and about 1940 (Figures 14.9 and 14.10).

The degree to which agricultural land use increased magnitudes and frequencies of floods and soil erosion probably peaked in the 1940s. Figure 14.9 shows that the right bank at the Doyle Site has eroded only a few metres since 1940 at a rate of about 12 cm year^{-1}. Furthermore, a slight reduction in the channel cross-section capacity of the meander belt shows that net sediment storage is occurring here as is occurring elsewhere in other headwater tributaries of the region. Other direct evidence of decreasing surface runoff since the 1940s is indicated by the magnitudes of annual maximum floods. Six south-western Wisconsin watersheds with drainage areas that average 560 km^2 show remarkably similar trends of decreasing flood magnitudes over the last half century (Figure 14.11). Climate change does not explain the trend because an analysis of time series of the area's storm rainfalls equal to or greater than 2.5 cm showed no statistically significant trend for the period 1940–1986 (Baker, 1990). The changes in flooding appear to reflect the shift to contour ploughing, strip cropping, and other land conservation measures that began to be introduced to south-western Wisconsin in the 1940s. The introduction of herbicides and pesticides in the 1950s probably had an even greater impact on reducing magnitudes and frequencies of floods because the use of chemicals along with heavy fertiliser applications allowed for abandonment of the "grid system" of corn planting. Up to this time corn was planted

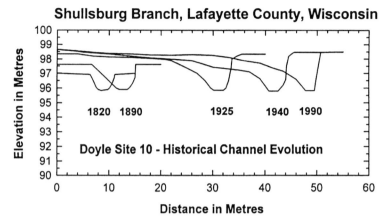

Figure 14.9 Agricultural settlement beginning about 1820 in the 25 km² Shullsburg Branch watershed in south-western Wisconsin led to increased flooding and soil erosion. The increased frequency of overbank floods transported heavy loads of silty sediment from upland fields and a large fraction of this sediment was deposited on floodplains during overbank flows. Increasing bank heights above a stable channel bed ultimately increased the erosive energies of floods, causing rapid bank erosion and expansion of the meander belt. Rates of bank erosion were especially rapid from about 1890 to 1940, but have slowed since then in response to better upland land use, followed by a reduction in flood magnitudes and frequencies

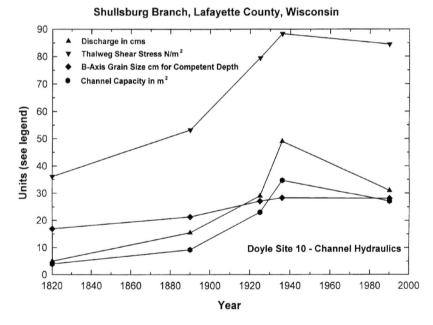

Figure 14.10 Hydraulic adjustments associated with the historical channel evolution shown in Figure 14.9

in clusters of three or four plants on a grid with spacings of about 107 cm in each direction. The grid arrangement allowed for mechanical cultivation in two directions to remove weeds from between the corn plants. The wide spacing exposed much bare soil to raindrop impact and erosion for a large part of the growing season. Furthermore, if rain followed cultivation up- and downslope, then surface runoff and rill erosion were extreme. During the 1950s corn began to be planted on the contour in rows that were more closely spaced, with plants also more closely and uniformly spaced in the rows. The trend to increasing corn plant densities continues to the present. The introduction of chisel ploughing, minimum tillage practices, and other land conservation measures in the 1980s and 1990s have contributed further benefits toward reducing floods and soil erosion.

The division of the agricultural period into three episodes is quite evident in the overbank alluvium of the lower reaches of the Little Platte and Grant Rivers in south-western Wisconsin (Figures 14.12 and 14.13). These downstream low energy reaches have not experienced the type of historical bank erosion typical of the headwater tributaries as shown in Figure 14.9. The 3.5 m of culturally accelerated overbank sedimentation for these sites is typical of the amount that has occurred on many of the low gradient lowermost reaches of small Mississippi River tributaries in this sector of the UMV. Variations in grain size, organic carbon and trace metals from historical mining provide a framework for age and sedimentation rate determinations

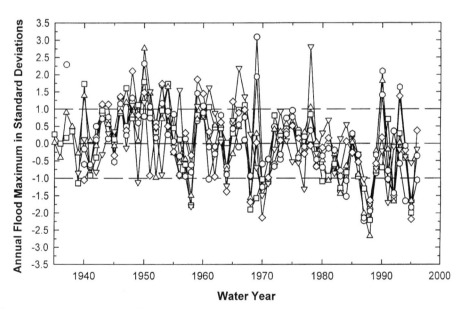

Figure 14.11 Respective flood series of the six south-western Wisconsin rivers were first converted to logarithms to normalise distributions and then individual floods were expressed as standardised deviations from the mean value of the respective flood series. The general similarity of time series variability for the six rivers shows a gradual decline in magnitudes of high frequency floods since about 1950. This long-term trend is related to improved land management practices. Calculations by the author; flood data source: US Geological Survey

of stratigraphic horizons (Figure 14.12). Prior to agricultural disturbance, the average rate of vertical accretion for south-western Wisconsin Holocene floodplains was only 0.02 cm year^{-1} in watersheds ranging from less than 10 km^2 to about 600 km^2 (Knox, 1985). The long-term historical overbank sedimentation rate at the Platte and Grant River sites has averaged about 2 cm year^{-1}, but the averages poorly reflect short-term variations in rates. The historical rates greatly exceed natural background rates. For example, the radiocarbon age of a log buried in late Holocene point bar sediment at the Little Platte River site shows that the rate of accumulation of 1.6 m of floodplain sedimentation between about AD 735 and AD 1820 was only about 0.15 cm year^{-1}. However, the overlying historical overbank sedimentation averaged 1.0–1.5 cm year^{-1} between 1820 and about 1880. Between about 1880 and 1950 the average overbank sedimentation rate increased to 3.1 cm year^{-1}, but since then has averaged 2.5 cm year^{-1} (Figure 14.12). The overbank sedimentation rate in the last several decades has averaged about 1.6 cm year^{-1}, and the higher average for post-1950 reflects the extreme sedimentation from a few very large floods that occurred during the years 1950–1954 (Figure 14.12). At the Grant River Yager Site between 1950 and 1954 a series of large floods deposited about 25 cm of sediment on the floodplain (Figure 14.13). At the same time, floods on the Little Platte River deposited about 40 cm of overbank sediment across the floodplain. These deposits represent short-term average annual rates of 5–8 cm year^{-1}. However, recognising that duration of overbank flow usually does not exceed 20 hours for a given flood at these sites, and commonly is one-half of this duration, the minimum rate of deposition for the five 1950–1954 floods was 0.25–0.40 cm h^{-1}. These high magnitude sedimentation rates show that changes in the recurrence frequencies of large floods, as occurs with changes of climate and changes of land use, greatly influence the nature of overbank sedimentation on floodplains. As large floods in the UMV tributaries of south-western Wisconsin have generally declined in magnitude and recurrence frequency during the last half of the 20th century, the average long-term floodplain sedimentation rate has greatly declined and stratigraphic bedding has become less evident in the overbank alluvium (Figure 14.13).

CONCLUDING SUMMARY

Four distinct major environmental episodes characterise the magnitudes and frequencies of floods in the upper Mississippi River valley (UMV) since about 25 000 ^{14}C years BP. Between 25 000 and about 14 000 ^{14}C years BP the UMV was under the influence of the continental ice sheet. The Mississippi River and many of its principal tributaries were dominated by seasonal meltwater floods and aggradation with sand and gravel outwash sediment. Extreme floods, which were probably produced by rapid drainage of glacial or proglacial lakes, occasionally interrupted the seasonal floods. Hydraulic modelling based on relict sedimentological structures indicated that one of these floods experienced a magnitude that was about twice the magnitude of the largest flood recorded since 1828 for a river reach along south-western Wisconsin. After about 14 000 ^{14}C years BP the continental ice sheet began a general period of retreat that resulted in the formation of several large proglacial lakes located between the retreat-

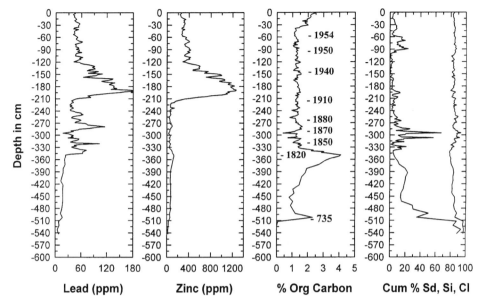

Figure 14.12 Vertical accretion on the floodplain of the 253 km² south-western Wisconsin Little Platte River watershed at the Hinderman Site has averaged about 2 cm year⁻¹ since the beginning of agricultural settlement around 1820. The natural rate of floodplain buildup from AD 735 (1270 ± 50 ¹⁴C years BP, WIS-2132) to 1820 was only 0.15 cm year⁻¹, which includes a small point bar component. The rate is 0.11 cm year⁻¹ if only former overbank sedimentation is applied. The high level of organic carbon at 350 cm depth marks a mollisol that formed on a relatively stable floodplain surface prior to agricultural settlement. The sudden upward increases in lead and zinc concentrations correspond with mine openings of known ages. Large increases in sand denote deposition from large floods of high energy such as occurred during 1950–1954

ing ice front and moraines of former ice advance positions. These lakes functioned as sediment traps and discharged waters of relatively low sediment concentration which then initiated an episode of major downcutting in the upper Mississippi River. These large lakes also occasionally experienced catastrophic outlet failures which produced exceptionally large and erosive floods on the upper Mississippi River. Hydraulic modelling indicates these floods were as small as 20 000–30 000 m³ s⁻¹, but some might have been as large as 60 000–125 000 m³ s⁻¹. The large range of estimates reflects the uncertainty associated with relict cross-section dimensions associated with downcutting.

No reliable estimates of Holocene floods on the upper Mississippi River are available because remobilisation of colluvial and alluvial sediment from tributaries has aggraded the Mississippi River during the Holocene and buried much of the early Holocene record. Furthermore, there is a general paucity of relict Holocene channels in the upper valley, and sediment textures are not sufficiently coarse to apply competent depth methodologies. Side valley tributaries of south-western Wisconsin and adjacent areas, however, preserve excellent records of Holocene floods. The tributary records indicate that relatively small-magnitude climate changes involving shifts in

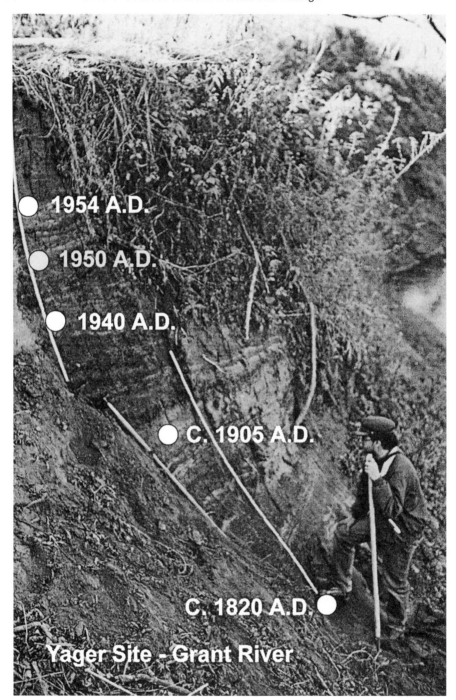

Figure 14.13 Vertical accretion of the Yager Site floodplain of the 695 km² Grant River watershed in south-western Wisconsin. Post-1940 flood depositional units were correlated to the historical discharge gauging record at a site 2 km downstream. The age of the 1905 horizon was established by correlation with mining-related trace metals as in Figure 14.11. The 1820 horizon is the top of the organic-rich mollisol which was the stable floodplain surface prior to beginning of settlement. Photo by the author

mean annual temperature of 1–2 °C and shifts in mean annual precipitation of about 15% have been associated with large changes in flooding characteristics. Long episodes occur in which magnitudes of high frequency floods of 1–2 year recurrence probability have varied from about 20% less to about 15% more than present-day floods of the same probability. Large overbank floods were equally sensitive. During a relatively warm and dry period, 5000–3000 ^{14}C years BP, the largest flood indicated in the stratigraphic record was no bigger than a present-day flood of 25–50 years recurrence probability. However, following a shift to a cooler and wetter climate after 3000 ^{14}C years BP, large overbank floods that equate with modern floods of 100–500 years recurrence probability became common.

The fourth major environmental episode that resulted in a distinct change in magnitudes and frequencies of floods in the UMV was initiated by the introduction of Euro-American agriculture in the early decades of the 19th century. The replacement of the prairie and forest vegetation with corn, wheat and hay crops and pasture by the late 19th and early 20th centuries increased magnitudes of high frequency floods five to six times above previous natural magnitudes in small tributary watersheds. The accelerated flooding was associated with greatly accelerated erosion and sedimentation. Long-term overbank sedimentation rates during the agricultural period have ranged from 1 to 3 cm year^{-1}, but the pre-agriculture long-term average Holocene rate was only 0.02 cm year^{-1}. Since the 1940s, improved land conservation practices and changes in the way crops are planted and grown have reduced magnitudes and frequencies of floods and erosion and sedimentation from their peak values attained in the late 19th and early 20th centuries.

Alluvial sediments and their associated landforms provide a proxy record of magnitudes and recurrence frequencies of past floods. The examination of the proxy record of floods that occurred in the UMV since 25 000 ^{14}C years BP shows strong non-stationary behaviour in the mean and variance of long-term flood series. While non-stationary behaviour is anticipated for geological timescales, the results of the present analyses demonstrate that even minor shifts in mean climate have been associated with major shifts in magnitudes and frequencies of floods. This suggests that shifts in mean climate conditions probably involve important adjustments in the recurrence frequencies of specific atmosphere/ocean circulation patterns that trigger floods. Changes in land use also resulted in strong non-stationary behaviour in flood series of the UMV. The high level sensitivity of floods to even small-magnitude environmental changes underscores the concern that modern short-duration instrumental records inadequately represent the past. The results also imply that past proxy records of floods may not necessarily be good analogues of potential future flood behaviour, unless the record is adjusted for human influences.

ACKNOWLEDGEMENTS

This research was supported by the US National Science Foundation, Grants EAR-8108721, EAR-8511280, EAR-8707504, EAR-9206854 and EAR-9409778; and University of Wisconsin Vilas Awards AK15 and BD52. Frank Magilligan, David Leigh, Peter Church and David May assisted with various components of field and laboratory efforts.

REFERENCES

Allen, J. R. L. 1984. *Sedimentary Structures: Their Character and Physical Basis*. Developments in Sedimentology, 30.

Anderson, D. B. and Burmeister, I. L. 1970. *Floods of March–May 1965 in the Upper Mississippi River Basin*. US Geological Survey Water Supply Paper 1850A.

Baker, J. P. 1990. An analysis of the surface hydrology in the Galena Watershed: 1940–1987. MS Thesis, Department of Geography, University of Wisconsin, Madison, WI.

Baker, R. G., Bettis, E. A., Schwert, D. P., Horton, D. G., Chumbley, C. A., Gonzàlez, L. A. and Reagan, M. K. 1996. Holocene paleoenvironments of northeast Iowa. *Ecological Monographs*, **66**, 203–234.

Bartlein, P. J., Webb, T. III and Fleri, E. 1984. Holocene climatic change in the northern Midwest: pollen-derived estimates. *Quaternary Research*, **22**, 361–374.

Beach, T. 1994. The fate of eroded soil: sediment sinks and sediment budgets of agrarian landscapes in southern Minnesota, 1851–1988. *Annals of the Association of American Geographers*, **84**, 5–28.

Becker, W. M. 1995. Reconstruction of Late Glacial discharges in the upper Mississippi Valley. MS Thesis, University of Wisconsin, Madison, WI.

Benedetti, M. M. 1993. Sediment budget response to land use changes: Big Jack Watershed, Grant County, Wisconsin. MS Thesis, Department of Geography, University of Wisconsin, Madison, WI.

Bettis, E. A., III, Baker, R. G., Green, W. R., Whelan, M. K. and Benn, D. W. 1992. *Late Wisconsinan and Holocene Alluvial Stratigraphy, Paleocology, and Archaeological Geology of East-central Iowa*. Iowa Department of Natural Resources, Geological Survey Bureau, Iowa City, Guidebook Series No. 12.

Blanchard, W. O. 1924. *The Geography of Southwestern Wisconsin*. Wisconsin Geological and Natural History Survey, Bulletin 65.

Borchert, J. R. 1950. The climate of the central North American grassland. *Annals of the Association of American Geographers*, **40**, 1–39.

Brown Survey 1931. Upper Mississippi River, Hastings, Minnesota to Grafton, Illinois Survey 1929–1930, US Corps of Engineers. Williams and Heintz Company, Washington, DC, Sheets 23–59.

Bryson, R. A. 1966. Air masses, streamlines, and the boreal forest. *Geographical Bulletin*, **8**, 228–269.

Bryson, R. A. and Wendland, W. M. 1967. Tentative climatic patterns for some late-glacial and post-glacial episodes in central North America. In Mayer-Oakes, W. J. (Ed.) *Life, Land, and Water*. University of Manitoba Press, Winnipeg, 271–298.

Carney, S. J. 1996. Paleohydrology of the western outlets of Glacial Lake Duluth. MS Thesis, University of Minnesota, Duluth.

Church, P. E. 1984. The archaeological potential of Pool No. 10, upper Mississippi River: a geomorphological perspective. US Army Corps of Engineers, Waterways Experiment Station, Vicksburg, Mississippi.

Clayton, L. 1983. Chronology of Lake Agassiz drainage to Lake Superior. In Teller, J. T. and Clayton, L. (Eds) *Glacial Lake Agassiz*. Geological Association of Canada, Special Paper 26, 291–307.

Clayton, L. and Attig, J. W. 1989. *Glacial Lake Wisconsin*. Geological Society of America, Memoir 173.

Dorale, J. A., Gonzàlez, L. A., Reagan, M. K., Pickett, D. A., Merrell, M. T. and Baker, R. G. 1992. A high-resolution record of Holocene climate change in speleothem calcite from Coldwater Cave, northeast Iowa. *Science*, **258**, 1626–1630.

Dunne, T. and Leopold, L. B. 1978. *Water in Environmental Planning*. W. H. Freeman, San Francisco.

Dyke, A. S. and Prest, V. K. 1987. Late Wisconsinan and Holocene retreat of the Laurentide Ice Sheet. Geological Survey of Canada, Map 1702A, scale 1:5 000 000.

Flock, M. A. 1983. The Late Wisconsinan Savanna Terrace in tributaries to the upper Mississippi River. *Quaternary Research*, **20**, 165–176.

Hajic, E. R. 1991. Terraces in the central Mississippi Valley. In Hajic, E. R., Johnson, W. H. and Follmer, L. R. (Eds) *Quaternary Deposits and Landforms, Confluence Region of the Mississippi, Missouri, and Illinois Rivers, Missouri and Illinois: Terraces and Terrace Problems*. Midwest Friends of the Pleistocene, 38th Field Conference Guidebook, 1–30.

Hallberg, G. R. and Kemmis, T. J. 1986. Stratigraphy and correlation of the glacial deposits of the Des Moines and James Lobes and adjacent areas in North Dakota, South Dakota, Minnesota, and Iowa. In Sibrava, V., Bowen, D. Q. and Richmond, G. M. (Eds) *Quaternary Glaciations in the Northern Hemisphere. Quaternary Science Review*, **5**, 65–68.

Hoggan, D. H. 1989. *Computer-Assisted Floodplain Hydrology and Hydraulics*. McGraw-Hill, New York.

Johnson, W. H. 1986. Stratigraphy and correlation of the glacial deposits of the Lake Michigan Lobe prior to 14 ka BP. In Sibrava, V., Bowen, D. Q. and Richmond, G. M. (Eds) *Quaternary Glaciations in the Northern Hemisphere. Quaternary Science Review*, **5**, 17–22.

Keen, K. L. and Shane, L. C. K. 1990. A continuous record of Holocene eolian activity and vegetation change at Lake Ann, east-central Minnesota. *Geological Society of America Bulletin*, **102**, 1646–1657.

Kehew, A. E. 1993. Glacial lake outburst erosion of the Grand Valley, Michigan, and impacts on glacial lakes in the Lake Michigan Basin. *Quaternary Research*, **39**, 36–44.

Knox, J. C. 1977. Human impacts on Wisconsin stream channels. *Annals of the Association of American Geographers*, **67**, 323–342.

Knox, J. C. 1983. Responses of river systems to Holocene climates. In Wright, H. E. Jr. (Ed.) *Late Quaternary Environments of the United States, Volume 2, The Holocene*. University of Minnesota Press, Minneapolis, 26–41.

Knox, J. C. 1985. Responses of floods to Holocene climatic change in the upper Mississippi Valley. *Quaternary Research*, **23**, 287–300.

Knox, J. C. 1993. Large increases in flood magnitude in response to modest changes in climate. *Nature*, **361**, 430–432.

Knox, J. C. 1996. Late Quaternary upper Mississippi River alluvial episodes and their significance to the lower Mississippi River system. *Engineering Geology*, **45**, 263–285.

Laird, K. R., Fritz, S. C., Maasch, K. A. and Cumming, B. F. 1996. Greater drought intensity and frequency before AD 1200 in the Northern Great Plains, USA. *Nature*, **384**, 552–554.

Leigh, D. S. and Knox, J. C. 1994. Loess of the Upper Mississippi Valley Driftless Area. *Quaternary Research*, **42**, 30–40.

Mason, J. A. and Knox, J. C. 1997. Age of colluvium indicates accelerated late Wisconsin hillslope erosion in the Upper Mississippi Valley. *Geology*, **25**, 267–270.

Matsch, C. L. 1983. River Warren, the southern outlet of Glacial Lake Agassiz. In Teller, J. T. and Clayton, L. (Eds) *Glacial Lake Agassiz*. Geological Association of Canada, Special Paper 26, 231–244.

Matsch, C. L. and Wright, H. E. Jr. 1967. The southern outlet of Lake Agassiz. In Mayer-Oakes, W. J. (Eds) *Life, Land, and Water*. University of Manitoba Press, Winnipeg, 121–140.

McCorvie, M. R. and Lant, C. L. 1993. Drainage district formation and the loss of Midwestern wetlands, 1850–1930. *Agricultural History*, **67**, 13–39.

Mickelson, D. M., Clayton, L., Fullerton, D. S. and Borns, H. W. Jr. 1983. The Late Wisconsin glacial record of the Laurentide Ice Sheet in the United States. In Wright, H. E. Jr. and Porter, S. C. (Eds) *Late-Quaternary Environments of the United States, Volume 1, The Late Pleistocene*. University of Minnesota Press, Minneapolis, 3–37.

Ojakangas, R. W. and Matsch, C. L. 1982. *Minnesota's Geology*. University of Minnesota Press, Minneapolis.

Porter, D. A. and Guccione, M. J. 1994. Deglacial flood origin of the Charleston Alluvial Fan, lower Mississippi alluvial valley. *Quaternary Research*, **41**, 278–284.

Saucier, R. T. 1994. *Geomorphology and Quaternary Geologic History of the Lower Mississippi Valley*. US Corps of Engineers, Mississippi River Commission, Vicksburg, Mississippi.

Schafer, J. 1932. *The Wisconsin Lead Region*. Wisconsin Domesday Book General Studies III,

State Historical Society of Wisconsin, Madison, WI.
Sorenson, C. J. 1977. Reconstructed Holocene bioclimates. *Annals of the Association of American Geographers*, **67**, 214–222.
Stuiver, M. and Reimer, P. 1993. Extended ^{14}C data base and revised CALIB3.0 ^{14}C age calibration program. *Radiocarbon*, **35**, 215–230.
Teller, J. T. 1987. Proglacial lakes and the southern margin of the Laurentide Ice Sheet. In Ruddiman, W. F. and Wright, H. E. Jr. (Eds) *North America and Adjacent Oceans during the Last Deglaciation, The Geology of North America*, K-3, Geological Society of America, 39–69.
Teller, J. T. 1990. Volume and routing of late-glacial runoff from the southern Laurentide Ice Sheet. *Quaternary Research*, **34**, 12–23.
Trimble, S. W. 1983. A sediment budget for Coon Creek basin in the Driftless Area, Wisconsin, 1853–1977. *American Journal of Science*, **283**, 454–474.
Wiele, S. and Moores, H. D. 1989. Glacial River Warren: steady state and peak discharge. *Geological Society of America Abstracts with Programs*, **21**, A60.
Willman, H. B. and Frye, J. C. 1969. *Pleistocene stratigraphy of Illinois*. Illinois State Geological Survey, Bulletin 94.
Winkler, M. G., Swain, A. M. and Kutzbach, J. E. 1986. Middle Holocene dry period in the northern midwestern United States: lake levels and pollen stratigraphy. *Quaternary Research*, **25**, 235–250.
Wright, H. E. Jr. 1987. Synthesis: the land south of the ice sheets. In Ruddiman, W. F. and Wright, H. E. Jr. (Eds) *North America and Adjacent Oceans during the Last Deglaciation, The Geology of North America*, K-3, Geological Society of America, 479–488.
Wright, H. E. Jr. 1992. Patterns of Holocene climatic change in the midwestern United States. *Quaternary Research*, **38**, 129–134.

15 High Resolution Palaeochannel Records of Holocene Valley Floor Environments in the North Tyne Basin, Northern England

ANDREW J. MOORES, DAVID G. PASSMORE and
ANTHONY C. STEVENSON
Department of Geography, University of Newcastle upon Tyne, UK

INTRODUCTION

In recent years there has been widespread study of the lithostratigraphy, biostratigraphy and chronology of Holocene alluvial valley floors in upland regions of the British Isles. In particular, geomorphological and geoarchaeological investigations have sought to elucidate the dynamics, timing and controls of post-glacial channel and floodplain development (e.g. Harvey *et al.*, 1984; Macklin and Lewin, 1986; Robertson-Rintoul, 1986; Hooke *et al.*, 1990; Macklin *et al.*, 1992a, 1992b, 1998; Passmore *et al.*, 1993; Passmore and Macklin, 1994, 1997; Rumsby and Macklin, 1994; Tipping, 1994, 1995, 1996; Taylor and Lewin, 1996, 1997), patterns of valley floor vegetation and land-use change (Passmore *et al.*, 1992; Tipping and Halliday, 1994; Tipping, 1996) and the archaeological implications of valley floor development (Macklin *et al.*, 1992a, 1992c; Brown, 1997; Passmore and Macklin, 1997). These studies have demonstrated that Holocene fluvial activity in upland regions has responded in a sensitive and complex manner to Holocene environmental changes, most notably over historic timescales under the combined influence of climate change and accelerated human activity.

Assessment of valley floor development and its palaeoenvironmental significance over extended Holocene timescales in these environments is typically constrained, however, by the combination of two factors. First, older Holocene terraces are liable to have been partially or wholly eroded by subsequent fluvial activity. This is especially problematic in upper and middle valley reaches, lying upstream of low-lying alluvial basins (Macklin *et al.*, 1992a; Tipping 1994, 1996) and perimarine locations (Passmore *et al.*, 1992), where fluvial activity has typically been associated with high rates of postglacial channel migration and a tendency towards net incision of valley floors (Macklin *et al.*, 1992a). Secondly, alluvial sequences in upper and middle valley reaches

Fluvial Processes and Environmental Change. Edited by A. G. Brown and T. A. Quine.
© 1999 John Wiley & Sons Ltd.

tend to be dominated by coarse-grained sediments with relatively thin fine members and little or no preservation of organic materials. This has tended to inhibit both development of alluvial chronologies via radiocarbon assay, and also palaeoecological analyses of valley floor vegetation histories (Passmore and Macklin, 1997).

These difficulties have constrained investigations of Holocene alluvial histories over a range of research issues. For example, the frequent lack of chronological control has ensured that discerning the respective roles of long-term climate and land-use changes in controlling geomorphological activity remains a contentious issue (e.g. Bell, 1992; Macklin et al., 1992a, 1992c; Tipping, 1992, 1995; Macklin and Lewin, 1993; Passmore, 1994; Brown, 1997). Similarly, investigations of the role of rare, high-magnitude flood events in conditioning valley floors (e.g. Newson, 1989; Newson and Lewin, 1991; Rumsby and Macklin, 1994) are proving difficult to extend over Holocene timescales that lie beyond the (recent historic) dating range of lichenometry and (in catchments that have experienced mining of metalliferous ores) trace metal analyses.

Recent investigations in the Rivers Tweed (Tipping and Halliday, 1994) and North Tyne (Passmore, 1994; Passmore and Macklin, 1997) have demonstrated, however, that upland valley floors may locally have experienced conditions conducive to the development and subsequent preservation of well-developed, organic-rich Holocene alluvial sequences. This chapter presents some preliminary results from geomorphological and palaeoecological investigations in middle reaches of the River North Tyne and its principal tributary the River Rede, northern England, that demonstrate the potential of these sites to extend our knowledge of valley floor development from the 5th millennium BC. Particular attention is drawn to palaeoecological analyses of channel fill deposits that are offering new high-resolution insights into long-term valley floor vegetation histories and human activity. Consideration is also extended to the geoarchaeology of these sites and their potential for preserving high-resolution geomorphological records of Holocene flood histories. Finally we briefly consider the potential contribution of these reaches to elucidation of regional chronologies and controls of Holocene fluvial activity.

BACKGROUND TO THE STUDY AREA AND METHODOLOGY

The Rivers North Tyne and Rede drain the southern margins of the Cheviot Hills and the Kielder Forest uplands with a catchment area of $1118\,km^2$ (Figure 15.1). The underlying geology is dominated by Lower Carboniferous rocks, which include coal measures, limestones and Fell sandstones. Hillslopes and valley floors throughout the basin are mantled with glacial, periglacial and glaciofluvial sediments deposited during the late Devensian (c. 14 000–10 000 BP), while present river channels are inset within these Pleistocene deposits, Holocene alluvium or bedrock (Passmore and Macklin, 1997).

Published palaeoecological studies of Holocene environments in the Tyne Basin are comparatively rare by comparison with the extensive history of regional archaeological survey and excavation (e.g. Higham, 1986). Indeed, dated multi-period pollen diagrams specific to the catchment are limited to upland mire and bog sites at Steng Moss (NY 965913, located on the interfluve between the catchments of the River Rede

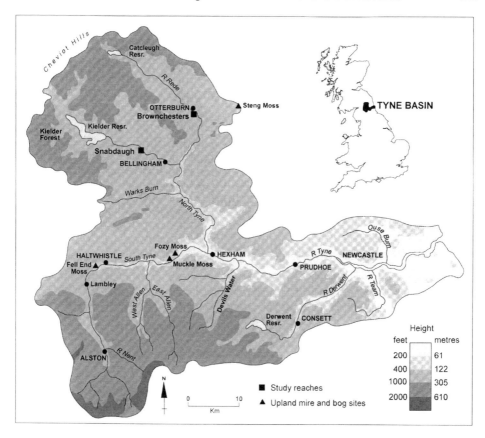

Figure 15.1 Location and relief of the Tyne basin showing study sites and upland mire and bog sites mentioned in text

and River Wansbeck; Davies and Turner, 1979) and near Hadrian's Wall in the lower South Tyne valley at Fellend Moss (NY 679658; Davies and Turner, 1979) and Fozy Moss (NY 830714; Dumayne, 1992, 1993a, 1993b; Barber et al., 1993; Dumayne and Barber, 1994) (Figure 15.1). Pollen evidence from these sites suggests that human impact on vegetation and soil cover in the basin prior to c. 2000 BC was minimal. Small-scale and temporary forest clearance, associated with pastoral agriculture and some arable cultivation, is evident during the Bronze Age and Iron Age, but widespread and long-term deforestation and subsistence activity is not recorded until the period of Roman occupation.

Recent archaeological survey suggests, however, that levels of tillage in upland areas during the Iron Age may have been underestimated (Topping, 1989), while extensive Iron Age and Romano-British settlement has been recorded by archaeological survey on the lower valley slopes and valley floors of the Rivers North Tyne and Rede (e.g. Higham, 1986; Jobey and Jobey, 1988) which are likely to have been the most favourable locations for pastoral agriculture and arable cultivation. Thus, although the North Tyne Basin has an archaeological record extending back to the Mesolithic

Table 15.1 Radiocarbon dates and calibrated calendar ages (after Stuiver *et al.*, 1993) for sites described in the text

Site and palaeochannel/ terrace	Sample depth (cm)	Lab. no. (beta)	Radiocarbon date (uncal. BP; 1 sigma)	Calibrated calendar age (2 sigma)
Snabdaugh T3	70–90	45553	6110 ± 60	5220–4850 BC
Snabdaugh T5	240	45555	2330 ± 70	760–210 BC
Snabdaugh T5	220	45556	1720 ± 50	AD 130–430
Snabdaugh T5	140	45557	920 ± 70	AD 950–1260
Brownchesters T4	224–241	96127	6110 ± 80	5240–4825 BC
Brownchesters T5	140–142	96124	3900 ± 70	2570–2145 BC
Brownchesters T5	340–350	96125	5110 ± 80	4055–3725 BC
Brownchesters T6	158–166	96122	3410 ± 80	1890–1520 BC
Brownchesters T6	337–340	96123	3910 ± 80	2585–2140 BC
Brownchesters T7	130–135	90753	650 ± 80	AD 1245–1430

period (Higham, 1986; Adams, 1996), palaeoecological investigations of valley floors are regarded as essential in order to ascertain the chronology and scale of prehistoric and historic deforestation and sedentary agriculture (Passmore and Macklin, 1997).

Methodology

Alluvial valley floors in both study reaches have been subject to geomorphological field mapping (using enlarged aerial photographs as base maps), survey and a combination of sediment coring and/or machine trenching through selected palaeochannel fills. Exposed sedimentary sequences were logged and sampled for pollen and ^{14}C analyses. All ^{14}C dates have been calibrated according to calibration curves after Stuiver *et al.* (1993) (Table 15.1) and are quoted in the text as age spans of the 2-sigma calibration range. In addition, some inorganic fine-grained alluvial deposits have been dated using palaeomagnetic techniques (cf. Clark, 1992). Detrital remanent magnetism of sampled sediment columns were measured in a Molspin fluxgate spinner magnetometer, with remanence directions corrected for field orientations using geomagnetic data for the UK published by the British Geological Survey. Mean palaeomagnetic directions for sediment columns were then compared to the UK master curve (Clark *et al.*, 1988).

Laboratory preparation of pollen slides followed standard methods (Moore *et al.*, 1991), with the exception of an additional stage prior to staining and mounting. This method, termed the "Amsterdam technique" (Munsterman and Kersholt, 1996), uses the heavy liquid sodium polytungstate to separate excess fine minerogenic material from the sample. Lycopodium tablets were added to each of the samples prior to preparation to allow the determination of pollen concentration (Stockmarr, 1971).

Pollen analysis was carried out using an Olympus CH-2 microscope at ×400 magnification (×1000 oil immersion/phase contrast for problematic grains) and counting of each slide continued until a minimum of 200 arboreal and shrub pollen grains were identified. In addition, all counted pollen was assigned to one of four deterioration classes (based upon the work of Cushing, 1967) that differentiate degraded, corroded and crumpled/broken grains. Degraded pollen in alluvial contexts may, in particular, be associated with allochthonous material deposited during flooding, and/or local reworking of pollen during flood episodes. A number of studies, for example, have found a correlation between inwashed minerogenic material and degraded pollen grains within peat bogs (Birks, 1970; Waller, 1993).

HOLOCENE ALLUVIAL SEQUENCES IN THE RIVERS NORTH TYNE AND REDE

Investigations have focused on middle reaches of the River North Tyne at Snabdaugh Farm, located 6 km upstream of Bellingham (catchment area 429 km^2), and the River Rede (the principal tributary of the North Tyne) at Brownchesters Farm (catchment area 214 km^2) which lies 1 km south-west of Otterburn village (Figure 15.1). Modern river channels in these reaches occupy valley floors up to 750 m wide and exhibit gently meandering, single-thread planforms (Figures 15.2 and 15.4) with slopes ranging between 0.002 and 0.006 m m^{-1}. Historical maps (OS County Series) indicate that channels have been laterally stable since the mid-19th century, although Holocene channel shifts are attested in both reaches by palaeochannels developed on alluvial terrace surfaces. Alluvial histories for each reach are outlined below.

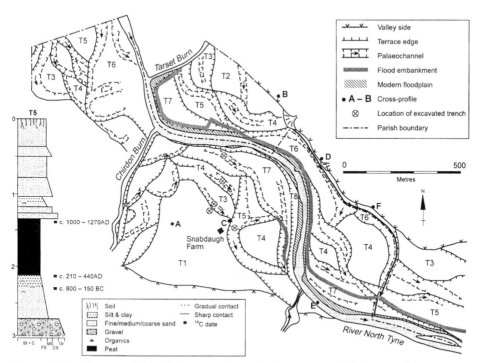

Figure 15.2 Geomorphological map of late Pleistocene and Holocene river terraces and palaeochannels at Snabdaugh Farm, showing the locations of cross-profiles and excavated trenches and the simplified lithostratigraphy of the investigated T5 palaeochannel fill

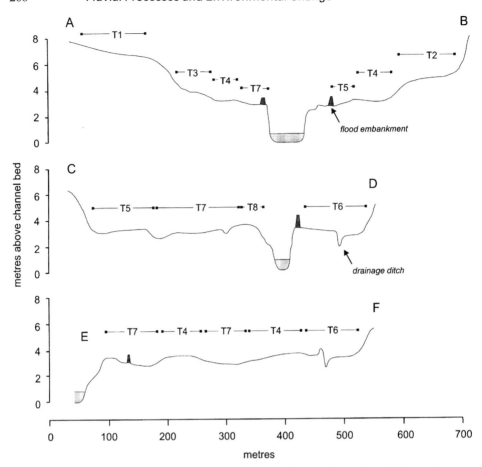

Figure 15.3 Surveyed cross-profiles of the valley floor at Snabdaugh Farm. For profile locations, see Figure 15.2

River North Tyne at Snabdaugh Farm (NY 787846)

Preliminary analyses of the Holocene alluvial history at Snabdaugh are documented in Passmore (1994) and Passmore and Macklin (1997), while an extended and detailed revised account will be published elsewhere; here we present a brief outline of valley floor development that focuses on the later prehistoric period. A total of eight alluvial terrace assemblages, designated T1–T8, have been differentiated at Snabdaugh on the basis of morphostratigraphic relationships, elevation above the current river and a variety of dating controls. Inevitably many discrete terrace units lack direct dating control and are classified on a provisional basis, while unsurveyed terrace units have not been classified. The morphology, lithostratigraphy and chronology of these alluvial units are summarised in Figures 15.2 and 15.3 and Table 15.2.

The highest elevation terraces, T1 and T2 (respectively lying 7–8 m and 5–6 m above the current river bed, Figures 15.2 and 15.3), represent the oldest alluvial surfaces

Table 15.2 Depth, lithostratigraphy and dating controls of late Pleistocene and Holocene alluvial fills exposed in machine-excavated trenches at Snabdaugh; for trench locations see Figure 15.2. Details of dating controls and radiocarbon calibrations are given in the text and Table 15.1

Terrace	Depth (cm)	Description	Dating control
T1	0–45	Dark brown silty sand with well-developed topsoil	—
	45–120 +	Coarse sandy gravel with occasional 20 cm thick sand lenses	—
T3	0–30	Dark brown silty sand topsoil	—
	30–70	Orange-brown unstructured fine silty sands	—
	70–90	Dark grey peaty silts with wood fragments	Wood and plant fragments ^{14}C dated to c. 5220–4850 BC
	90–220	Unstructured, fining-upward brown silty sands	—
	220 +	Coarse sandy gravel	—
T4	0–30	Dark brown silty sand topsoil	—
	30–150	Orange-brown poorly bedded silty sands with occasional mottled manganese staining	Palaeomagnetic date of c. 1500 BC between 56 and 115 cm
	150–200 +	Fining-upward medium–fine sandy gravels, rising towards channel edge in northern end of trench and interdigitating with fine channel fill sediments described above	—
T5	0–40	Dark brown silty sand topsoil	—
	40–50	Fining upward brown medium–fine silty sand	—
	50–105	Fining upward orange-brown medium–coarse sand with occasional laminations and erosional lower contact	—
	105–135	Laminated, fining upward light-brown coarse–fine sands with erosional lower contact	—
	135–210	Unstructured dark grey fine sandy silt peat with occasional wood and plant fragments, becoming less peaty up-profile between 135 and 160 cm	Wood fragment ^{14}C dated to c. AD 950–1260
	210–240	Fining-upward dark grey peaty fine sands and silts with occasional wood and plant fragments	Wood fragment ^{14}C dated to c. AD 130–430
	240–270	Grey medium-fine sands with occasional laminations	Wood fragment ^{14}C dated to c. 760–210 BC
	270 +	Coarse sandy gravels	—
T7	0–30	Dark brown silty sand topsoil	—
	30–75	Unstructured orange-brown fine silty sand	—
	75–160	Poorly bedded, fining-upward coarse–fine silty sands with occasional fine coal flecks	Palaeomagnetic date of c. 1350 between 140 and 160 cm
	160 +	Fine sandy gravels	—

evident in the study reach although these currently lack dating control. Trenching of T1 near Snabdaugh Farm has revealed this terrace to comprise a thin (up to 0.45 m) sandy fine member overlying sandy coarse gravels (Table 15.2). Progressively inset below these fills are terraces T3 (4 m) and T4 (3–2.5 m, Figures 15.2 and 15.3) which have been investigated by machine trenching of associated palaeochannels north of Snabdaugh Farm (Figure 15.2).

Infilling the T3 palaeochannel are 2.2 m of largely unstructured sands and silts, interbedded with peaty silts between 0.7 and 0.9 m; radiocarbon assay on wood and plant fragments from the organic horizon has returned a date of c. 5220–4850 BC (Table 15.2) and suggests that alluviation of T3 and subsequent abandonment of the channel and floodplain occurred between late Pleistocene times and the later Mesolithic period (Passmore, 1994). Sediments infilling the T4 palaeochannel comprise up to 1.5 m of poorly bedded inorganic silty sands (Table 15.2). Palaeomagnetic dating of sediments between 0.56 and 1.15 m indicates that infilling of the upper levels of this channel was occurring around c. 1500 BC (Noel, 1991; Passmore, 1994), and suggests incision of T3 sediments and subsequent alluviation of the T4 unit occurred sometime between the later Mesolithic and the mid-second millennium BC.

T5 alluvial units also lie between 2–3 m above the present river bed, but are differentiated from earlier T4 fills by well-defined terrace scarps and, immediately east of Snabdaugh Farm, a dated palaeochannel fill sequence up to 2.7 m thick (Figure 15.2, Table 15.2). Infilling the base of the channel are 0.3 m of fining-upward grey sandy silty clays that grade up-profile into 0.25 m of peaty sandy silt (Figure 15.2). A large piece of timber, clearly bearing toolmarks and resembling a truncated plank, was recovered lying at the transition between these fills and has been ^{14}C dated to c. 760–210 BC (Passmore, 1994; Table 15.2). This date indicates that local abandonment of the T5 channel occurred sometime during the late Bronze Age and Iron Age periods. Overlying these sediments are 0.75 m of silty peats with frequent wood and plant inclusions and interbedded fine silty lenses. Radiocarbon assays on wood and plant fragments extracted from the base (depth 2.2 m) and top (1.4 m) of the peats bracket this period of organic-rich sedimentation to between c. AD 130–430 and c. AD 950–1260 respectively (Figure 15.2, Table 15.2). Overlying sediments largely comprise up to 1.3 m of inorganic coarse-fine sands and silts (Figure 15.2). Organic-rich sediments within this palaeochannel fill are the focus of the palaeoecological analyses described below.

Fluvial activity dating to the medieval and later periods in the study reach is manifested by terrace assemblages T6–T8 (Figure 15.2). These fills have been differentiated on the basis of morphostratigraphic relationships, cartographic evidence and an inorganic palaeochannel fill (associated with T7) that has been dated using palaeomagnetic techniques to a period centred on c. AD 1350 (Figure 15.2, Table 15.2; Noel, 1991; Passmore, 1994). Medieval and later fills truncate and locally encircle older alluvial surfaces (Figure 15.2) although associated terraces are typically also developed at elevations between 2–3 m above the present river bed (Figure 15.3); indeed, the upper levels of T7 and T8 are locally perched up to 1 m *above* T5 palaeochannels. Alluviation to these levels is likely to have been promoted during the recent historic period through confinement of the active channel zone by flood protection measures (Passmore, 1994).

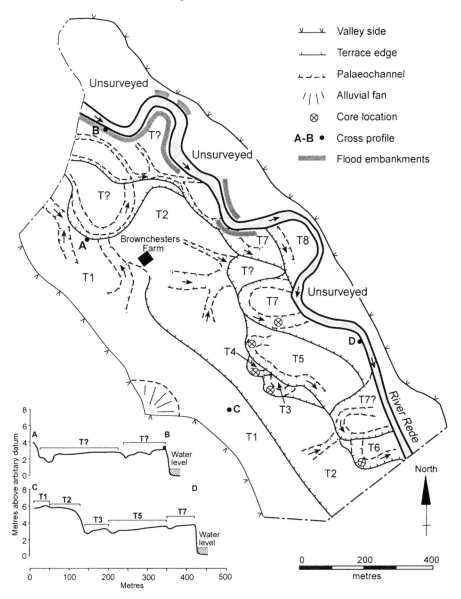

Figure 15.4 Geomorphological map of late Pleistocene and Holocene river terraces and palaeochannels at Brownchesters Farm showing sediment core locations and surveyed valley cross-profiles

River Rede at Brownchesters Farm (NY 889922)

Brownchesters Farm lies 1 km south-west of Otterburn village on the valley floor of the River Rede (Figure 15.1). The study reach occupies a small alluvial basin up to 600 m wide and extends over a valley length of 1.2 km (Figure 15.4). The modern river channel is deeply entrenched on the east side of the valley floor and, like the North Tyne, exhibits a meandering planform that is locally confined by flood embankments constructed sometime before the mid-19th century (Figure 15.4). Investigations have focused on alluvial terraces developed between the present river and the west valley side. Here the valley floor is dominated by two extensive terraces, designated T1 and T2, which lie up to 6 m and 5 m respectively above present river levels (Figure 15.4). No lithostratigraphic data or dating control are yet available for these units.

Inset below T1 and T2 are a series of Holocene alluvial terraces that are characterised by well-developed palaeochannels and surface elevations (where surveyed) restricted to a range between 2.5 and 3 m above present river levels (Figures 15.4 and 15.5). Surveyed terraces have asymmetric surface profiles that progressively rise in elevation towards their point of truncation by younger fills, while maximum heights are attained immediately adjacent to the present active channel belt (Figure 15.4). Detailed reconstruction of long-term fluvial sedimentation styles is hindered by lithostratigraphic data that are currently restricted to sediment cores from selected channel fills (Figures 15.4 and 15.6). However, all Holocene terraces, bar one, feature single-thread palaeochannels of moderate–high sinuosity and no evidence of ridge-and-swale topography diagnostic of migrating point bars. Holocene floodplain development appears therefore to have been dominated by overbank alluviation in the context of laterally stable meanders, although one terrace unit in the northern part of the study reach features relatively narrow palaeochannels that divide around one or more mid-channel bars (Figure 15.4).

In the context of previously documented upland British river valleys, palaeochannel environments at Brownchesters are unusual in that they exhibit thick deposits of peat and organic-rich clayey silt alluvium (Figure 15.6). These units are frequently finely laminated with silt and fine sand lenses (c. 1–2 mm thick), and are occasionally interbedded with thicker (> 1 cm) fine–medium sand units with laminations of organic-rich sandy silts. Discrete inorganic silt and sand horizons throughout these sequences are interpreted as representing deposits of moderate–large floods postdating channel abandonment. Deposits of this type that are greater than 1 cm in thickness are depicted in Figure 15.6.

The restricted range of terrace elevations at Brownchesters precludes long-distance correlation of discrete alluvial units throughout the study reach on the basis of relative height. Abundant *in situ* peaty deposits within palaeochannels are, however, readily datable by radiometric techniques. To date, a combination of seven ^{14}C assays and morphostratigraphic relationships for adjacent alluvial surfaces have established a provisional terrace sequence comprising a minimum of five Holocene alluvial fills, respectively designated T3–T8. Details of ^{14}C ages, calibrations and depths are presented in Table 15.1. Four dates have been obtained at or near the base of organic-rich sediments within palaeochannel cores T3, T5, T6 and T7 (Figure 15.6). Assuming these represent the latest date at which each associated

Figure 15.5 View from high T2 gravel terrace towards Otterburn village. T4, T5 and T7 palaeochannels can be seen clearly due to ponding of floodwaters. Note also rig and furrow cultivation just visible on T5 terrace. (Photo: L.E. Ellis)

terrace fill was deposited, episodes of alluviation at Brownchesters can be bracketed to the periods between the early Holocene and c. 5030 BC (terraces T3 and T4), c. 5030–3950 BC (terrace T5), c. 3950–2440 BC (terrace T6), c. 2440 BC–AD 650 (terrace T7) and sometime after AD 650 (terrace T8) (all dates are mid-points of the calibrated 2-sigma age span). It is noted, however, that the upper levels of each terrace may have been subject to overbank alluviation associated with later periods of fluvial activity. In this chapter we focus on palaeoecological records from two sediment cores extracted from channels associated with T5 and T6, respectively, dated to the periods between c. 4055–3715 and 2570–2145 BC (T5), and c. 2585–2140 BC and 1890–1520 BC (T6).

Summary: Holocene Valley Floor Development in the Middle Reaches of the North Tyne Basin

Extensively developed and upstanding T1 and T2 terraces that lack direct dating control are a feature of both study reaches, although the relative elevation and (at Snabdaugh) predominantly coarse-grained composition of T1, as has been argued elsewhere in upland British river valleys, suggests this terrace is of late Devensian age (e.g. Harvey, 1985; Tipping, 1995; Passmore and Macklin, 1997).

Inset below these surfaces are sequences of Holocene alluvial fills that, for the most part, appear to have been constructed by laterally stable, meandering channels prone to episodic cut-off and/or avulsion. The available chronology and approximate thick-

294 Fluvial Processes and Environmental Change

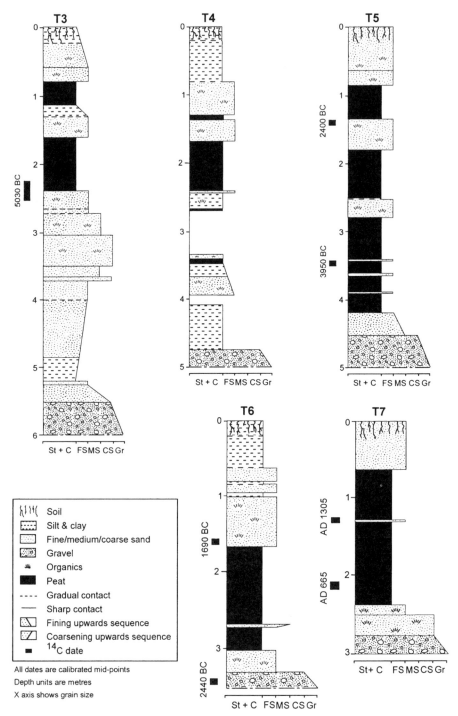

Figure 15.6 Simplified lithostratigraphy and ^{14}C dates for palaeochannel sediment cores at Brownchesters Farm. For core locations, see Figure 15.4

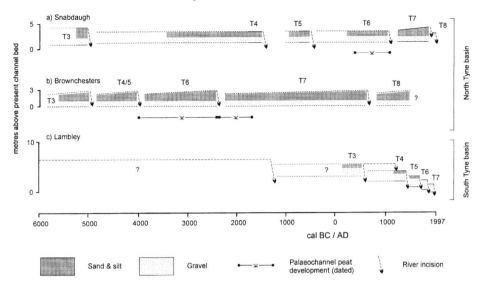

Figure 15.7 Age–depth diagram showing Holocene alluvial units and incision episodes in (a) the River North Tyne at Snabdaugh Farm, (b) the River Rede at Brownchesters Farm, and (c) the River South Tyne at Lambley. For the sake of clarity, ^{14}C dates bracketing alluviation and incision episodes are plotted at the mid-point of age spans (2-sigma error spans) obtained from calibration (cf. Stuiver et al., 1993)

ness of Holocene alluvial terraces at Snabdaugh and Brownchesters, together with episodes of channel incision and/or floodplain reworking, are schematically represented as height–age diagrams in Figure 15.7(a) and (b) respectively. For purposes of comparison these diagrams include the alluvial history of the River South Tyne in its middle reaches at Lambley, near Haltwhistle (Figures 15.1 and 15.7(c); Passmore, 1994; Passmore and Macklin, 1997). Here the South Tyne has a catchment area of 276 km^2, higher gradients than the North Tyne study reaches (0.01–0.006 m m^{-1}), and a present river channel that is frequently divided and subject to locally high rates of lateral migration.

The propensity of valley floors in the middle North Tyne Basin to preserve alluvial fills of prehistoric age contrasts markedly with the South Tyne at Lambley (Figure 15.7) and suggests these reaches have experienced relatively limited rates of lateral channel migration over Holocene timescales. Figures 15.3 and 15.4 also illustrates a long-term tendency of the North Tyne and Rede to develop successive alluvial fills at elevations similar to, or even slightly higher than, early Holocene terraces. Net aggradation of later Holocene valley floors contrasts with contemporary alluvial sequences described elsewhere in upper and middle reaches of northern British rivers, including the South Tyne at Lambley (Figure 15.7(c)); these are typically characterised by net *incision* of valley floors and the development of younger terraces at progressively lower elevations (e.g. Harvey, 1985; Macklin et al., 1992a, 1992b; Passmore, 1994; Passmore and Macklin, 1994, 1997; Tipping, 1995).

Differences in valley morphology and gradient, allied to contrasting catchment land-use histories and coarse sediment supply regimes, are likely to have played a significant role in conditioning the variable character and chronology of Holocene valley floor development across the Tyne Basin (Passmore, 1994; Passmore and Macklin, 1997; Macklin et al., 1998). An updated review of these issues will be published elsewhere; here we draw attention to the unusual long-term preservation of organic-rich channel fills in middle reaches of the North Tyne and Rede which is believed to have been promoted by the net tendency of these rivers to aggrade their valley floors and thereby experience persistently high groundwater levels (cf. Brown and Keough, 1992; Brown, 1996).

PALAEOECOLOGICAL ANALYSES OF HOLOCENE CHANNEL FILL DEPOSITS

Background

Organic-rich alluvial palaeochannel fills provide comparatively little utilised sedimentary sequences yet constitute a potentially important means of evaluating valley floor palaeoenvironments for a number of reasons (Brown, 1997). First, models of pollen transport indicate that pollen rain in these spatially restricted environments is taphonomically derived from predominantly local floodplain and adjacent terrace sources (Tauber, 1965; Janssen, 1973, 1981; Jacobsen and Bradshaw, 1981; Prentice, 1988), and hence primarily reflect valley floor vegetation (Brown and Keough, 1992; Brown, 1996).

Secondly, palaeochannels are often proximal to fertile and productive alluvial soils which are ideal for agriculture (Brown, 1996), although not necessarily settlement (Evans et al., 1988). Past agricultural activity can be discerned within the pollen record by the presence of certain anthropogenic indicator species (Behre, 1981), especially cereal grains (Edwards and McIntosh, 1988; Tipping, 1996). Previous work has demonstrated that cereal-type pollen is more likely to be recorded if samples are taken close to edges of wetlands where arable crops may have been growing in the immediate vicinity (Turner, 1975; Edwards and McIntosh, 1988; Edwards, 1991a, 1991b). Indeed, Janssen (1986) has found cereal values of up to 30% in palaeochannel sequences dating from medieval times in South Limburg, the Netherlands.

Thirdly, palaeochannels provide low-lying sediment sinks within the floodplain environment which may experience relatively rapid accumulation of material, and hence offer potentially high-resolution palaeoenvironmental records. Lewis and Lewin (1983), for example, have recorded fine sediment accumulation rates of 0.003–0.071 m year^{-1} within cut-offs in Wales and the borderlands. However, the precise nature of alluvial lithofacies type (Miall, 1983) is controlled primarily by the mode of formation. Palaeochannels formed by chute and neck cut-offs tend to contain fine-grained and organic sediments (Bridge, 1985; Waters, 1992) whereas those created by avulsion and mobile bar cut-off will have variable stratigraphic characteristics dependent on the cut-off angle and the relative elevation of the abandoned channel.

A number of potential problems need to be taken into consideration, however, when examining pollen records from palaeochannel contexts (Brown, 1997). In particular, the re-mobilisation and secondary deposition of polleniferous sediment, notably during flood events, may introduce both temporal and spatial error into samples (Martin, 1963; Tschudy, 1969; Fall, 1987; Burrin and Scaife, 1988). Assessment of the preservation state of pollen grains may offer a means of identifying reworked materials and is being approached in this study via an adaptation of Cushing's (1967) methodology described above.

A further source of potential confusion arises when interpreting the significance of the appearance of certain taxa within the pollen record. A number of opportunistic herbaceous taxa invade disturbed ground associated with river banks and can be interpreted as indicators of anthropogenic impact. However, the dynamics of the palaeochannel sedimentary environment may lead to natural fluctuations in these taxa, for instance, due to bank collapse and other natural processes. Further statistical analyses of the vegetation with respect to sedimentary logs and other proxy data sources may assist in establishing causal mechanisms for certain species fluctuations.

Summarised pollen diagrams are presented here in Figures 15.8, 15.9 and 15.10, with composite curves generated to highlight anthropochorous taxa. The constituent species of these curves have been based upon classifications of cultivated plants and indicator species from Behre (1981) and Birks (1990). Full details of the taxa which contribute to each of the summary curves can be found within the Appendix to this chapter.

Pollen Records from the River Rede at Brownchesters, T5 Palaeochannel

Figure 15.8 spans channel fill sediments between 60 and 440 cm in depth. ^{14}C assays at depths of 345 cm and 140 cm returned dates of 4055–3715 BC and 2570–2145 BC respectively (Table 15.1). Figure 15.8 shows a high but fluctuating quantity of arboreal taxa, including the thermophilous tree *Quercus* and *Ulmus* which attain a combined percentage of around 30%. *Alnus* percentages show the greatest degree of variability, with high values frequently associated with large peaks in pollen concentration which reflect the local incidence of this taxa. *Pinus* pollen attains values approaching 20% in the lower levels in the core, an amount usually indicative of regional as opposed to local presence (Huntley and Birks, 1983). However, owing to the probable small pollen source area for this sedimentary context it is likely that *Pinus* was growing somewhere relatively proximal, although not in significant quantities at this time. The peak in *Betula* at the top of the diagram is somewhat anomalous, resulting from an extremely local presence which has caused a large increase in pollen concentration. Attention is drawn here to the presence of anthropogenic indicator species, allied with incidences of cultivated taxa throughout the core (Figure 15.8). This suggests a continuous anthropogenic impact upon the valley floor environment since late Mesolithic times.

Pollen concentrations vary quite markedly throughout the sediment core, correlating reasonably well with percentage loss on ignition (LOI) values. This indicates that the pollen concentration is higher in the more organic sediment – a function of the slower accumulation rate for this type of sediment and also possibly the better preservation of palynomorphs. The classification of pollen deterioration types also appears to be correlated to the LOI values although further statistical work is required to validate this relationship.

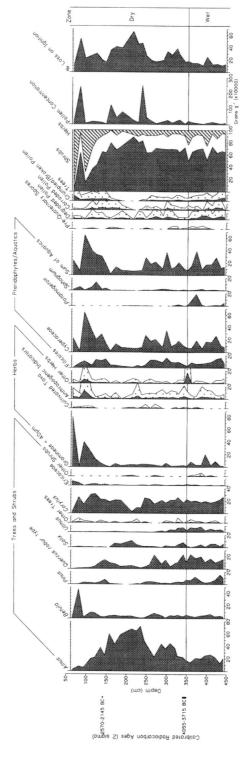

Figure 15.8 Pollen diagram for the T5 palaeochannel, Brownchesters Farm, Rede valley. Exaggeration × 5

Pollen Records from the River Rede at Brownchesters, T6 Palaeochannel

Figure 15.9 spans channel fill sediments between 80 and 330 cm in depth. ^{14}C assays at depths of 345 cm and 160 cm have returned dates of 2585–2140 BC and 1890–1520 BC respectively (Table 15.1), and hence the basal part of this sequence broadly overlaps with the uppermost part of the T5 core. Overall, Figure 15.9 shows a very similar pattern to the T5 diagram (Figure 15.6), with a large peak in *Alnus* pollen in the middle of the sequence. Other tree species do not attain significantly large percentages with the exception of *Quercus* in the lower, mesic zone of Figure 15.9 and *Corylus*, which remains fairly constant at a level of 20%. Significant quantities of anthropogenic indicator species and cultivated plants again occur throughout the diagram and, in combination with a large quantity (> 60%) of grasses in the top of the core, indicates a significant human impact upon valley floor vegetation during the period approximately 2500–1500 BC.

Pollen concentration values in Figure 15.9 reach extremely high levels associated with the peak in *Alnus* values. This can be attributed to pollen deposition from the *in situ* vegetation as opposed to a large decrease in accumulation rates, as the whole diagram spans a period of only *c.* 1000 years. Pollen concentration values also correlate extremely well with LOI values and inversely with Filicales percentages, a proxy indicator of post-depositional pollen preservation (Cundill and Whittington, 1987; Tipping *et al.*, 1994).

Pollen Records from the River North Tyne at Snabdaugh, T5 Palaeochannel

The pollen record from Snabdaugh spans a period ^{14}C dated to between 760–210 BC and AD 950–1260. Figure 15.10 shows extremely low percentages of arboreal taxa, with the exception of peaks in *Alnus* and *Salix*, which can be attributed to *in situ* sources. The quantity of anthropogenic indicator species reaches over 40% in this diagram, which, allied with a continuous presence of cultivated plants, is suggestive of fairly intensive utilisation of the valley floor environment for agricultural purposes. Late Iron Age human activity at this site is also attested by the recovery of a worked wooden plank, dated to 760–210 BC, from near the base of this core (see above).

Preliminary comparison of the pollen deterioration index with sediment logs for all palaeochannel sequences described here suggests that high proportions of degraded pollen broadly correlate with inorganic sediments, reflecting relatively high energy depositional environments associated with flood episodes. This would indicate that a relatively small percentage of the pollen within the palaeochannels has undergone fluvial transport and subsequent secondary deposition. Further statistical work is to be undertaken to test this potential relationship.

DISCUSSION

Anthropogenic Impact on Valley Floor Environments

The results of pollen analysis from these sites has demonstrated the influence of anthropogenic factors on the vegetation of valley floors in the North Tyne Basin over

300

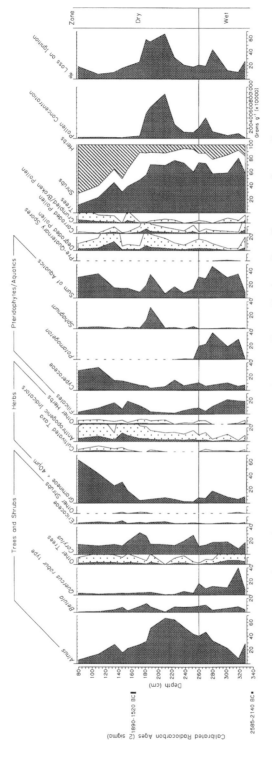

Figure 15.9 Pollen diagram for the T6 palaeochannel, Brownchesters Farm, Rede valley. Exaggeration × 5

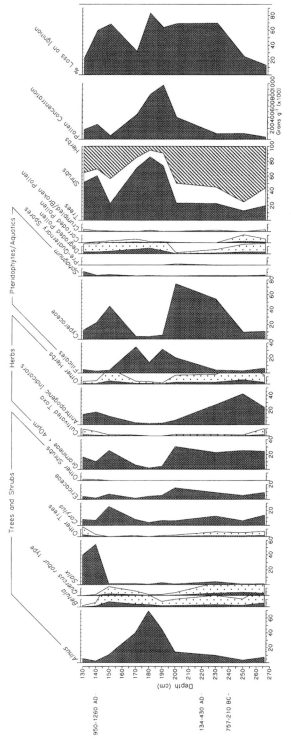

Figure 15.10 Pollen diagram for the T5 palaeochannel, Snabdaugh Farm, North Tyne valley. Exaggeration × 5

an almost continuous period between the late 5th millennium BC and late medieval times. Indeed, all the pollen diagrams presented show markedly higher percentages of non-arboreal and anthropochorous taxa than published regional diagrams (Davies and Turner, 1979; Barber *et al.*, 1993; Dumayne, 1993a, 1993b; Dumayne and Barber, 1994) and investigations carried out by one of the authors (Moores, unpublished data). This suggests that alluvial valley floors have been an important focus for human activity, including both arable and pastoral agriculture, since the late Mesolithic period.

Evidence for comparatively early human activity is particularly important given that archaeological records of a Mesolithic presence comprise only a few lithic scatters on regional hillslopes and higher river terraces of the lower Tyne (Tolan-Smith, 1996, 1997), although various authors have suggested that river and lake edges are likely to have been favoured by Mesolithic communities (Weyman, 1984). However, it remains unclear whether these patterns reflect preferential utilisation of the valley floor by past societies or rather the insensitivity of regional pollen diagrams to local anthropogenic changes in the vegetation.

It is also interesting to note that the long record of valley floor vegetation disturbance at Brownchesters does not seem to have been associated with especially high rates of lateral channel migration and reworking of older Holocene alluvial fills, particularly in comparison to documented sites elsewhere in the Tyne Basin (Passmore and Macklin, 1997; Macklin *et al.*, 1998). This runs counter to widespread expectations of such a geomorphological response (e.g. Brown, 1997; Passmore and Macklin, 1997), although it is possible that *riparian* vegetation may have been less affected by clearance. Relatively low valley floor gradients, allied possibly to relatively cohesive fine-grained channel banks, may also have acted to restrict lateral channel erosion.

Vegetation Succession in Palaeochannel Contexts

The pollen records from the sites studied at Brownchesters and Snabdaugh also allow a reconstruction of the hydroseral vegetation succession within the palaeochannels (Janssen *et al.*, 1995; Brown, 1996). For instance, the *Potamageton* curves from the Brownchesters cores have been used for the wet/dry zonation of the pollen diagrams (Figures 15.8 and 15.9). This is due to the ecological requirements of the species for standing water and the fact that the pollen which is produced within the water body is extremely fragile and unlikely to survive transportation. Therefore, the presence of *Potamageton* can be used to infer that the palaeochannel existed as a cut-off oxbow feature during the period where this taxa occurs in the record. This is supported by the sedimentological evidence since when the *Potamageton* curve ends, the sediment becomes more consistently organic, with minerogenic laminae only present where floods have encroached the former channel.

In addition, the point at which the *Potamageton* record terminates corresponds to an increase in *Alnus* percentages. This can be used to model the succession of alder into the newly terrestrialised palaeochannel, the progressive drying out and slight succession of other tree species and then the removal of woodland from the palaeochannel and the increase in grass and herbaceous species. Janssen *et al.* (1995) have related hydroseral succession to groundwater-level fluctuations, and have highlighted periods of vegetation regression characterised by swamp vegetation replacing alder carr. At Brownchesters T5 and T6, alder carr is replaced in the uppermost levels of the analysed sedimentary sequences (Figures 15.8 and 15.9), but the species associated with this indicate an anthropogenic cause rather than one associated solely with groundwater-level fluctuation.

Geoarchaeology of Valley Floor Environments in the North Tyne Basin

As well as providing extended palaeoecological records of valley floor human activity, these investigations are extending geoarchaeological models of landform, sedimentary and archaeological associations in regional alluvial environments (Macklin et al., 1992c; Passmore and Macklin, 1997). As is the case throughout upland river valleys, the fragmentary nature of Holocene fluvial terrace survival will have led to partial or complete erosion of prehistoric and historic archaeological landscapes in the North Tyne Basin. However, comparatively restricted rates of lateral reworking in the middle reaches of these valley floors point to the greater likelihood of older Holocene surfaces remaining at least partially intact. Preservation of archaeological landscapes in these environmental settings will also have been promoted by the susceptibility of palaeochannels, floodplains and possibly also fluvial terrace surfaces to burial beneath vertically accreted, fine-grained overbank alluvium. Persistent waterlogging of alluvial deposits will have ensured long-term survival of organic archaeological materials such as wood and leather, as has already been demonstrated by the recovery of an Iron Age worked wooden plank from alluvial sediments at Snabdaugh Farm (see above). However, deeply buried archaeological contexts are invisible to surface survey and require evaluation by geoarchaeological techniques.

Utility of Palaeochannel Fills for Reconstruction of Holocene Flood Histories

Floods, and particularly moderate to large-scale events, play a critical role in fluvial sediment transfers and the development of channel and floodplain environments, and in recent years investigations of palaeoflood hydrology have gained impetus with the recognition that long-term palaeohydrological data will assist the forecasting and management of river channel and floodplain environmental change (Knox, 1995). Particular attention has focused on slackwater facies in arid environments (e.g. Baker et al., 1983; Baker and Pickup, 1987; Webb et al., 1988; Enzel et al., 1993) where, in the context of vertically stable bedrock channels, it has proved possible to reconstruct former flood magnitude and frequency. In temperate regions such as north-west Europe, by contrast, estimates of flood frequency and magnitude over Holocene timescales are typically hindered by the paucity of well-dated Holocene alluvial sequences that preserve well-defined sedimentary evidence of individual flood events (Macklin and Needham, 1992; Macklin et al., 1992b). Here, episodes of enhanced Holocene flooding tend to be indirectly inferred from broadly synchronous periods of increased fluvial activity reflected, for example, in meander avulsions (cf. Starkel and Kalicki, 1996) and episodes of basin-wide alluviation and/or channel incision (Macklin et al., 1998).

Lithostratigraphic analysis of organic-rich palaeochannel fills does, however, offer a potential means of accessing readily datable sedimentary records of flood frequency for moderate to large-scale events that temporarily re-occupy former watercourses. These will be manifested as inorganic textural reversals in fining-upward and/or peaty channel-fill sequences (Erskine *et al.*, 1992; Malik and Khadkikar, 1996). At Brownchesters, thick deposits of interbedded and often finely laminated peats, organic-rich silts and inorganic sands and silts potentially offer high-resolution flood frequency records for the River Rede from the early to mid-Holocene period. Preliminary work suggests peaty deposits in palaeochannel core T5 include, for example, at least three major flood events (reflected as discrete 1–5 cm thick beds of fine sand) in the period shortly before and during *c.* 4055–3715 BC, and two phases of enhanced flood frequency (reflected as laminated minerogenic fine sands and silts and organic-rich silts) between this time and *c.* 2570–2145 BC (Figure 15.4). A systematic evaluation of these and other palaeochannel deposits at Brownchesters is the focus of ongoing research.

Chronology and Controls of Holocene Fluvial Activity

By comparison to documented Holocene alluvial histories elsewhere in northern and western Britain (Macklin and Lewin, 1993), the extended chronology of alluvial sequences in middle reaches of the North Tyne Basin afforded by well-dated palaeochannels offers an unusually detailed perspective on prehistoric and early historic valley floor development in an upland British river. Episodes of major channel abandonment appear for the most part to be non-synchronous between the study reaches, although during the mid-1st millennium AD (Brownchesters) and mid-1st millennium BC (Snabdaugh) (Figure 15.7(a) and (b)) these discontinuities are broadly coincident with phases of fluvial adjustment recorded elsewhere in the Tyne Basin. These periods have been previously correlated with independent climatic records (e.g. derived from regional peat stratigraphy; cf. Barber, 1981, 1982, 1985, 1994; Barber *et al.*, 1994) indicating shifts to cooler and/or wetter conditions (Macklin *et al.*, 1992b, 1998; Passmore *et al.*, 1993; Passmore, 1994). Furthermore, channel abandonment and incision evident at both Snabdaugh and Brownchesters shortly before *c.* 5000 cal BC (Figure 15.7(a) and (b)) corresponds with an early neoglacial phase centred on *c.* 5000 BC identified by Meese *et al.* (1994) and high precipitation in the Cairngorms, Scotland recorded by Dubois and Ferguson (1985). In general, however, the explanation of non-synchronous fluvial activity between the study reaches may lie in reach-specific geomorphological factors or complex-response mechanisms (cf. Schumm, 1973) that have the potential to invoke autogenic channel adjustments (e.g. meander avulsion) or to render valley floors insensitive to environmental perturbations.

At basin scales a striking contrast is also evident between long-term alluvial chronologies in the North Tyne and South Tyne valleys (Figure 15.7); the latter shows comparatively little evidence of fluvial activity for much of the prehistoric period, but is characterised by a greater frequency of vertical and lateral channel and floodplain adjustments over the last 2500 years (Figure 15.4). Destabilisation of the South Tyne catchment during the later prehistoric and early historic period has also been recorded in several upland tributaries (Macklin *et al.*, 1992c, 1998; Passmore *et al.*, 1993) and, while in part reflecting the greater likelihood of later Holocene valley fills to be preserved in alluvial sedimentary records, most probably reflects the relatively late large-scale anthropogenic disturbance of catchment vegetation and soil cover evidenced by regional palynological and archaeological records (e.g. Davies and Turner, 1979; Higham, 1986; Passmore and Macklin, 1997). Longer and more intensive records of later Mesolithic and subsequent human activity exist for the North Tyne catchment (see above) and are currently being augmented by palynological investigations of channel fills in both study reaches described here. Disturbance of catchment and valley floor vegetation cover in the North Tyne Basin may therefore have promoted autogenic adjustment of channels and floodplains, as well as rendering regional river environments more sensitive to prehistoric climate change (Passmore and Macklin, 1997).

CONCLUSIONS

Investigations of Holocene fluvial activity and vegetation change in the middle reaches of the Rivers North Tyne and Rede, northern England, are yielding unusually high-resolution records of post-glacial environmental change. Although investigations are still at an early stage, available dating controls and morphostratigraphic relationships of alluvial fills indicate that these reaches, especially in Redesdale, will offer insights into rates and patterns of fluvial activity, including the frequency and impact of moderate–large flood events, over extended Holocene timescales. Long-term preservation of older Holocene terraces and their associated organic-rich palaeochannel fills most probably reflects the combination of relatively restricted rates of lateral reworking and a long-term tendency towards slight aggradation of valley floors.

Palaeoecological analyses of Holocene channel fill sequences are significantly enhancing our knowledge of prehistoric human activity in these valley floors and, especially in terms of the timing of this activity, have highlighted discrepancies between vegetation reconstructions on regional and local scales. Valley floors appear to have been subject in particular to clearance and possibly cultivation from late Mesolithic times, which is significantly earlier than previously indicated by existing palynological or archaeological records. Alluvial pollen records are also elucidating patterns of vegetation succession in abandoned river channels. It is anticipated that these and ongoing studies (including analyses of plant macrofossils and microscopic charcoal particles) will extend this research with a view to modelling long-term interactions between vegetation changes, human activity and channel and floodplain geomorphology.

ACKNOWLEDGEMENTS

The co-operation and the enthusiasm of the Allgood family at Snabdaugh Farm and the Davison family at Brownchesters Farm are gratefully acknowledged. The Bill Bishop Memorial Trust provided a contribution towards radiocarbon dating. Thanks also to Ann Rooke for drawing some of the figures and Lucy Ellis for permission to use her photos. Finally, thanks are extended to all the people who have assisted with fieldwork at the study sites, particularly Lucy Ellis, Mark Williams and Vanessa Thornes.

REFERENCES

Adams, M. 1996. Setting the scene: the Mesolithic in northern England. *Northern Archaeology* (Special Edition: Neolithic Studies in No-Mans Land), **13/14**, 1–5.
Baker, V. R. and Pickup, G. 1987. Flood geomorphology of the Katherine Gorge, Northern Territory, Australia. *Geological Society of America Bulletin*, **98**, 635–646.
Baker, V. R., Kochel, R. C., Patton, P. C. and Pickup, G. 1983. Palaeohydrologic analysis of Holocene flood sequence slack-water sediments. In Collinson, J. P. and Lewin, J. (Eds) *Modern and Ancient Fluvial Systems*. International Association of Sedimentologists, 229–239.
Barber, K. E. 1981. *Peat Stratigraphy and Climatic Change: A Palaeoecological Test of the Theory of Cyclic Peat Bog Regeneration*. Balkema, Rotterdam.
Barber, K. E. 1982. Peat-bog stratigraphy as a proxy climate record. In Harding, A. F. (Ed.) *Climatic Change in Later Prehistory*. Edinburgh University Press, 103–113.
Barber, K. E. 1985. Peat stratigraphy and climatic change: some speculations. In Tooley, M. J. and Sheail, G. M. (Eds) *The Climatic Scene*. Allen and Unwin, London, 175–185.
Barber, K. E. 1994. Deriving Holocene palaeoclimates from peat stratigraphy: some misconceptions regarding the sensitivity and continuity of the record. *Quaternary Newsletter*, **72**, 1–9.
Barber, K. E., Dumayne, L. and Stoneman, R. 1993. Climatic change and human impact during the late Holocene in northern Britain. In Chambers, F. M. (Ed.) *Climate Change and Human Impact on the Landscape*. Chapman and Hall, London, 225–236.
Barber, K. E., Chambers, F. M., Maddy, D., Stoneman, R. and Brew, J. S. 1994. A sensitive high resolution record of late Holocene climatic change from a raised bog in northern England. *The Holocene*, **4**(2), 198–205.
Behre, K. E. 1981. The interpretation of anthropogenic indicators in polled diagrams. *Pollen et Spores*, **23**, 225–245.
Bell, M. 1992. Archaeology under alluvium: human agency and environmental processes. Some concluding thoughts. In Needham, S. and Macklin, M. G. (Eds) *Alluvial Archaeology in Britain*. Oxbow Monograph No. 27, Oxbow Books, Oxford, 271–276.
Birks, H. J. B. 1970. Inwashed pollen spectra at Loch Fada, Isle of Skye. *New Phytologist*, **69**, 807–820.
Birks, H. J. B. 1990. Indicator values of pollen types from post-6000 BP pollen assemblages from southern England and southern Sweden. *Quaternary Studies in Poland*, **10**, 21–31.
Bridge, J. S. 1985. Palaeochannel patterns inferred from alluvial deposits: a critical evaluation. *Journal of Sedimentary Petrology*, **55**, 579–589.
Brown, A. G. 1996. Floodplain palaeoenvironments. In Anderson, M. G., Walling, D.E. and Bates, P. D. (Eds) *Floodplain Processes*. John Wiley, Chichester, 95–138.
Brown, A. G. 1997. *Alluvial Geoarchaeology: Floodplain Archaeology and Environmental Change*. Cambridge Manuals in Archaeology, Cambridge University Press.
Brown, A. G. and Keough, M. 1992. Holocene floodplain metamorphosis in the Midlands, United Kingdom. *Geomorphology*, **4**, 433–445.
Burrin, P. J. and Scaife, R. G. 1988. Environmental thresholds, catastrophe theory and landscape sensitivity: their relevance to the impact of man on valley alluviations. In Bintlif, J. L., Davidson, D. A. and Grant, E. G. (Eds) *Conceptual Issues in Environmental Archaeology*. Edinburgh University Press, 211–232.
Clark, A. J. 1992. Magnetic dating of alluvial deposits. In Needham, S. and Macklin, M. G. (Eds) *Alluvial Archaeology in Britain*. Oxbow Monograph 27, Oxford, 37–42.
Clark, A. J., Tarling, D. H. and Noel, M. 1988. Developments in archaeomagnetic dating in Britain. *Journal of Archaeological Science*, **15**, 645–667.
Cundill, P. and Whittington, G. 1987. Flood-plain sedimentation in the Upper Axe valley, Mendip England: a comment. *Transactions of the Institute of British Geographers*, **12**, 360–362.
Cushing, E. J. 1967. Evidence for differential pollen preservation in late Quaternary sediments in Minnesota. *Review of Palaeobotany and Palynology*, **4**, 87–101.

Davies, G. and Turner, J. 1979. Pollen diagrams from Northumberland. *New Phytologist*, **82**, 783–804.

Dubois, A. D. and Ferguson, D. K. 1985. The climate history of pine in the Cairngorms based on radiocarbon dates and stable isotope analysis, with an account of the events leading up to its colonisation. *Review of Palaeobotany and Palynology*, **46**, 55–80.

Dumayne, L. 1992. Late Holocene palaeoecology and human impact on the environment of northern Britain. PhD Thesis, Southampton.

Dumayne, L. 1993a. Invader or native-vegetation clearance in northern Britain during Romano British time. *Vegetation History and Archaeobotany*, **2**, 29–36.

Dumayne, L. 1993b. Iron Age and Roman vegetation clearance in northern Britain: further evidence. *Botanical Journal of Scotland*, **46**, 385–392.

Dumayne, L. and Barber, K. E. 1994. The impact of the Romans on the environment of northern England: pollen data from three sites close to Hadrian's Wall. *The Holocene*, **4**, 165–173.

Edwards, K. J. 1991a. Spatial scale and palynology: a commentary on Bradshaw. In Harris, D. R. and Thomas, K. D. (Eds) *Modelling Ecological Change: Perspectives from Neoecology, Palaeoecology and Environmental Archaeology*. Institute of Archaeology, 53–59.

Edwards, K. J. 1991b. Using space in cultural palynology: the value of the of-site pollen record. In Harris, D. R. and Thomas, K.D. (Eds) *Modelling Ecological Change: Perspectives from Neoecology, Palaeoecology and Environmental Archaeology*. Institute of Archaeology, 61–74.

Edwards, K. J. and McIntosh, C. J. 1988. Improving the detection rate of cereal-type pollen grains from Ulmus decline deposits from Scotland. *Pollen et Spores*, **30**, 179–188.

Enzel, Y., Ely, L., House, P. K., Baker, V. R. and Webb, R. H. 1993. Palaeoflood evidence for a natural upper bound to flood magnitudes in the Colorado River basin. *Water Resources Research*, **29**, 2287–2297.

Erskine, W., McFadden, C. and Bishop, P. 1992. Alluvial cut-offs as indicators of former channel conditions. *Earth Surface Processes and Landforms*, **17**, 23–37.

Evans, J. G., Limbrey, S., Mate, I. and Mount, R. J. 1988. Environmental change and land-use history in a Wiltshire river valley in the last 14000 years. In Barrett, J. C. and Kinnes, I. A. (Eds) *The Archaeology of Context in the Neolithic and Bronze Age: Recent Trends*. Department of Archaeology and Prehistory, University of Sheffield, 97–103.

Fall, P. L. 1987. Pollen taphonomy in a canyon stream. *Quaternary Research*, **28**, 393–406.

Harvey, A. M. 1985. The river systems of North-west England. In Johnson, R. H. (Ed.) *The Geomorphology of North-West England*. Manchester University Press, Manchester, 122–142.

Harvey, A. M., Alexander, R. W. and James, P. A. 1984. Lichens, soil development and the age of Holocene valley floor landforms: Howgill Fells, Cumbria. *Geografiska Annaler*, **66A**, 353–366.

Higham, N. J. 1986. *The Northern Counties to AD 1000*. Longman, Harlow.

Hooke, J. M., Harvey, A. M., Miller, S. Y. and Redmond, C. E. 1990. The chronology and stratigraphy of the alluvial terraces of the River Dane valley, Cheshire, N.W. England. *Earth Surface Processes and Landforms*, **15**, 717–737.

Huntley, B. and Birks, H. J. B. 1983. *An Atlas of Past and Present Pollen Maps for Europe: 0–13,000 Years Ago*. Cambridge University Press, Cambridge.

Jacobsen, G. L. and Bradshaw, R. H. W. 1981. The selection of sites for palaeoecological studies. *Quaternary Research*, **16**, 80–96.

Janssen, C. R. 1973. Local and regional pollen deposition. In Birks, H. J. B. and West, R. G. (Eds) *Quaternary Plant Ecology*. Blackwell Scientific, Oxford, 31–42.

Janssen, C. R. 1981. On the reconstruction of past vegetation by pollen analysis. *Botany Proceedings C*, **84**, 197–209.

Janssen, C. R. 1986. The use of local pollen indicators and of the contrast between regional and local pollen values in the assessment of the human impact on the vegetation. In Behre, K. E. (Ed.) *Anthropogenic Indicators in Pollen Diagrams*. A.A. Balkema, Rotterdam, 203–208.

Janssen, C. R., Berendsen, H. J. A. and van Broekhuizen, A. J. D. 1995. Fluvial activity and vegetation development 4000–2000 BP in south-western Utrecht, the Netherlands. *Mededelingen Rijkf Geologische Dienst*, **52**, 357–367

Jobey, I. and Jobey, G. 1988. Gowanburn River Camp: an Iron Age, Romano-British and more recent settlement in North Tynedale, Northumberland. *Archaeologia Aeliana*, Series 5, **16**.

Knox, J. C. 1995. Fluvial systems since 20,000 years BP. In Gregory, K. J., Starkel, L. and Baker, V. R. (Eds) *Global Continental Palaeohydrology*. John Wiley, Chichester, 87–108.

Lewis, G. W. and Lewin, J. 1983. Alluvial cut-offs in Wales and Borderlands. In Collinson, J. P. and Lewin, J. (Eds) *Modern and Ancient Fluvial Systems*. International Association of Sedimentologists, 145–154.

Macklin, M. G. and Lewin, J. 1986. Terraced fills of Pleistocene and Holocene age in the Rheidol Valley, Wales. *Journal of Quaternary Science*, **1**, 21–34.

Macklin, M. G. and Lewin, J. 1993. Holocene river alluviation in Britain. *Zeitschrift für Geomorphologie* (Supplement), **88**, 109–122.

Macklin, M. G. and Needham, S. 1992. Studies in British alluvial archaeology: potential and prospect. In Needham, S. and Macklin, M. G. (Eds) *Alluvial Archaeology in Britain*. Oxbow Monograph, Oxford, 9–23.

Macklin, M. G., Passmore, D. G. and Rumsby, B. T. 1992a. Climatic and cultural signals in Holocene alluvial sequences: the Tyne Basin, Northern England. In Needham, S. and Macklin, M. G. (Eds) *Alluvial Archaeology in Britain*. Oxbow Monograph, Oxford, 123–140.

Macklin, M. G., Rumsby, B.T. and Newson, M. D. 1992b. Historical floods and vertical accretion of fine grained alluvium in the Lower Tyne Valley, Northeast England. In Billi, P., Hey, R. D., Thorne, C. R. and Tacconi, P. (Eds) *Dynamics of Gravel-Bed Rivers*. John Wiley, Chichester, 573–589.

Macklin, M. G., Passmore, D. G., Cowley, D. C., Stevenson, A. C. and O'Brien, C. F. 1992c. Geoarchaeological enhancement of river valley archaeology in North East England. In Spoerry, P. (Ed.) *Geoprospection in the Archaeological Landscape*. Oxbow Monograph No.18, Oxford, 43–58.

Macklin, M. G., Passmore, D. G. and Newson, M. D. 1998. Controls of short and long term river instability: processes and patterns in gravel-bed rivers, the Tyne basin, northern England. *Proceedings of the 4th International Gravel Bed Rivers Conference*, Washington State, in press.

Malik, J. N. and Khadkikar, A. S. 1996. Palaeoflood analysis of channel-fill deposits, central Tapti river basin, India. *Zeitschrift für Geomorphologie*, **106**, 99–110.

Martin, P. S. 1963. *The Last 10,000 Years: A Fossil Pollen Record of the American South-west*. University of Arizona Press.

Meese, D. A., Gow, A. J., Grootes, P., Mayewski, P. A., Ram, M., Stuiver, M., Taylor, K. C., Waddington, E. D. and Zielinski, G. A. 1994. The accumulation record from the GISP2 Core as an indicator of climate change throughout the Holocene. *Science*, **266**, 1680–1682.

Miall, A. D. 1983. Basin analysis of fluvial sediments. In Collinson, J. P. and Lewin, J. (Eds) *Modern and Ancient Fluvial Systems*. International Association of Sedimentologists, 279–286.

Moore, P. D., Webb, J. A. and Collinson, M. E. 1991. *Pollen Analysis*. Blackwell Scientific, Oxford, 216.

Munsterman, D. and Kerstholt, S. 1996. Sodium polytungstate, a new non-toxic alternative to bromoform in heavy liquid separation. *Review of Palaeobotany and Palynology*, **91**, 417–422.

Newson, M. D. 1989. Flood effectiveness in river basins: progress in Britain in a decade of drought. In Beven, K. and Carling, P. A. (Eds) *Floods: Geomorphological, Hydrological and Sedimentological Implications*. John Wiley, Chichester, 151–169.

Newson, M. D. and Lewin, J. 1991. Climatic change, river flow extremes and fluvial erosion – scenarios for England and Wales. *Progress in Physical Geography*, **15**, 1–17.

Noel, M. 1991. An archaeomagnetic study of floodplain sediments from the North and South Tyne. Unpublished report by Archaeometrics, Archaeological Scientific Services, Wolsingham, County Durham.

Passmore, D. G. 1994. River response to Holocene environmental change: the Tyne Basin, northern England. Unpublished PhD Thesis, University of Newcastle Upon Tyne.

Passmore, D. G. and Macklin, M. G. 1994. Provenance of fine-grained alluvium and late Holocene land-use change in the Tyne basin, northern England. *Geomorphology*, **9**, 127–142.

Passmore, D. G. and Macklin, M. G. 1997. Geoarchaeology of the Tyne basin: Holocene river valley environments and the archaeological record. In Tolan-Smith, C. (Ed.) *Landscape*

Archaeology in Tynedale. Department of Archaeology, University of Newcastle upon Tyne, 11–27.
Passmore, D. G., Macklin, M. G., Stevenson, A. C., O'Brien, C. F. and Davis, B. A. S. 1992. A Holocene alluvial sequence in the lower Tyne valley, northern Britain: a record of river response to environmental change. *The Holocene*, **2**, 138–147.
Passmore, D. G., Macklin, M. G., Brewer, P. A., Lewin, J., Rumsby, B. T. and Newson, M. D. 1993. Variability of late Holocene braiding in Britain. In Best, J. L. and Bristow, C. S. (Eds) *Braided Rivers.* Geological Society, London, 205–232.
Prentice, I. C. 1988. Records of vegetation in time and space: the principles of pollen analysis. In Huntley, B. and Webb, T. III. (Eds) *Vegetation History. Handbook of Vegetation Science.* Kluwer Academic, Amsterdam, 17–42.
Robertson-Rintoul, M. S. E. 1986. A quantitative soil-stratigraphic approach to the correlation and dating of post-glacial river terraces in Glen Feshie, western Cairngorms. *Earth Surface Processes and Landforms*, **11**, 605–617.
Rumsby, B. T. and Macklin, M. G. 1994. Channel and floodplain response to recent abrupt climate change: the Tyne Basin, northern England. *Earth Surface Processes and Landforms*, **191**, 499–515.
Schumm, S. A. 1973. Geomorphic thresholds and complex response of drainage systems. In Morisawa, M. (Ed.) *Fluvial Geomorphology.* SUNY, Binghampton Publications in Geomorphology, 299–309.
Starkel, L. and Kalicki, T. (Eds) 1996. *Evolution of the Vistula River Valley during the Last 15,000 Years. Part VI.* Polish Academy of Sciences, Geographical Studies Special Issue No. 9.
Stockmarr, J. 1971. Tablets with spores used in absolute pollen analysis. *Pollen et Spores*, **13**, 614–621.
Stuiver, M., Long, A., Kra, R. S. and Devine, J. M. 1993. Calibration-1993. *Radiocarbon*, **35**.
Tauber, H. 1965. Differential pollen dispersion and the interpretation of pollen diagrams. *Danm Geols Unders*, **89**.
Taylor, M. P. and Lewin, J. 1996. River behaviour and Holocene alluviation: the River Severn at Welshpool, Mid-Wales, UK. *Earth Surface Processes and Landforms*, **21**, 77–91.
Taylor, M. P. and Lewin, J. 1997. Non-synchronous response of adjacent floodplain systems to Holocene environmental change. *Geomorphology*, **18**, 251–264.
Tipping, R. M. 1992. The determination of cause in the generation of major prehistoric valley fills in the Cheviot Hills, Anglo-Scottish Border. In Needham, S. and Macklin, M. G. (Eds) *Alluvial Archaeology in Britain.* Oxbow Monograph, Oxford, 111–121.
Tipping, R. M. 1994. Late prehistoric fluvial activity on the River Breamish at Powburn. In Harrison, S. and Tipping, R. (Ed.) *The Geomorphology and Late Quaternary Evolution of the Cheviot Hills.* British Geomorphological Research Group Field Guide, 38–50.
Tipping, R. M. 1995. Holocene evolution of a lowland Scottish landscape: Kirkpatrick Fleming. Part III, fluvial history. *The Holocene*, **5**, 184–195.
Tipping, R. M. 1996. The Neolithic landscapes of the Cheviot Hills and hinterland: palaeoenvironmental evidence. *Northern Archaeology* (Special Edition: Neolithic Studies in No-Mans Land), **13/14**, 17–33.
Tipping, R. M. and Halliday, S. P. 1994. The age of alluvial fan deposition at a site in the southern uplands of Scotland. *Earth Surface Processes and Landforms*, **19**, 333–348.
Tipping, R. M., Carter, S. and Johnston, D. 1994. Soil pollen and soil micromorphological analyses of old ground surfaces on Biggar Common, Borders region, Scotland. *Journal of Archaeological Science*, **21**, 387–401.
Tolan-Smith, C. 1996. The Mesolithic–Neolithic transition in the lower Tyne valley: a landscape approach. *Northern Archaeology* (Special Edition: Neolithic Studies in No-Mans Land), **13/14**, 7–15.
Tolan-Smith, C. 1997. The Stone Age landscape: the contribution of fieldwalking. In Tolan-Smith, C. (Ed.) *Landscape Archaeology in Tynedale.* Department of Archaeology, University of Newcastle upon Tyne, 79–90.
Topping, P. 1989. Early cultivation in Northumberland and the Borders. *Proceedings of the Prehistoric Society*, **55**, 161–179.

Tschudy, R. H. 1969. Relationship of palynomorphs to sedimentation. In Tschudy, R. H. and Scott, R. A. (Eds) *Aspects of Palynology*. John Wiley, Chichester, 79–96.

Turner, J. 1975. The evidence for land use by prehistoric farming communities: the use of three-dimensional pollen diagrams. In Evans, J. G., Limbrey, S. and Cleere, H. (Eds) *The Effect of Man on the Landscape: The Highland Zone*. Council for British Archaeology, 86–95.

Waller, M. P. 1993. Flandrian vegetational history of south-eastern England. Pollen data from Pannel bridge, East Sussex. *New Phytologist*, **124**, 345–369.

Waters, M. 1992. *Principles of Geoarchaeology*. University of Arizona Press.

Webb, R. H., O'Connor, J. E. and Baker, V. R. 1988. Palaeohydrologic reconstruction of flood frequency on the Escalante River, South Central Utah. In Baker, V. R., Kochel, R. C. and Patton, P. C. (Eds) *Flood Geomorphology*. John Wiley, New York, 403–418.

Weyman, J. 1984. The Mesolithic in North-East England. In Miket, R. and Burgess, C. (Eds) *Between and Beyond the Walls: Essays on the Prehistory and History of North Britain in Honour of George Jobey*. John Donald, Edinburgh, 38–51.

APPENDIX

Pollen taxa contributing to summary curves in Figures 15.8, 15.9 and 15.10

Cultivated plants (Behre, 1981)

Secale, Avena/Triticum, Hordeum, Vicia

Anthropogenic indicators (Birks, 1990)

Artemisia, Bidens, Cannabis, Centaurea, Chenopodiaceae, *Cirsium,* Cruciferae, *Filipendula,* Liguliflorae, *Malus, Malva, Papaver, Plantago lanceolata, Polygonum persicaria, Potentilla,* Ranunculaceae (excluding trichophyllus type), *Rumex, Spergula, Stellaria, Urtica, Valerianella.*

16 Fluvial Processes, Land Use and Climate Change 2000 Years Ago in Upper Annandale, Southern Scotland

RICHARD TIPPING,[1] PAULA MILBURN,[1] and STRATFORD HALLIDAY[2]
[1] *Department of Environmental Science, University of Stirling, UK*
[2] *RCAHMS, Edinburgh, UK*

INTRODUCTION

The recognition that river systems in north-west Europe have been transformed by substantial environmental changes within the present interglacial, the Holocene Stage (last 11 000 cal. years), is now firmly established (Burrin and Scaife, 1988; Starkel *et al.*, 1991; Needham and Macklin, 1992; Macklin and Lewin, 1994; Gregory *et al.*, 1995; Brown, 1997). What is far from established are the driving forces behind the transformations. The "competing" hypotheses of human activity – mediated, for example, through deforestation or ploughing (Shotton, 1977; Bell and Boardman, 1992) and climate change (Vita-Finzi, 1969; Macklin and Lewin, 1994) – need not be mutually exclusive. Macklin and Needham (1992) suggested a compromise position by recognising that anthropogenic influences can modulate the effects of primary climatic forcing mechanisms. Nevertheless, this position, whilst possibly helpful as a generalism, does little to clarify the cause(s) of individual fluvial "events", and begs questions, for instance, as to whether human influence is always secondary to climate as the prime driver.

One profound difficulty in this debate is the inappropriateness of some proxy data analysed to understand the cause of past fluvial changes. Working on the belief that "small is understandable", Tipping (1995a) chose to examine a small catchment, the Kirtle Water in south-west Scotland (Figure 16.1(b)), where proxy data on climate change could be made entirely relevant to the fluvial record by being derived from the catchment itself (Tipping, 1995b), as could data on human activities through archaeological survey (Mercer, 1997) and pollen analyses depicting anthropogenic change within the catchment (Tipping, 1995c). There are difficulties with this "holistic" or integrated approach, not least with regard to temporal correlation between different data sets (Tipping and Halliday, 1994), but it does offer interpretations from data

---

Fluvial Processes and Environmental Change. Edited by A. G. Brown and T. A. Quine.
© 1999 John Wiley & Sons Ltd.

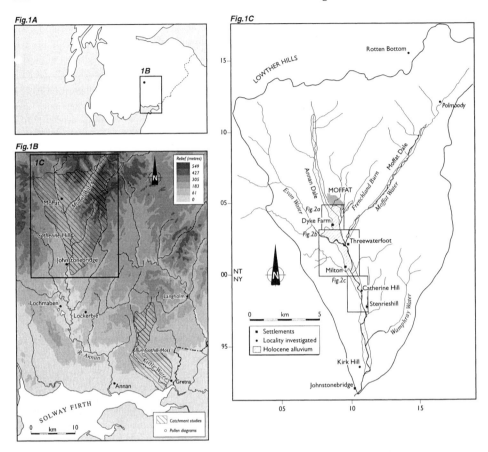

Figure 16.1 (a) Location of the study area in southern Scotland; (b) location of upper Annandale within the catchment of the River Annan, showing catchments and localities outwith the study area mentioned in the text (modified with permission from RCAHMS 1997); and (c) the study area in detail, showing the major rivers, the extent of Holocene alluvium, localities within the study area mentioned in the text and the outlines of the areas depicted in Figure 16.2

which are all collected at comparable and relevant scales, temporal and spatial. This approach is pursued in this chapter by analysing the fluvial, archaeological, land-use, vegetation and climatic records of a catchment to the north of the Kirtle Water, in upper Annandale (Figure 16.1(b) and (c)).

Over the last five years comprehensive records of each of these aspects have been generated (Tipping, 1997a; RCAHMS, 1997; Milburn and Tipping, unpublished). These cover the entire duration of the Holocene, and it is not intended to combine these analyses for this period here; this will be developed in other publications. The late Iron Age and Romano-British periods (c. 200 cal. BC to AD 410) will be analysed here because (a) there is current interest in the degree of native and Roman influence on the landscape in this region (Dumayne, 1993; Dumayne and Barber, 1994; Tipping, 1997b); (b) evidence for accelerated fluvial activity in this region over this period is common

(Bishop, 1963; Walker, 1966; Potter, 1976, 1979; Macklin *et al*., 1992; Tipping, 1992, 1995a; Dark and Dark, 1997); and (c) there are apparent difficulties in relating both climate and land-use proxy records to alluviation in these periods (Tipping, 1995a) which need further resolution and which directly address the issues raised by this volume.

UPPER ANNANDALE

The area investigated extends from the headwaters of the River Annan north and north-east of Moffat to Johnstonebridge in the south (Figure 16.1(c)), where a gorge cut in late Devensian fluvioglacial sediments inhibits the development of Holocene sediments. The main western tributary, the Evan Water, has been disrupted by reconstruction of the M74 road and has not been mapped above its confluence with the River Annan (Figures 16.1(c) and 16.2(b)). The Annan catchment drains the highest ground of the Southern Uplands, above 700–800 m OD. These hills are carved in Ordovician and Silurian shales and mudstones, with younger rocks found on the valley floor north and south of Moffat. The slopes from the Southern Uplands to the valley floors are steep, and the valley floors are deeply incised. The fall in the valley floor from north of Moffat to Johnstonebridge, however, is only around 100 m in 20 km. This is broadly a piedmont reach, where steep-gradient mountain streams can dump sediment at the abrupt break-in-slope to the valley floor.

The lower slopes of the valleys are draped in Dimlington Stadial-age till, and these are overlain on valley floors by kamiform fluvioglacial sands and gravels (May, 1981) in excess of 75 m thick. These outwash deposits present undulating surfaces with a large number of late Devensian kettle-holes. Lower and flatter fluvial/fan terraces in reworked sands and gravels are thought to date to the Loch Lomond Stadial because of intra-formational frost-wedging in an exposure in the Frenchland Burn. The Evan Water fan (Figure 16.2(b)) was active in this period (unpublished data).

The present growing season within these deep but south-facing valleys, determined by temperature, is around 220 days per year, but in the higher parts of the hills it falls to 140 days: precipitation is very high, probably in excess of 3000 mm year^{-1} in the headwaters of the Annan (Harrison, 1997). The primary Holocene vegetation of the valley floors and lower slopes was an oak–elm–hazel woodland (Tipping, 1994a, 1995c; Milburn and Tipping, unpublished). On slopes above *c*. 300 m OD the woods became thinner, with elm and oak giving way to birch and hazel (Tight, 1987), but even on the high plateau at Rotten Bottom (610 m OD; Figure 16.1(c)) an open mixed deciduous woodland probably grew at the mid-Holocene "hypsithermal" between 6500 and 5000 cal. BP (Tipping 1997a, 1997c). These woodlands were probably altered by human communities in the Mesolithic (Tipping, 1995c), and were certainly modified from the Neolithic onwards. The archaeological record (RCAHMS, 1997) has a very limited number of Mesolithic and Neolithic findspots, although this can be attributed to the difficulties of recovering such material in the uplands, and may not reflect a real absence of settlement in these periods. Bronze Age activity on lower slopes appears very dense, but many sites are burnt mounds which probably do not represent permanent settlement and agriculture. The environmental impact of these

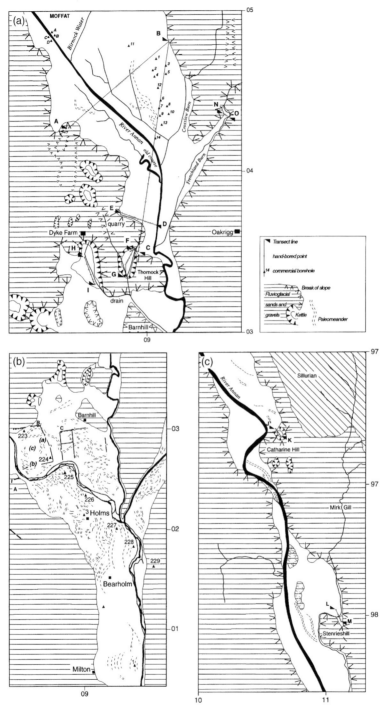

Figure 16.2 Geomorphological maps of (a) the Moffat Basin, (b) the Evan Water and (c) the River Annan south of the Threewatersfoot confluence (see Figure 16.1(c) for locations) showing transects and localities mentioned in the text. Holocene alluvium is left blank

possibly ephemeral sites might have been less than appears from the density of sites. Late Iron Age and Romano-British settlement patterns are considered later.

Holocene fluvial sediments are incised below the latest Devensian terrace. This incision, commonly of around 2.5–3 m, predates the earliest Holocene fluvial-depositional event, when the Annan eroded the north edge of a kettle-hole at Thornock Hill (Transect F–G, Figure 16.2(a)) and deposited minerogenic sediment across the site after 8323–7909 cal. BP (GU-4268). Although a very early event, pollen analyses (McKinnon, 1997) show this to be stratigraphically secure. Peat grew over the fluvial sediments after 5575–5156 cal. BP (GU-4269). Pre-Neolithic terrace formation is found at Kirkhill (Tipping, unpublished), and here alluviation also accompanied the use of a burnt mound on the Annan valley floor within the Neolithic (Pollard, 1993). In the Moffat Basin the truncated fluvioglacial surface is overlain by peat dated to 4995–4835 cal. BP (SRR-5083); therefore valley-widening occurred to some extent prior to this. Peat was replaced by overbank sedimentation after 4871–4651 cal. BP (SRR-5084), but this fluvial event had ceased by 4083–3837 cal. BP (SRR-5085). The full extent of valley-widening along transect D–E (Figure 16.2(a)) may have occurred during this phase (unpublished data). These events will be discussed in detail elsewhere.

METHODS

Fluvial Deposits and Chronology

Holocene fluvial features in this area occur as erosional channels on late Devensian sand and gravel fan surfaces (e.g. Evan Water), as low gravel terraces in upland valley floors like Moffatdale, as thin veneers of overbank silts and palaeochannels on lag gravel surfaces (e.g. the valley floors above Moffat and between Threewatersfoot and Johnstonebridge), and as thick accumulations of fine-grained overbank sediments, stacked fills (Burrin and Scaife, 1984; Macklin et al., 1991) representing periodic vertical accretion of flood sediments (e.g. in the basin south of Moffat). Terraces are present only at Kirkhill, just north of the incised gorge at Johnstonebridge (Figure 16.1(c)), but archaeological and ^{14}C dating (Pollard, 1993) show these to predate the late prehistoric period (Tipping, 1997a). The controls on these contrasting alluvial architectural styles will be evaluated elsewhere.

These different alluvial styles require different methods of analysis. Figure 16.2 is a composite of aerial-photographic interpretation and field-mapping. Holocene features are eroded by incision or valley-widening into the bordering fluvioglacial sediments, and many kettle-holes at the edges of the present valley floor show partial lateral erosion. These have been investigated by sediment-stratigraphic transects levelled to OD (Figure 16.2(a) and (c), transects F–G, H–I, J–K and L–M; unpublished data) to establish the age of erosion of kettle-hole edges by fluvial erosion and the occurrence of fluvially derived minerogenic inwash bands. The Moffat Basin sequence was examined by sediment-stratigraphic transects, levelled to OD, along transects A–B, B–C and D–E, and that at Frenchland Burn by transect N–O (Figure 16.2(a)). Commercial borehole data were combined with hand-bored Eijelkamp gouge transects.

Dating by stratigraphic association with archaeological features and ^{14}C assays was possible at Kirkhill (above) and at the Evan Water (Speller and Leslie, 1995) but the chronology in this chapter is largely constructed from a suite of 28 radiometric ^{14}C dates. These have been sampled by hand-operated large-diameter Russian corers and by power-driven Stitz corers. All assays were derived from laterally continuous stratigraphic contexts such as boundaries between peats and overlying fluvial sediments in peat-rich kettle-holes, and also from peat–overbank sediment boundaries in the stacked fill within the Moffat Basin. Peat continues to form on the floodplain around, for example, transect point E (Figure 16.2(a)), and when flooding frequencies and overbank sediment deposition were negligible peat was able to expand onto the floodplain surface. Peats overlying mineral sediments provide more secure contexts; the assumption here is that peat re-establishment occurred very soon after cessation of fluvial deposition, which seems reasonable in these sedimentary environments. Unless indicated, dates are given in cal. BP; calibrated age-ranges are expressed at 2σ.

Archaeology

The study area represents part of an intensive survey by the Royal Commission on Ancient and Historical Monuments of Scotland (RCAHMS, 1997). Methods include detailed aerial-photographic interpretation, analysis of cropmarks, intensive field-walking, surveying of upstanding monuments and documentary search. The recovery of archaeological monuments can be considered as complete as monument survival (Stevenson, 1975) and antiquarian and current research activities allow.

Vegetation History

Detailed and nearly continuous Holocene pollen records have been produced from a kettle-hole fill at Catharine Hill and from a high-level blanket peat at Rotten Bottom above Moffatdale (Tipping, 1997a). The analyses are supported by nine ^{14}C assays at Catharine Hill and 18 at Rotten Bottom. Methods of preparation and analysis were standard (see Tipping, 1995b).

Climate Reconstruction

Barber *et al.* (1994) provide a proxy record of groundwater changes (and so, it is argued, of effective precipitation) in a raised moss at Bolton Fell, near Carlisle, over the last 6000 years from plant macrofossil and colorimetric analyses. Tipping (1995b) used different techniques to obtain the same sort of record over the last 9500 years at a comparable ombrotrophic bog at Burnfoothill Moss, in lower Annandale. Colorimetric and pollen preservation data (cf. Tipping, 1995b) have been used to interpret similar changes in the water-shedding blanket peat at Rotten Bottom (unpublished data).

LATE PREHISTORIC AND EARLY HISTORIC ENVIRONMENTAL CHANGE IN UPPER ANNANDALE

Fluvial Activity

On transect D–E (Figure 16.2(a)) the deposition of minerogenic sediments began to exceed peat growth at 2718–2346 cal. BP (SRR-5086). By 659–540 cal. BP (SRR-5088) minerogenic sediment had extended across the entire transect to point E; no major peat growth has occurred on this transect since then. Within the kettle-hole F–G (Figure 16.2(a)) peat that had formed since the late Neolithic (above) was covered by a final capping of fluvial sediment after 3049–2789 cal. BP (GU-4270). At and around transect point C the peat of a completely destroyed kettle-hole was overlain by minerogenic overbank sediments prior to 2331–2112 cal. BP (SRR-5082). This assay represents the re-growth of peat over minerogenic sediments in the late Iron Age, a fluctuation in flood frequency/intensity not seen at the adjacent transect F–G. Flood-plain stability at point C was, however, short-lived since a gradual change to further overbank deposition is dated to 1603–1406 cal. BP (SRR-5090). This assay dates the last appearance of peat on this part of the floodplain.

The Frenchland Burn (Figure 16.2(a)) contains a stacked fill very similar to that in the Moffat Basin. The earliest sediments above fluvioglacial gravels are peats, and this marshy valley floor persisted from 3393–3167 cal. BP (SRR-5569) until 715–576 cal. BP (SRR-5095), with an apparently only limited amount of fluvial deposition at $c.$ 1500–1270 cal. BP (SRR-5568 and -5567; two dates bracketing the minerogenic band but statistically inseparable). Fine-grained overbank sediments have dominated this floodplain since 715–576 cal. BP.

Further south, the peat within an eroded kettle-hole at Stenrieshill (transect L–M; Figure 16.2(c)) contains some mid-Holocene, poorly dated, fluvial sediment, indicating that erosion of the kettle-hole allowed access by the River Annan before the late prehistoric. After 1687–1411 cal. BP (SRR-5094) fluvially derived sediments replaced peat, and continue to do so. Such a sequence could simply relate to the proximity of the river channel to the kettle-hole through channel migration. To test this a very similar sediment stratigraphy at neighbouring Catharine Hill was ^{14}C dated. An initial assay on the contact between peat and overlying minerogenic sediment, of 1520–1300 cal. BP (GU-4272), appears to confirm the quite precise synchroneity of increasing flood frequency/intensity in this reach, but a further date on this contact obtained later on sediments sampled for pollen analysis (below), of 660–510 cal. BP (GU-4642), causes interpretative problems. Although both assays are internally consistent, the extrapolation of age from underlying assays is better if GU-4272 is assumed to be correct. The adoption of GU-4642 provides a very low peat accumulation rate (Milburn and Tipping, unpublished), and the closely comparable sediment stratigraphies at Stenrieshill and Catharine Hill (with a comparable ^{14}C assay at Stenriehill to GU-4272), all argue for the retention of assay GU-4272 and the rejection of GU-4642.

In upland reaches of the study area, high channel mobility (cf. Tipping, 1994b) and gravel deposition restrict organic matter accumulation, and the significance of the one ^{14}C assay on the complex multi-channel sequence at Polmoody in upper Moffatdale (Figure 16.1(c)) is uncertain, but gravel terrace aggradation occurred after 1173–

962 cal. BP. On the Evan Water a much more detailed reconstruction of events is possible not through ^{14}C dating but through relations between the fluvial sequence and a series of Roman temporary camps on and next to a late Devensian fan surface. Figure 16.2(b) shows the pattern of palaeochannels on the fan surface, many of which could relate to the late Devensian. Overlying most of these are the cropmarked outlines of Roman temporary camps mapped from aerial photographs by Maxwell and Wilson (1987). The relations are complex but rewarding. Camp A on Figure 16.2(b) is represented only by fragmentary ditch-sections, but the age and nature of this camp have been demonstrated by excavation (GUARD, unpublished). The camp occupies the surface of the late Devensian kamiform fluvioglacial terrace. However, the north angle of the camp has been eroded since the 1st century AD by southward shift of the Evan Water. At the time of its construction, Camp A would have extended onto what is now the left bank of the Evan Water, and it is likely that the course of the Evan Water contemporaneous with Camp A lay to the north of an "island" of older terrace ((c) on Figure 16.2(b)), perhaps occupying the sinuous channel (a). Temporary camps B and C both postdate nearly all the palaeochannels that cover the fan surface, including channel (a). The north surface of the fan has been stable since before the 1st century AD; Camp B in particular overlies the set of meander scroll bars marked (b). The southward migration of the Evan Water at this point commenced prior to construction of Camp B, but after the construction of Camp A, at the end of the 1st century AD. This earliest historic period was clearly one of considerable channel mobility and valley-widening on the Evan Water.

In summary, the extensive floodplain of the Moffat Basin records accelerated alluviation within the Iron Age, possibly as early as 2900–3000 cal. BP. However, most localities show that alluviation was most extensive within or after the 1st century AD, and that except for Frenchland Burn (above), all reaches within upper Annandale were receiving overbank flood sediments after 1500 cal. BP.

Archaeology and Land Use

Figure 16.3 depicts the archaeological record for the study area for three periods: (a) the late Bronze/early Iron Age (c. 3000–2600 cal. BP); (b) the mid to late Iron Age (c. 2600–1900 cal. BP); (c) Romano-British native settlements (c. 1900–1500 cal. BP); and (d) Roman sites, all assumed to be of either Flavian (1st century AD) or Antonine (2nd century AD) age. These temporal divisions are not hard-and-fast, and the chronology of these distributions is fully discussed elsewhere (RCAHMS, 1997). Broadly, the divisions represent early timber-built settlements, both unenclosed and enclosed, the movement after the early Iron Age to defended sites and earthwork enclosures, and then in Figure 16.3(c) the construction of various types of lightly enclosed settlements, many of them with scooped interiors (the so-called scooped settlements). The cultivation trace called cord rig (Figure 16.3(a)) has a long chronology (Carter, 1995) but a good case may be made for a late prehistoric age in the Borders (RCAHMS, 1997, p. 44–47).

Biases in site preservation, such as afforestation, mean the distribution patterns may be distorted. On the west of the Annan, above the Evan Water, the RCAHMS identified an expanse of upland settlement that remains largely unaffected by medieval

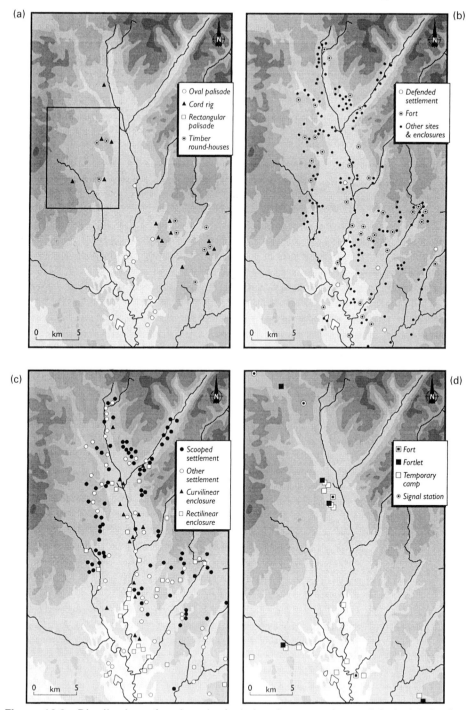

Figure 16.3 Distribution of archaeological sites in upper Annandale in (a) the late Bronze/early Iron Ages, (b) the later Iron Age, (c) the Romano-British period (native sites) and (d) the Romano-British period (Roman sites), modified with permission from RCAHMS (1997). The box in Figure 16.3(a) indicates the area described in detail in RCAHMS (1997)

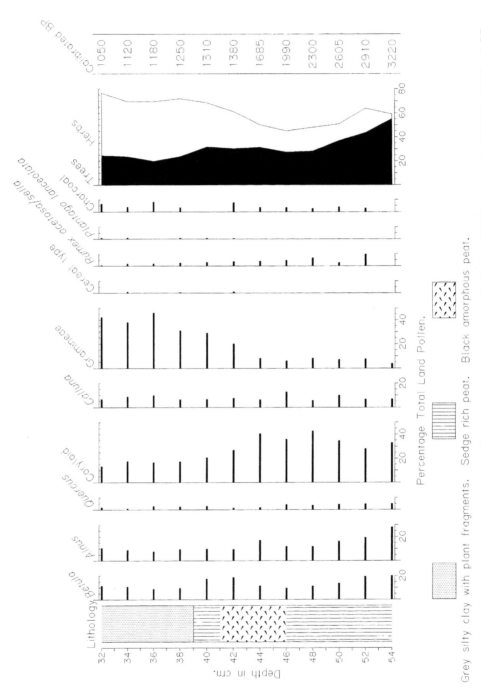

Figure 16.4 Sediment stratigraphy and pollen diagram of selected taxa at Catharine Hill for the period 3000 to 1000 cal. BP

and later activity, and implications concerning changes in population within the study area are made from this detailed study (RCAHMS, 1997; see box in Figure 16.3(a)), probably typical of the valley. Upper Annandale does not appear to have been highly populated in the early Iron Age compared to some other areas to the south-east (RCAHMS, 1997). The rise of the defended sites need not represent a significant change in population but the mass of small settlements that succeeded them, including the scooped settlements (Figure 16.3(c)) are, however, argued to represent a significant expansion of small farmsteads, apparently accompanied by extensive enclosures. Poorly dated, some of these settlements evidently belong in the Romano-British period (Burgess, 1984), but many workers prefer to see their origin in the late Iron Age (Hill, 1982; Mercer and Tipping, 1994; RCAHMS, 1997). Of these settlements, 28 are found within the detailed survey area above Beattock alone (Figure 16.3(a)). Similar arguments concerning a population expansion related to scooped settlements were developed by Mercer and Tipping (1994) for the northern Cheviot Hills, their significance in this regard increasing when several examples suggest each settlement contained three or four "houses". In addition, there is excavated evidence from one hill-fort, at Castle O'er in upper Eskdale, for its continued use into the 1st millennium AD (RCAHMS, 1997), so that the pattern of small settlements may accompany some of the larger defended sites rather than replace them in eastern Dumfriesshire.

Despite the problem of dating the settlements, those broadly attributable to the Romano-British period are concentrated on the lower slopes of the hills below 250 m OD. Stock-farming appears from archaeological data to have played a major role in the agricultural economy throughout the later prehistoric and Romano-British period (Halliday, 1982, 1993). The Roman sites (Figure 16.3(d)) are seen as only military fortifications, and to have no direct relation to land use; they may have promoted a market or tax system, but the periods of occupation in upper Annandale were probably too brief (RCAHMS, 1997) to have substantially influenced the landscape. Pollen analyses that should typify the landscape occupied by such settlements come from Catharine Hill (Figures 16.1(c) and 16.2(c): Milburn and Tipping, unpublished). Figure 16.4 shows selected taxa and a summary of the major ecological groupings plotted against a calibrated timescale. It shows that prior to $c.$ 1500 cal. BP (GU-4272) woodland clearance was small-scale and of limited duration. Major clearance is recorded after this time, and very shortly after this the pollen sequence is distorted by fluvially derived minerogenic sediment. This clearance, by far the largest event in the Holocene, was for both arable and pastoral agriculture. High above the farmed lowland landscape, human impact was also felt on the 600 m OD plateau around Rotten Bottom (Tipping, unpublished), but much earlier, at $c.$ 2700 cal. BP, when proportions of grazing indicator taxa such as *Plantago lanceolata* and grasses increased. This interest in the uplands, potentially linked to Iron Age transhumance, persisted through the Romano-British period without seeming to intensify, but reached a peak at around 1300 cal. BP.

Climate Change

In the last 3000 years there appears to have been only one significant or detectable shift in peat humification within the blanket peat of Rotten Bottom, at $c.$ 2600 cal. BP. This is a shift to wetter ground conditions from a dry peat between c. 3700 and 2600 cal. BP, but since this is broadly coincident with the increase in grazing pressures (above) this may not be an unambiguous climatic signal (cf. Blackford and Chambers, 1991). In the lowlands at the raised moss of Burnfoothill Moss, near lower Annandale (Tipping, 1995b), interpretative difficulties affect the understanding of effective precipitation between c. 3000 and 1900 cal. BP. By 1900 cal. BP the peat surface is thought to have been generally dry; a shift to wetter conditions occurred at c. 1200 cal. BP. Further south still, at Bolton Fell Moss, near Carlisle, Barber et al. (1994) identified shifts in increased bog surface wetness at c. 2800–2500 cal. BP (in apparent agreement with Rotten Bottom), 2000–1800 cal. BP and at c. 1000 cal. BP (possibly that recorded at Burnfoothill Moss). Separating these wet phases are periods of drier bog surface conditions. More extensive though less well dated analyses of plant macrofossil stratigraphies from raised mosses throughout southern Scotland by Stoneman (1993) suggest broadly synchronous wet "shifts" at c. 2800–2700 cal. BP, c. 2500–2350 cal. BP, c. 2100–1900 cal. BP and at c. 1050 cal. BP.

DISCUSSION AND SYNTHESIS

In rivers draining the Cheviot Hills in south-east Scotland, a major phase of accelerated fluvial aggradation was closely ^{14}C dated to the mid to late Iron Age, and continuing through the Romano-British period (Tipping, 1992). This has been associated with strong palynological and archaeological evidence for large-scale woodland clearance and land-use intensification (Mercer and Tipping, 1994). Comparable very rapid and extensive woodland clearances have been identified throughout southern Scotland (e.g. Dumayne, 1993; Tipping, 1994a), and away from Hadrian's Wall they were the work of native late Iron Age communities (Tipping, 1997b). Within the Kirtle Water in lower Annandale, Tipping (1995c) identified a very substantial and very abrupt woodland clearance in either the latest Iron Age or the beginning of the Romano-British period. However, the fluvial chronology for the Kirtle Water (Tipping, 1995a) failed to endorse the relation established in the Cheviot Hills between anthropogenic deforestation and fluvial aggradation; instead, the period of woodland clearance has no evidence for accelerated fluvial activity, and aggradation occurred some 300–400 years later at around 1500 cal. BP.

Within upper Annandale the abundance and variety of deposits datable by ^{14}C allows the relation between causal agents and fluvial response to be viewed afresh. First, the date of initiation and consequent span of overbank deposition varies depending on the location within the floodplain of the dated deposit. Deposits on the floodplain near the channel, which appears stable (unpublished data), suggest sustained aggradation from much earlier than localities at the floodplain edge such as kettle-hole sequences. Sites adjacent to the channel are likely, of course, to receive flood sediment more frequently than those farther away, but it is of critical importance to know what palaeo-fluvial signal to correlate with "causal" proxy data. In the analysis presented above greater significance has been placed on evidence for "distal to the channel" deposition, because this is taken to represent some sort of peak in intensity or frequency of flooding within the basin. Such localities in upper Annandale show that accelerated fluvial activity across the floodplain was broadly synchronous at c. 1900–1500 cal. BP.

Secondly, although synchroneity remains one of the major tests for climatic causation in palaeo-fluvial records (Macklin and Lewin, 1994), in most British river systems phases of fluvial activity are rarely dated by more than one control. The abundance of secure datable contexts in upper Annandale allows the recognition that different contexts yield different ages for the same fluvial event; this makes comparisons between less adequately dated systems (i.e. most systems) more difficult and weakens the argument for synchroneity between catchments.

Within upper Annandale, despite the demonstration of a major synchronous aggradational event at c. 1900–1500 cal. BP, a climatic origin for it can be questioned because, although a climatic deterioration is now recognised in southern Scotland in the early Romano-British period (above), one well-dated tributary, the Frenchland Burn, did not respond to this climate change. Again, the sheer number of datable contexts in upper Annandale allows a more satisfactory test of synchroneity in fluvial behaviour than elsewhere. It is true that complex-response models (Schumm, 1977) allow for parts of the fluvial system to remain in balance during environmental change, but there is a danger in too readily resorting to the non-testable outcomes of such models. In the reconstruction preferred here it is considered that the Frenchland Burn and the River Annan should have exhibited comparable fluvial responses to the same environmental change; they did not. At a wider scale, the onset of this characteristic late prehistoric/early historic alluviation is non-synchronous between south-east Scotland (Tipping, 1992), the North Tyne catchment (Macklin et al. 1992) and south-west Scotland (Tipping, 1995a; this study).

The idea that the sequence in upper Annandale is nevertheless "climatically driven" and merely "anthropogenically blurred" (Macklin and Needham, 1992) could still be maintained; the late Bronze Age/early Iron Age climatic deterioration at around 2800–2500 cal. BP is a signal recognised throughout southern Scotland (above), and one interpreted by van Geel et al. (1996) as of major significance (see also Burgess, 1985, 1995). It may have been that slope instability was accelerated during this apparent increase in precipitation, priming the temporary sediment storage system which was then reworked by anthropogenic activity 700–1000 years later, but we have no evidence for this; we need to understand more about slope–channel linkages. One difficulty at present with Macklin and Needham's (1992) model is that it is not parsimonious, and therefore less satisfying than a hypothesis which attributes cause to one event.

In the sediment stratigraphy at Catharine Hill (Figure 16.4) there is a very good relation between woodland clearance and the appearance of overbank flood sediment. This relation is not direct, because sediment movement from deforested areas into streams still requires "natural" events like rainstorm events, but a causal relation is argued for here, and the geomorphological effects of this striking woodland clearance phase are further endorsed in the Cheviot Hills (Tipping, 1992) and sites in northern England (Bartley et al., 1976). Despite the comparatively comprehensive data set in upper Annandale, there are interpretative issues. One is that the ^{14}C chronology for this event at Catharine Hill is not secure (above). A second is that channel mobilisation on the Evan Water predates woodland clearance by at least 200 years. A third is that sediment supplied to the Annan at Catharine Hill may have been derived from upstream sources, as well as also deriving from reworked older fluvial sediments, and we know nothing of the timing or scale of clearance on the slopes in and around, for instance, lower Moffatdale. A still greater density of pollen analyses is required.

Correlation with the archaeological record is relatively poor. Figure 16.3 and its analysis (above) indicates the likelihood of a population expansion approximately coincident with vegetational and geomorphological events, but the date of origin and life-span of scooped settlements, the farms argued here and in Mercer and Tipping (1994) to represent the population expansion responsible for environmental change, are inadequately known. It would be valuable to demonstrate, for example, that scooped settlements in the Cheviot Hills were of late Iron Age, but of Romano-British age in eastern Dumfriesshire, because the environmental changes in the Cheviots occurred at an earlier time. However, the archaeological record is inadequate to answer this, and will remain so as long as archaeological chronologies are based on so few excavated settlements.

In summary then, studies such as this in upper Annandale are characterised by the assembling of large and complex data sets, geomorphological, archaeological, palaeobotanical and palaeoclimatic. These data sets are constructed at appropriate spatial scales; that is, they are all related to the one relatively small catchment. There are conflicts between data sets, such as the weakness in comparing environmental change with archaeological change, but this combining of data at the catchment scale allows more precise questions to be asked of the data. Using these data we argue that in this example from the late prehistoric/early historic periods, there is little requirement for complex hypotheses to explain accelerated alluviation. Anthropogenic impacts appear at present to explain the very major environmental changes wrought in this landscape in the early 1st millennium AD.

ACKNOWLEDGEMENTS

We would like to thank Historic Scotland (HS), RCAHMS, Scottish Natural Heritage, the British Geomorphological Research Group, Edinburgh University Trust Fund and Glasgow University Archaeological Research Division (GUARD) for funding. Gordon Cook at SURRC, and Douglas Harkness and Brian Miller at NERC, provided the ^{14}C dates, and Patrick Ashmore (HS) and the NERC ^{14}C Steering Committee are thanked for their support.

We thank all the landowners for site access, and Gordon Maxwell, Tony Pollard, Keith Speller and Alan Leslie (GUARD) for access to archaeological data. Bob McCulloch, Anthony Newton, Malcolm Murray, John Lowe, Francis Mayle, Charles Sheldrick, Ingrid Robson, Andy Haggart, Andy Moir, Angela Wardell and Ian Parker kindly assisted with fieldwork. Bill Jamieson produced the figures at short notice. We thank Tony Brown and Tim Quine for inviting this contribution.

REFERENCES

Barber, K. E., Chambers, F. M. and Maddy, D. 1994. Sensitive high-resolution records of Holocene palaeoclimate from ombrotrophic bogs. In Funnell, B. M. and Kay, R. L. F. (Eds) *Palaeoclimate of the Last Glacial/Interglacial Cycle*. NERC Earth Science Directorate Special Publication No. 94/2, London, 57–60.

Bartley, D. D., Chambers, C. and Hart-Jones, B. 1976. The vegetational history of parts of south and east Durham. *New Phytologist*, **77**, 437–468.
Bell, M. G. and Boardman, J. 1992. *Past and Present Soil Erosion*. Oxbow, Oxford.
Bishop, W. W. 1963. Late-glacial deposits near Lockerbie, Dumfriesshire. *Transactions of the Dumfries & Galloway Natural History and Antiquarian Society*, **40**, 117–132.
Blackford, J. J. and Chambers, F. M. 1991. Proxy records of climate from blanket mires: evidence for a Dark Age (1400 BP) climatic deterioration in the British Isles. *The Holocene*, **1**, 63–67.
Brown, A. G. 1997. *Alluvial Environments: Geoarchaeology and Environmental Change*. Cambridge University Press, Cambridge.
Burgess, C. 1984. The prehistoric settlement of Northumberland: a speculative survey. In Miket, R. and Burgess, C. (Eds) *Between and Beyond the Walls: Essays on the Prehistory and History of North Britain in Honour of George Jobey*. John Donald, Edinburgh, 126–175.
Burgess, C. 1985. Population, climate and upland settlement. In Spratt, D. and Burgess, C. (Eds) *Upland Settlement in Britain. The 2nd Millennium BC and After*. BAR British Series 143, Oxford, 195–219.
Burgess, C. 1995. Bronze Age settlements and domestic pottery in northern Britain: some suggestions. In Kinnes, I. and Varndell, G. (Eds) *Unbaked Urns of Rudely Shape – Essays on British and Irish Pottery for Ian Longworth*. Oxbow, Oxford, 145–158.
Burrin, P. J. and Scaife, R. G. 1984. Aspects of Holocene valley sedimentation and floodplain development in southern England. *Proceedings of the Geologists' Association*, **95**, 81–96.
Burrin, P. J. and Scaife, R. G. 1988. Environmental thresholds, catastrophe theory and landscape sensitivity: their relevance to the impact of man on valley alluviations. In Bintliff, J. L., Davidson, D. A. and Grant, E. H. (Eds) *Conceptual Issues in Environmental Archaeology*. Edinburgh University Press, Edinburgh, 211–232.
Carter, S. P. 1995. Radiocarbon dating evidence for the age of narrow cultivation ridges in Scotland. *Tools & Tillage*, **7**, 83–91.
Dark, K. and Dark, P. 1997. *The Landscape of Roman Britain*. Sutton Publishing, Stroud.
Dumayne, L. 1993. Iron Age and Roman vegetation clearance in northern Britain: further evidence. *Botanical Journal of Scotland*, **46**, 385–392.
Dumayne, L. and Barber, K. E. 1994. The impact of the Romans on the environment of northern England: pollen data from three sites close to Hadrian's Wall. *The Holocene*, **4**, 165–173.
Gregory, K. J., Starkel, L. and Baker, V. R. 1995. *Global Continental Palaeohydrology*. John Wiley, Chichester.
Halliday, S. P. 1982. Later prehistoric farming in south-east Scotland. In Harding, D. W. (Ed.) *Later Prehistoric Settlement in South-east Scotland*. Department of Archaeology, Edinburgh University, Edinburgh, 75–91.
Halliday, S. P. 1993. Marginal agriculture in Scotland. In Smout, T. C. (Ed.) *Scotland Since Prehistory. Natural Change and Human Impact*. Scottish Cultural Press, Aberdeen, 64–78.
Harrison, J. 1997. Central and southern Scotland. In Wheeler, D. and Mayes, J. (Eds) *Regional Climates of the British Isles*. Routledge, London, 205–227.
Hill, P. H. 1982. Settlement and chronology. In Harding, D. W. (Ed.) *Late Prehistoric Settlement in South-east Scotland*. Department of Archaeology, Edinburgh University, Edinburgh, 141–188.
Macklin, M. G. and Lewin, J. 1994. Holocene river alluviation in Britain. *Zeitschrift für Geomorphologie* (Supplement), **88**, 109–22.
Macklin, M. G. and Needham, S. 1992. Studies in British alluvial archaeology: potential and prospect. In Needham, S. and Macklin, M. G. (Eds) *Alluvial Archaeology in Britain*. Oxbow, Oxford, 9–23.
Macklin, M. G., Passmore, D. G., Cowley, D. C., Stevenson, A. C. and O'Brien, C. F. 1991. Geoarchaeological enhancement of river valley archaeology in north east England. In Spoerry, P. (Ed.) *Geoprospection in the Archaeological Landscape*. Oxbow, Oxford, 43–58.
Macklin, M. G., Passmore, D. G. and Rumsby, B. T. 1992. Climatic and cultural signals in Holocene alluvial sequences: the Tyne basin, northern England. In Needham, S. and Macklin,

M. G. (Eds) *Alluvial Archaeology in Britain.* Oxbow, Oxford, 123–140.

Maxwell, G. S. and Wilson, D. R. 1987. Air reconnaissance in Roman Britain 1977–84. *Britannia,* **18**, 1–48.

May, J. A. 1981. The glaciation and deglaciation of upper Nithsdale and Annandale. Unpublished PhD Thesis, University of Glasgow.

McKinnon, H. 1997. Vegetation change and accelerated fluvial activity in the Moffat Basin. Unpublished BSc Thesis, University of Stirling.

Mercer, R. 1997. *Kirkpatrick Fleming. An Anatomy of a Parish in South West Scotland.* Dumfries and Galloway Natural History and Antiquarian Society, Dumfries.

Mercer, R. and Tipping, R. 1994. The prehistory of soil erosion in the northern and eastern Cheviot Hills, Anglo-Scottish Borders. In Foster, T. and Smout, T. C. (Eds) *The History of Soils and Field Systems.* Scottish Cultural Press, Aberdeen, 1–25.

Needham, S. and Macklin, M. G. 1992 *Alluvial Archaeology in Britain.* Oxbow, Oxford.

Pollard, T. 1993. *Kirkhill Farm.* Glasgow University Archaeological Research Division, Glasgow.

Potter, T. W. 1976. Valleys and settlement: some new evidence. *World Archaeology,* **8**, 207–219.

Potter, T. W. 1979. *Romans in North West England: Excavations at the Roman Forts of Ravenglass, Waterbrook and Bowness on Solway.* Cumberland & Westmorland Antiquarian & Archaeological Society, Kendal.

RCAHMS 1997. *Eastern Dumfriesshire: An Archaeological Landscape.* HMSO, Edinburgh.

Schumm, S. A. 1977. *The Fluvial System.* John Wiley, Chichester.

Shotton, F. W. 1977. Archaeological inferences from the study of alluvium in the lower Severn–Avon valleys. In Limbrey, S. and Evans, J. G. (Eds) *The Effect of Man on the Landscape: The Lowland Zone.* Council for British Archaeology, London, 27–31.

Speller, K. and Leslie, A. 1995. *Beattock.* Glasgow University Archaeological Research Division, Glasgow.

Starkel, L., Gregory, K. J. and Thornes, J. B. 1991. *Temperate Palaeohydrology.* John Wiley, Chichester.

Stevenson, J. B. 1975. Survival and discovery. In Evans, J. G., Limbrey, S. and Cleere, H. (Eds) *The Effect of Man on the Landscape: The Highland Zone.* Council for British Archaeology, London, 104–108.

Stoneman, R. 1993. Holocene palaeoclimates from peat stratigraphy: extending and refining the model. Unpublished PhD Thesis, University of Southampton.

Tight, J. A. 1987. The Late Quaternary history of Wester Branxholme and Kingside Lochs, South-east Scotland. Unpublished PhD Thesis, University of Reading.

Tipping, R. 1992. The determination of cause in the generation of major prehistoric valley fills in the Cheviot Hills, Anglo-Scottish Border. In Needham, S. and Macklin, M. G. (Eds) *Alluvial Archaeology in Britain.* Oxbow, Oxford, 111–121.

Tipping, R. 1994a. The form and fate of Scotland's woodlands. *Proceedings of the Society of Antiquaries of Scotland,* **124**, 1–54.

Tipping, R. 1994b. Fluvial chronology and valley floor evolution of the upper Bowmont Valley, Borders Region, Scotland. *Earth Surface Processes and Landforms,* **19**, 641–657.

Tipping, R. 1995a Holocene evolution of a lowland Scottish landscape: Kirkpatrick Fleming. III Fluvial history. *The Holocene,* **5**, 184–195.

Tipping, R. 1995b. Holocene evolution of a lowland Scottish landscape: Kirkpatrick Fleming. I Peat- and pollen-stratigraphic evidence for raised moss development and climatic change. *The Holocene,* **5**, 69–81.

Tipping, R. 1995c. Holocene evolution of a lowland Scottish landscape: Kirkpatrick Fleming. II Regional vegetation and land-use change. *The Holocene,* **5**, 83–96.

Tipping, R. 1997a. Holocene landform evolution. In RCAHMS *Eastern Dumfriesshire: An Archaeological Landscape.* HMSO, Edinburgh, 30–68.

Tipping, R. 1997b. Pollen analysis, late Iron Age and Roman agriculture around Hadrian's Wall. In Gwilt, A. and Haselgrove, C. (Eds) *Reconstructing Iron Age Societies.* Oxbow, Oxford, 239–247.

Tipping, R. 1997c. Vegetational history of southern Scotland – a review. In Proctor, J. (Ed.)

Scottish Vegetation. Botanical Journal of Scotland, **49**, 151–162.

Tipping, R. 1998. Accelerated geomorphic activity and human causation: problems in proving the links in proxy records. In O'Connor, T. P. and Nicholson, R. (Eds) *Human Impact on the Landscape*. Oxbow, Oxford, in press.

Tipping, R. and Halliday, S. 1994. The age of alluvial fan deposition at Hopecarton in the upper Tweed valley, Scotland. *Earth Surface Processes and Landforms*, **19**, 333–348.

Van Geel, B., Buurman, J. and Waterbolk, S. T. 1996. Archaeological and palaeoecological indications of an abrupt climate change in the Netherlands, and evidence for climatological teleconnections around 2650 BP. *Journal of Quaternary Science*, **11**, 451–460.

Vita-Finzi, C. 1969. *The Mediterranean Valleys*. Cambridge University Press, Cambridge.

Walker, D. 1966. The late Quaternary history of the Cumberland Lowland. *Philosophical Transactions of the Royal Society of London*, **B251**, 1–210.

17 A 1000 Year Alluvial Sequence as an Indicator of Catchment/Floodplain Interaction: The Ruda Valley, Sub-Carpathians, Poland

K. KLIMEK
Earth Sciences Faculty, University of Silesia, Poland

INTRODUCTION

Human influence on alluvial channel dynamics and related erosional/aggradational tendencies has occurred since the first human interference with nature. In the mid-European temperate climate zone it was primarily caused by gatherers and hunters, and then by farmers clearing the fields and extending the farmlands. During the last few centuries the human impact noted in the fluvial system has been caused mainly by mineral resource exploitation, urbanisation, land drainage, dams, road and railway construction as well as engineering correction of river channels.

The scale of human impact on the intensity and results of geomorphological processes had not been well understood up to the beginning of the 20th century (Sherlock, 1923). Since then many original papers have been published which present the results of studies on different aspects of that problem. Several publications give us a retrospective review of these problems (Gregory, 1977, 1986, 1987; Knox, 1977, 1987; Lewin, 1977; Graf, 1979; Brookes, 1985; Douglas, 1985; Macklin, 1985, 1996; Lewin and Macklin, 1987; Petts, 1989; Anderson *et al.*, 1996; Brown, 1996).

In Poland, human influence on fluvial processes was not noticed before the 1960s (Falkowski, 1967). The significance of the agricultural colonisation of the Wisłoka catchment on the Carpathian foreland was assessed in the 1970s (Klimek and Starkel, 1974). Additional data were published in the special issues edited by Starkel (1990, 1991, 1995, 1996), Starkel and Kalicki (1996), or in review publications (Klimek, 1987). The influence of the hard-coal mining industry and urbanisation on the valley systems was manifested in waste material supply, reflected in heavy metals recorded in the alluvium of the Upper Vistula. The problems mentioned above have been carefully studied since the late 1980s (Klimek and Zawilińska, 1985; Macklin and Klimek, 1992; Klimek, 1994).

In the Upper Odra catchment in the Carpathian and Sudetic foreland, the results of

Fluvial Processes and Environmental Change. Edited by A. G. Brown and T. A. Quine.
© 1999 John Wiley & Sons Ltd.

hard-coal exploitation and industrial development are superimposed on those of early medieval agricultural colonisation. This chapter presents the consequences of human activity recorded in the structure and geochemical features of the alluvium of a medium size river draining an area filled with Quaternary sediments, the Ruda Valley – a typical right-bank tributary of the River Odra.

THE STUDY AREA AND HISTORY OF HUMAN IMPACT DURING THE LAST MILLENNIUM

The Odra River basin drains the northern slopes of the metamorphic eastern Sudetic massif (1000–1500 m), the northern slopes of the flysch Western Carpathians (1000–1300 m), the Sudety foreland and a part of the Carpathian foredeep (200–350 m) (Figure 17.1).

The Ruda River drainage basin (190–310 m a.s.l.) is a part of the Carpathian

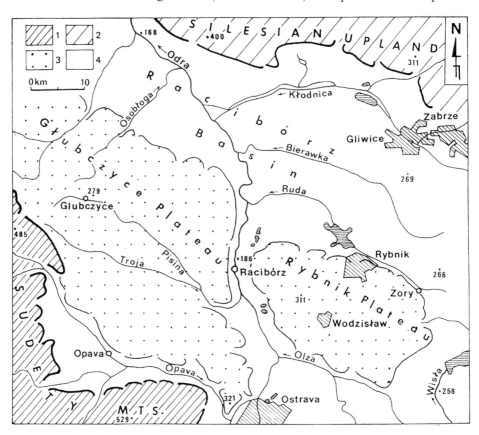

Figure 17.1 Location of the study area within the upper Odra drainage basin. 1, mountains; 2, uplands; 3, elevated plateaux; 4, plains

foredeep. The Scandinavian ice sheet twice covered the region (the South Polish and the Middle Polish Glaciations) and left a genetically differentiated sedimentary complex with a gravelly–sandy series up to tens of metres in thickness (Klimek, 1972; Lewandowski, 1987). During the last Pleistocene cold stage, the Vistulian (Weichselian), the area was within the periglacial zone of Central Europe. At that time the elevated part of the Ruda catchment – the Rybnik Plateau – was covered by silty aeolian deposits (Maruszczak, 1986). Both the slopes of hills and valleys were smoothed by wash-out and solifluction processes.

The climatic warming of the Pleistocene/Holocene transition caused the development of forest communities, which stabilised the valley slopes. The river valleys of the northern Carpathian foredeep were incised, mainly in unconsolidated Quaternary sediments, by a process of river valley floor widening by meandering channels (Klimek, 1995) which started 13 000–12 000 years BP. Wide alluvial plains with well-preserved palaeochannels are the typical scenery of the Ruda Valley and its main tributaries (Figure 17.2). The Ruda drainage basin covers an area of $515\,km^2$ and receives 700–800 mm annual precipitation. The Ruda River average discharge reaches more than $2\,m^3\,s^{-1}$ in the lowermost course (Absalon, 1996).

Favourable meso- and microclimatic conditions, and the dense network of the Ruda tributaries, have facilitated the colonisation of the Rybnik Plateau since prehistoric times. Recent investigations into the colonisation history during the early Middle Ages and later times (Panic, 1992) enable the reconstruction of the human impact on the natural environments of river valleys since this time (Klimek, 1996).

During the early Middle Ages, a time of feudal partition in Poland, the study area was a part of the Opole Duchy, whose history and national status changed many times. At first it became a fiefdom of the Bohemian Kingdom, then in 1526 it became a part of the Habsburg State (Austria), and from 1742 a part of the Kingdom of Prussia. Since 1945 this area has again become a part of Poland. From the 13th and 14th centuries the area of the upper Odra drainage area, together with the southern part of the Ruda catchment, has been densely populated as a result of economic development. About 20 new settlements were established up to the end of the 13th century on the northern part of the Rybnik Plateau over an area of 140–$160\,km^2$. At the beginning of the 14th century one village and its fields covered $9\,km^2$. Extremely low-yielding crops of that historic period forced one family to farm one feud of land, i.e. 15–30 ha. Hence one small village used an excessively large area of land (Panic, 1992). It led directly to a deforestation of the more fertile lands. As a result, the overland flow of rain and snowmelt water increased flood wave amplitudes. Most probably, on the loess-like silty deposits covering the deforested slopes of the Rybnik Plateau soil erosion increased greatly, having a far-reaching effect. In all probability at the same time as this erosion, a network of gullies developed dissecting the Rybnik Plateau slopes as in other regions of southern Poland (Buraczyński, 1989/90). They are no longer active (Dwucet, 1986).

The Ruda as well as its tributaries have low-angle longitudinal slopes and broad floors which create favourable conditions for artificial pond construction. At first they were commonly used as fish-farms. Then they were used for damming water for powering water-mills and ironworks using local siderite deposits. At the end of the 18th century there were about 260 ponds in the Ruda drainage basin (Harnisch,

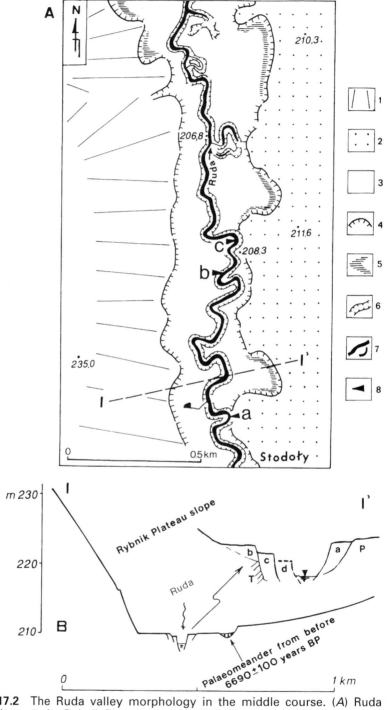

Figure 17.2 The Ruda valley morphology in the middle course. (*A*) Ruda valley morphology: 1, the Rybnik Plateau slope; 2, Pleistocene sandy plain/terrace; 3, Holocene valley floor; 4, meandering river undercuts; 5, palaeomeanders; 6, zone of inset terraces; 7, actual channel; 8, investigated alluvial sequences: a, Stodoły Bend; b, Stodoły Forest; c, Stodoły Border. (*B*) Ruda valley cross-section: T, Tertiary clays; P, Pleistocene deposits; a,b,c,d, inset terraces

1794–1801), which together stored about 20 million m³ of water (Kocel, 1995). This resulted in considerable equalisation of the Ruda River discharges and lowering of the flood waves.

The unchanging pattern of fields and baulks since medieval times resulted in an equilibrium of sediment transfer from the slopes to the valley floor. However, the land-use changes occurring here after the Napoleonic Wars in the mid-19th century transformed the farmland pattern (Korzeniowska, 1995). Around the same time, steam machines were introduced and the development of heavy industry began, which together, little by little, reduced the need for water energy supply. The ponds stopped being the source of energy, so they disappeared in a short time (Kocel, 1995). This caused increases in the discharge fluctuations in the Ruda River. Hard-coal exploitation on a large scale was initiated in the Ruda catchment during the second half of the 19th century (Jaros, 1984) and, together with the tailings, caused the delivery of large amounts of fine coal into the surface water (Klimek, 1995). At present the fine coal is easily identified in the Ruda alluvium.

VERTICAL SEQUENCES OF THE RUDA ALLUVIUM AS AN INDICATOR OF HUMAN IMPACT DURING THE LAST 1000 YEARS

The Valley Floor Topography

The Ruda valley floor in its middle and lower reaches is 0.3–0.8 km wide and has an average slope of 1.9 m km^{-1}. The valley floor is bordered with meandering undercuts of the Upper Pleistocene terrace (Figure 17.2(a)). Palaeomeanders with organic infills, 0.5–1 m thick, often exist at the base of these erosional slopes. The oldest radiocarbon dating to date, from basal parts of the infills, is 6690 ± 100 (Gd-10047) (Figure 17.2(b)) although palaeobotanic investigations of some other palaeomeanders put their age at the Pleistocene/Holocene boundary (D. Nalepka, pers. comm.). A system of younger palaeomeanders with a smaller radius of curvature occur close to the present-day Ruda channel (Figure 17.2(a)). Three (a, b and c) or possibly four (d) levels of narrow inset terraces have been noted on most river bends (Figure 17.2(b)). The highest reaches 2.4–2.6 m above the average water level. The deposition of alluvial sequences building up the system of inset terraces started about 1000 years ago and the valley floor topography from previous fluvial development has been partly fossilised. The structure, geochemical features and radiocarbon dating of the terraces enable us to reconstruct in more detail the direction and character, as well as intensity, of the fluvial processes during the last millennium.

The Internal Structure of Inset Terraces

At the Stodoły Bend site, fossil meandering channel sediments of the Ruda are overlain by overbank sediments (Figure 17.3(b)). Rhythmically bedded organic–clastic deposits, occurring in the lower part of the fossil channel fill, indicate vertical and horizontal accretion of organic matter interrupted by flood water inflows depositing sandy layers. Two samples taken from 5 cm and 24 cm above the bottom of the fossil

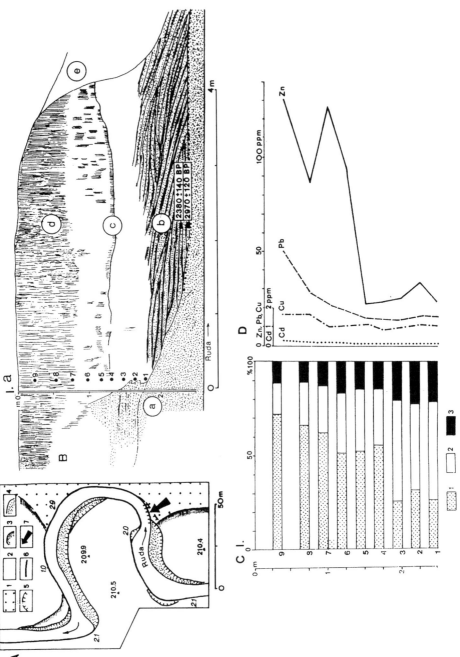

Figure 17.3 Ruda alluvial sequence at the Stodoły Bend exposure. (A) Ruda valley floor morphology: 1, Pleistocene level; 2, valley floor; 3, traces of palaeomeanders; 4, present-day active floodplain; 5, edges of inset terraces; 6, active undercuts; 7, exposure of inset terrace a. (B) Alluvial sequence: a, Pleistocene fluvial deposits; b, fossil channel organic fill; c, sandy-silty fossil channel fill; d, cambic

Table 17.1 Metal concentrations (ppm) in inset terrace vertical sequences, Ruda valley, Upper Silesia, Poland

Sampled sequences and points	Cadmium	Lead	Zinc
Stodoły Bend (see Figure 17.3, a, points 1–9 from the top)			
9	0.3	51	131
8	0.2	28	87
7	0.2	29	127
6	0.1	18	93
5	0.1	14	22
4	0.1	13	22
3	0.1	13	25
2	0.1	15	33
1	0.1	15	22
Stodoły Forest (see Figure 17.4(b), points 1–5 from the top)			
1	0.3	9	44
2	0.3	11	88
3	0.3	10	45
4	0.6	23	267
5	0.6	25	90
Stodoły Border (see Figure 17.4(c), points 1–5 from the top)			
1	1.1	43	98
2	2.8	52	220
3	3.4	48	207
4	1.4	63	310
5	3.1	102	490
Ruda Kozielska Forest (points 1–8 from the top)			
1	7.5	53	260
2	4.9	39	177
3	6.4	90	445
4	3.3	50	235
5	5.4	94	381
6	3.3	75	227
7	4.1	52	265
8	4.7	38	254

channel produced radiocarbon dates of 2970 ± 120 and 2380 ± 140 years BP (Gd-9413 and Gd 10229) respectively. Pilot palynological analyses of them showed a presence of cereals from an indistinct source (D. Nalepka, pers. comm.). Assuming that the sedimentation rate was stable, the estimated depositional rate of this sedimentary complex equals 0.3 mm year^{-1}, which suggests that undisturbed infilling of the cut-off lasted up to the beginning of the Middle Ages.

The upper section of the sequence has coarsening-upwards sandy silts (Figure 17.3(c)). It can be inferred that the flood amplitude would increase as soil erosion intensifed. Increased overbank deposition could be correlated with the first phase of early medieval agricultural colonisation and deforestation of the Rybnik Plateau. Especially supportive of such an interpretation are the Zn, Cd and Pb concentrations in the vertical profile of the alluvial sequence (Figure 17.3(d) and Table 17.1). Zinc, cadmium and lead contents within the fossil channel infill range as follows: Zn 25–30 ppm, Cd 0.1–0.2 ppm, Pb 13.5–15 ppm, and appear to represent the primary geochemical background of the drainage basin. Within the profile the concentration of Zn increases upwards, but variably, five to six times. There is no direct correlation between organic matter concentration and Zn concentration, although there is a degree of coincidence between the two in the upper part of the profile. This tendency coexists with uniform Cd and Pb concentrations in all the studied profiles. The present-day concentration of Zn in the near-surface soil layer reaches 240 ppm in the forested parts of the Ruda catchment, whereas it is much lower in the cultivated parts. Similarly the bottom deposits of former ponds in the Ruda River basin, supplied with water from forested catchments, have higher Zn concentrations than the deposits of ponds supplied with water from the agricultural catchments (Kocel, 1997). Therefore the increased concentration of Zn up-profile could be interpreted as the result of the erosion of the forest soils, rich in Zn, and their redeposition in the Ruda Valley as overbank deposits throughout the agricultural colonisation of the Rybnik Plateau. A well-developed profile of a cambic gleyo-eutric fluvisol in the upper part of the excavation suggests that it accumulated over several hundred years in conditions of temperate climate. It indicates that overbank sedimentation ended at least several hundred years ago, which seems to have been an effect of fluvial activity stabilisation or channel incision. Besides the general climatic reasons, the process was caused mainly by a stabilisation in agricultural expansion. The development of hammer ponds, which stored large amounts of water and lowered the flood amplitudes, have also influenced that process.

The morphogenetic impact of the Little Ice Age, with conditions "transitional" between oceanic and continental climate, are most clearly recorded in the highest part of the Carpathians, in the Tatry Mountains above 1400 m a.s.l. (Kotarba, 1993–94). In the low-lying Ruda catchment (310–190 m a.s.l.), filled with Quaternary sediments and forested for the greater part, the above-mentioned climate deterioration is not recorded in the fluvial processes and sediments.

At the Stodoły Forest site, 1.6 m of alluvium covers the older clay, of Tertiary age (Wydawnictwa Geologiczne, Geological Map of Poland, 1:50 000). Alluvial deposits here form a narrow zone of inset terraces, referred to here as inset terrace b (Figure 17.2(b) and 17.4). There are coarse- and medium-grained sands with horizontal lamination, cross-laminated in places, and with a clearly visible erosional surface. In the middle part of the profile, sandy co-sets are interlayered with sandy–silty laminae of several centimetres thickness, rich in organic detritus. Sedimentary structures indicate that these deposits accumulated due to bedload transport or to the migration of central/medial bars or side bars several centimetres thick in a straight or meandering channel. A change of sedimentary structures occurs in the top part of sequence. Here layers of fine-grained cross-laminated sands appear which are interlayered with silty–sandy laminae. The structure and evident decrease in bed thickness is thought to represent accretion on previously existing platforms and bars evolving to overbank accumulation.

In the lower part the of the sedimentary complex, within truncated sand packages, there are vertical remnants of *in situ* roots cutting through layers with organic detritus (Figure 17.4, terrace b). Their presence indicates an environment of sandy bars colonised by riparian communities and long inter-flood periods. Intercalations of sandy silt or fine sands with organic detritus indicate the existence of local accumulation zones,

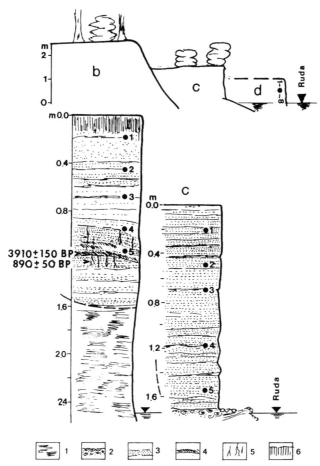

Figure 17.4 Medieval and 19th century inset terraces sequences at Stodoły Forest (terrace b) and Stodoły Border (terrace c): 1, Tertiary clays; 2, sandy-gravel channel deposits; 3, rhythmically bedding sands; 4, organic detritus laminae; 5, burie cut-off roots; 6, soil horizon; •1–5, sampling points

dominated by fine-grained load or fine organic matter settling downstream from sandy bars (Figure 17.4, terrace b).

The upper 0.4–0.5 m of the sedimentary profile represent an initial phase of overbank accumulation. Rhythmic deposits occurring here point to the close vicinity of a channel and high-velocity flowing water due to narrowing of the depositional area (Klimek, 1974; Macklin and Klimek, 1992). Radiocarbon dating of roots buried and cut-off in sandy deposits indicate a (non-calibrated) age of 890 ± 50 years BP (Gd-10045). This suggests early medieval sedimentation following an intensive agricultural colonisation period in the southern part of the Ruda catchment. Organic detritus from silty intercalations was dated to 3919 ± 190 years BP. It indicates that this organic matter has been derived from older organic infills of palaeochannels by a migrating channel and redeposited here.

The properties of the alluvial deposits in the lower and middle part of the 2.4 m inset terrace at the Stodofy Forest site suggest that a stable Ruda channel and narrow floodplain with presumably dense riparian vegetation has existed here for the last 1000 years. At the present time alluvial channels of this mainly meandering type are abundant on the Silesian Upland and on neighbouring lowlands filled with Quaternary deposits (Zieliński, 1993). They are commonly found in the forested areas of the temperate climate zone (Sundborg, 1956; Koutaniemi, 1987).

The lower step of the inset terrace c, 1.8 m thick and several metres wide, is noted mostly on the inner banks of river bends. Distinct rhythmic bedding exists in all the alluvial sequences (Figure 17.4, terrace c). The rhythmite thickness of 15–20 cm decreases upwards to 10 cm. Lower terrace sediments are more poorly sorted and contain distinct black intercalations of fine coal, in the form of several millimetre thick laminae capping ripples or 8–10 cm thick layers of quartz-coal sands. In the first case the hard-coal grains were deposited from suspension in the final phase of deposition as their specific weight is lower than the quartz. The Nacyna Stream, a local tributary joining the Ruda River several kilometres upstream from this site, is the source of the coal. The Nacyna is fed by waters derived from the coal mines. Part of the bedload and suspended load of the stream is caught in settling ponds. In the second case the dark laminae within the quartz-coal sand layers point to the intermittent supply of two petrographically different sand-sized grain types.

The geochemical properties of the lower inset terrace alluvium are different from those of the upper. Generally, in most cases, concentrations of trace metals (of $< 63 \mu m$ grain size) are near to double (in the case of Zn) or one order of magnitude (in the case of Cd) higher than in the lower terrace sediments (Table 17.1). Exploitation and coal processing (Jaros, 1984), as well as industrial development relating to local coal deposits at the beginning of the 20th century, were the source of these elements.

The petrographic, structural and geochemical features of the alluvium, as well as the lack of a soil horizon in the uppermost part of the sediments, clearly indicate a relatively young age for the lower inset terrace c. The lowermost layers of the sequence were deposited at the beginning of the period of increased hard-coal exploitation in the second half of the 19th century, although most of the sequence originated at the turn of the 20th century, when the surface water network was contaminated by waste material from the coal mines.

The lowermost inset terrace level d, about 1 m high (Figure 17.4), is occasionally noted in the middle Ruda reach on the outer banks of bends, but it is more frequently adjacent to a highly sinuous channel in the lower reach of the river. A typical sedimentary sequence of this 1.2 m thick terrace was found in the Ruda Kozielska Forest site, with very poorly sorted, horizontally stratified silty-sand. These were interpreted as the deposits of sandy point-bars in the middle part of the sequence and overbank sediments of a low natural levee in the upper part. They were characterised by a high content of coal grains uniformly interspersed throughout the sedimentary profile. This feature points to a redepositional origin of sediments, which underwent short-distance transport without fractional and petrographic sorting.

Water stage amplitudes, measured at the Hydrological Survey gauging station located 18 km downstream, showed that during 38 years (1955–1993) the bankfull stage was exceeded at least 17 times. This could account for the origin of the uppermost alluvial sequence.

The most characteristic feature of the youngest alluvial sediments is the high concentration of cadmium, zinc and lead (Table 17.1). In some levels the cadmium concentration is up to 7.5 ppm. This is an indicator of the very young age of this alluvial sequence, deposited in the last decades of rapid industrial development and urbanisation in the Ruda catchment area. Analogous relationships were noted in the alluvium of neighbouring catchments in Upper Silesia (Macklin and Klimek, 1992; Klimek, 1994).

SUMMARY

Sequences of inset terraces following the Ruda alluvial channel indicate a depositionally active floodplain with a relatively stable horizontal and vertical, meandering channel during the last millennium. The present-day Ruda channel bed lies 0.5–1.0 m below the base level of the last millennium alluvium. During the last 1000 years the average rate of incision did not exceed 0.5–1.0 mm year^{-1}. The sedimentary sequences of the inset terraces accumulated in short periods. Radiocarbon dates, well-developed soils capping the uppermost terrace, and the high variability of trace elements, indicate that the alluvium of the 2.4 m high terraces (a and b) were deposited over a few centuries, the alluvium of the lower terrace (c), 1.8 m high, was deposited during the last 100 years, and the alluvium of the lowermost terrace (d) accumulated over the last few decades.

RESULTS AND CONCLUSIONS

In the Ruda Valley, a right-bank tributary of the Odra River, deposition of an alluvial complex containing three or four vertical sedimentary sequences has occurred during the last millennium. The structural and geochemical features of these sediments are significantly different due to fluctuations in the intensity of human activity (Figure 17.5).

The impact of agricultural activity was connected with the medieval colonisation of the primarily forested southern parts of the catchment, and resulted in an evident increase in eroded soil transfer from the farmed slopes to the valley floor, and their deposition within the Stodoły Bend sequence (Figure 17.3).

Industrial human impact was connected with the beginning of intense hard-coal exploitation in the second half of the 19th century, and simultaneous development of the coal industry. It was characterised by the supply of fine coal particles and trace elements to the fluvial networks, which were derived from coal exploitation and processing. This phenomenon has been recorded in the geochemical properties of the alluvium: the zinc concentration doubled and cadmium concentrations increased by nearly one order of magnitude within the sediments of the younger inset terraces c and d. The petrographic characteristics of these deposits have changed as well. They contain admixtures or intercalations of fine-grained coal material that are absent in the deposits of the older inset terraces.

In spite of no evidence of a climate change during the period of sediment formation,

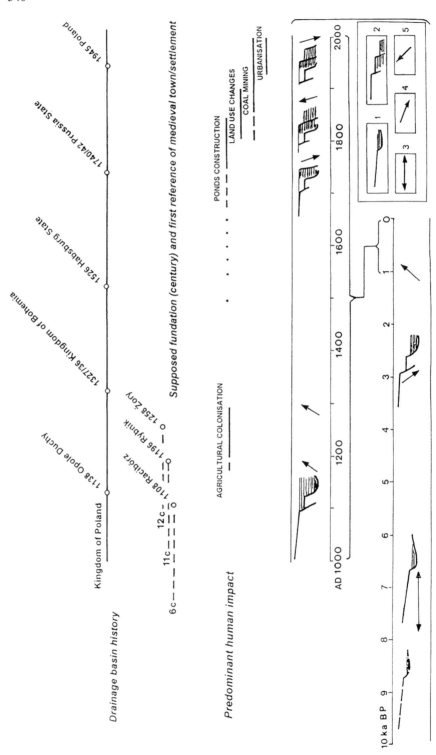

Figure 17.5 Ruda drainage basin during the last millennium. History of human impact and river aggradation/incision tendencies: 1, palaeomeanders; 2, inset terraces; 3, lateral migration; 4, incision; 5, accumulation

the process of insetting younger alluvium (terraces c and d) into incised older alluvium (terraces a and b) took place. This was presumably caused by progressive channel incision due to base-level (i.e. the Odra River) lowering. A second possible factor is changes in the Ruda River hydrological regime due to the retention of significant amounts of water in artificial ponds, which have existed here since the beginning of the 20th century.

REFERENCES

Absalon, D. 1996. Anthropogenic transformation of runoff from the Ruda drainage area. Unpublished Report, University of Silesia, Sosnowiec (in Polish).

Anderson, M. G., Bates, P. D. and Walling, D. E. 1996. The general context for floodplain process research. In Anderson, M. G., Walling, D. E. and Bates, P. D. (Eds) *Floodplain Processes*. John Wiley, Chichester, 1–13.

Brookes, A. 1985. River channelization. *Progress in Physical Geography*, **9**, 44–73.

Brown, A. G. 1996. Floodplain paleoenvironments. In Anderson, M. G., Walling, D. E. and Bates, P. D. (Eds) *Floodplain Processes*. John Wiley, Chichester, 95–138.

Buraczyński, J. 1989/90. Development of the gullies in the Goraj Roztocze during the last millennium. *Annales Universitatis Mariae Curie-Skłodowska*, Lublin, Series B, **44/45**, 95–104 (in Polish).

Douglas, I. 1985. Urban sedimentology. *Progress in Physical Geography*, **9**, 255–280.

Dwucet, K. 1986. Differentiation of morphology against the background of the lithology of the Rybnik Plateau dust formation. *Prace Naukowe Uniwersytetu Śląskiego w Katowicach*, **759**, Katowice (in Polish).

Falkowski, E. 1967. Evolution of the Holocene Vistula River at the Zawichost-Solec reach and an engineering–geological prognosis of its further development. *Instytut Geologiczny, Biuletyn*, Warszawa, **198**, 57–142 (in Polish).

Graf, W. L. 1979. Mining and channel response. *Annals of the Association of American Geographers*, **69**, 262–275.

Gregory, K. J. 1977. The context of river channel changes. In Gregory, K. J. (Ed.) *River Channel Changes*. John Wiley, Chichester, 1–12.

Gregory, K. J. 1986. Human impact on the fluvial environment. In *IGCP 158, Palaeohydrological Changes in the Temperate Zone in the Last 15 000 Years*. J. J. Gaillard, Lund.

Gregory, K. J. 1987. River channels. In Gregory, K. J. and Walling, D. E. (Eds) *Human Activity and Environmental Processes*. John Wiley, Chichester, 207–235.

Harnisch, J. 1794–1801. Situation Plan von einen Theile Ober Schlesien einder Oestreich und Neu Schleisischen Grentze, 1:80 000. *Archiwum. Katowice*.

Jaros, J. 1984. *Historical Dictionary of Coal Mines in Polish Territory*. Silesian Research Institute, Katowice (in Polish).

Klimek, K. 1972. The Racibórz–Oświęcim Basin. In Klimaszewski, M. (Ed.) *Geomorfologia Polski*, vol. 1, Warszawa, 116–138 (in Polish).

Klimek, K. 1974. The structure and mode of sedimentation of flood-plain deposits in the Wisłoka Valley (southern Poland). *Studia Geomorphologica Carpatho-Balcanica*, Kraków, **8**, 135–151.

Klimek, K. 1987. Man's impact on fluvial processes in the Polish Western Carpathians. *Geografiska Annaler*, Stockholm, **69A**, 221–226.

Klimek, K. 1994. River alluvial deposits. Contamination of the Vistula river basin overbank deposits by heavy metals. *Excursion Guide Book 2nd International Symposium on Environmental Geochemistry*. Kraków, Poland, 14–35.

Klimek, K. 1995. The role of orography in the river channel pattern transformation during the Late Vistulian, Subcarpathian Oświęcim Basin, Poland, *Questiones Geographicae*, Special Issue, **4** Poznań, 147–153.

Klimek, K. 1996. The Ruda river alluvia as indicator of 1000 years degradation of the Rybnik Plateau. In Kostrzewski, A. (Ed.) *Geneza, litologia i stratygrafia utworów czwartorzędowych.* UAM, Poznań, 155–166 (in Polish).

Klimek, K. and Starkel, L. 1974. History and actual tendency of floodplain development at the border of the Polish Carpathians. *Abhandlungen der Akademie der Wissenshaften in Gottingen, Mathematisch-Physikalische Klasse* 3, **29**, 185–196.

Klimek, K. and Zawilińska, L. 1985. Trace elements in alluvia of the Upper Vistula as indicators of palaeohydrology. *Earth Surface Processes and Landforms*, **10**, 273–280.

Knox, J. C. 1977. Human impacts on Wisconsin stream channels. *Annals of the Association of American Geographers*, **67**, 323–342.

Knox, J. C. 1987. Historical valley floor sedimentation in the upper Mississippi Valley. *Annals of the Association of American Geographers*, **77**, 224–244.

Kocel, K. 1995. Artificial ponds in the Ruda river valley (Upper Silesia) as indicators of anthropogenic landscape transformation. *Stanowisko 5. Rudy:stawy w dorzeczu Rudy. III Zjazd Geomorfologów Polskich, Przewodnik wycieczek.* Sosnowiec, 66–71 (in Polish).

Kocel, K. 1997. The bottom deposits of the hammer ponds as indicators of environment changes in the Ruda drainage basin. *Scripta Rudensia*, Rudy, **7**, 75–84 (in Polish).

Korzeniowska, W. 1995. Socio-economical situation of the Upper-Silesian village in the mid-19th century. *Scripta Rudensia.* Rudy, **4**, 29–36 (in Polish).

Kotarba, A. 1993–94. Record of Little Ice Age in lacustrine sediments of Morskie Oko Lake, High Tatra Mts. *Studia Geomorphologica Carpatho-Balcanica*, Kraków, **27/28**, 61–69 (in Polish).

Koutaniemi, L. 1987. Little Ice Age flooding in the Ivalojoki and Oulankajoki valleys, Finland. *Geografiska Annaler*, **69A**, 71–83.

Lewandowski, J. 1987. The Middle Pleistocene in the Upper Odra valley; Moravian Gate–Racibórz Basin. *Przegląd Geologiczny*, Warszawa, **36**, 465–474 (in Polish).

Lewin, J. 1977. Channel pattern changes. In Gregory, K. J. (Ed.) *River Channel Changes*. John Wiley, Chichester, 167–184.

Lewin, J. and Macklin, M. G. 1987. Metal mining and floodplain sedimentation. In Gardiner V. (Ed.) *International Geomorphology 1986, Part I.* John Wiley, Chichester, 1009–1027.

Macklin, M. G. 1985. Floodplain sedimentation in the upper Axe valley, Mendip, England. *Transactions of the Institute of British Geographers*, New Series, **10**, 235–244.

Macklin, M. G. 1996. Flux and storage of sediment – associated heavy metals in floodplain systems: assessment and river basin management issues at a time of rapid environmental change. In Anderson, M. G., Walling, D. E. and Bates, P. D. (Eds) *Floodplain Processes*. John Wiley, Chichester, 441–460.

Macklin, M. G. and Klimek, K. 1992. Dispersal, storage and transformation of metal contaminated alluvium in the upper Vistula basin, south-west Poland. *Applied Geography*, **12**, 7–30.

Maruszczak, H. 1986. Loesses in Poland, their stratigraphy and paleogeographical interpretation. *Annales Universitatis Mariae Curie-Skłodowska*, Lublin, Series B, **41**, 15–54.

Panic, I. 1992. Settling in Opole Duchy in the early Middle Ages. *Muzeum Śląskie*, Katowice (in Polish).

Petts, G. E. 1989. Historical analysis of fluvial hydrosystems. In Petts, G. E. Moller, H and Roux, A. L. (Eds) *Historical Change of Large Alluvial Rivers: Western Europe*. John Wiley, Chichester, 1–18.

Sherlock, R. L. 1923. The influence of man as an agent in geographical change. *Geographical Journal*, **61**, 258–273.

Starkel, L. 1990. *Evolution of the Vistula river valley during the last 15 000 years. Part III.* Geographical Studies, Special Issue No. 5, Wrocław.

Starkel, L. 1991. *Evolution of the Vistula river valley during the last 15 000 years. Part IV.* Geographical Studies, Special Issue No. 6, Wrocław.

Starkel, L. 1995. *Evolution of the Vistula river valley during the last 15 000 years. Part V.* Geographical Studies, Special Issue No. 8, Wrocław.

Starkel, L. and Kalicki, T. 1996. *Evolution of the Vistula river valley during the last* 15 000 *years. Part VI.* Geographical Studies, Special Issue No. 8, Wrocław.
Sundborg, A. 1956. The river Klaralven. A study of fluvial processes. *Geografiska Annaler*, **38**, 125–361.
Wydawnictwa Geologiczne 1959. *Racibórz*. Geological Map of Poland 1:50 000. Wydawnictwa Geologiczne, Warszawa.
Zieliński, T. 1993. Bed morphology and sediment of the present-day Biała Przemsza alluvial channel (S. Poland). *Geologia, University of Silesia*, Katowice, **12/13**, 199–230.

18 Historic River Response to Extreme Flooding in the Yorkshire Dales, Northern England

STEPHEN P. MERRETT and MARK G. MACKLIN
School of Geography, University of Leeds, UK

INTRODUCTION

Recent research in the British uplands has demonstrated much higher rates of fluvial activity and enhanced upland and hillslope erosion over the last 500 years (Innes, 1983; Robertson-Rintoul, 1986; Harvey and Renwick, 1987; Richard *et al*., 1987; Brazier *et al*., 1988; Brazier and Ballantyne, 1989; Ballantyne, 1991; Macklin *et al*., 1992; Evans, 1996) in comparison to earlier parts of the Holocene. A similar pattern has emerged in northern, western and central Europe (Rumsby and Macklin, 1996) where debate has centred on establishing whether these changes are a result of anthropogenic activity or natural climatic influences. In the UK the majority of studies have focused on human disturbance for causal links (e.g. Harvey *et al*., 1984; Ballantyne, 1991). However, such studies offer few indications as to the responses of upland fluvial environments to future climate change. This gap is important as the small- to medium-scale variations in hydroclimate which have characterised the late Holocene, particularly since the Little Ice Age (Lamb, 1977, 1982), are likely to be of the same order of magnitude as those predicted for the foreseeable future as a result of global warming (Newson and Lewin, 1991).

This study is concerned with the response of small upland catchments to large "geomorphologically effective" floods (Newson and Macklin, 1990) over the last 300 to 400 years, and the impact of short-term climate and land-use changes on basin hydrologies. Small upland catchments in the Yorkshire Dales, similar to other upland areas in northern Britain, are sensitive to both localised convective summer thunderstorms and predominantly winter cyclones that produce high-intensity rainfall events. This coupled with steep channels and flashy runoff regimes produces relatively frequent large-magnitude floods with high unit stream powers and high rates of coarse sediment transport. Such fluvial environments, whose valley floor morphologies are predominantly governed by extreme events, are likely to be particularly vulnerable to the decadal-scale changes in hydroclimate outlined above.

As understanding of the interactions between climate change, flood frequency and magnitude and river valley development has recently advanced, there has been relatively little work dealing with the influence of rates of sediment supply on river behaviour over timescales of a century or more. Work of this nature has largely been concentrated on the identification of thresholds in transformations between single- and multi-thread channel planforms (Harvey, 1987, 1991; Passmore *et al.*, 1993). In this chapter we investigate the influence of changes in flood frequency and magnitude, and changes in sediment supply, in the Yorkshire Dales region where pronounced variations in lithology, catchment morphometry, soils and land management practices exist. This spatial approach enables patterns of synchroneity and divergence in river response to flooding to be investigated.

The aims of this study are as follows:
(1) to date coarse flood sediments and evaluate their hydrodynamic depositional environments;
(2) to identify changes in the temporal and spatial frequencies and magnitudes of major floods, and to assess how these might relate to climate, vegetation, catchment morphometry and land-use changes;
(3) to examine the role of floods in longer-term valley floor development viewed in the context of Holocene environmental changes in the UK and European uplands.

STUDY AREA

Investigations have focused on the upland catchments of the rivers Swale, Ure, Wharfe, Nidd and Aire, all major tributaries of the Yorkshire Ouse, within the limits of the Yorkshire Dales National Park (Figure 18.1). These rivers drain the Askrigg Block, a dissected peneplain delineated by the Dent Fault in the west and the Craven Fault in the south (De Boer, 1974). The peneplain is composed of Carboniferous rocks (Figure 18.2) with extensive major faulting of Hercynian age (Pounder, 1989).

The 25 m km^{-1} eastwards dip of the Askrigg Block, coupled with uplift and warping, has resulted in the eastwards drainage pattern of the rivers Swale, Ure and Nidd with the Wharfe and Aire draining to the south (De Boer, 1974). Extensive drainage modification occurred during the late Pleistocene and earlier glaciations.

South-westerly air streams prevail, bringing up to 2000 mm of rain a year in the headwaters of the Ure and Wharfe, with a marked easterly decline and dependence on altitude (Yorkshire Ouse and Hull River Authority, 1969). There is also a considerable period of snow cover, with heavy rainfall on snow generating many of the largest floods (Radley and Simms, 1971).

Geomorphological investigations have been concentrated in small, high relief (average basin relief 350 m) tributary catchments with steep ($> 0.07 \text{ m m}^{-1}$) cobble and boulder-bed channels and occasional bedrock reaches inset within Pleistocene and Holocene terraced river gravels or bedrock with side slopes covered with glacial and periglacial deposits. Basin vegetation is heather (*Calluna vulgaris*) and bracken (*Pteridium aquilinum*) moorland.

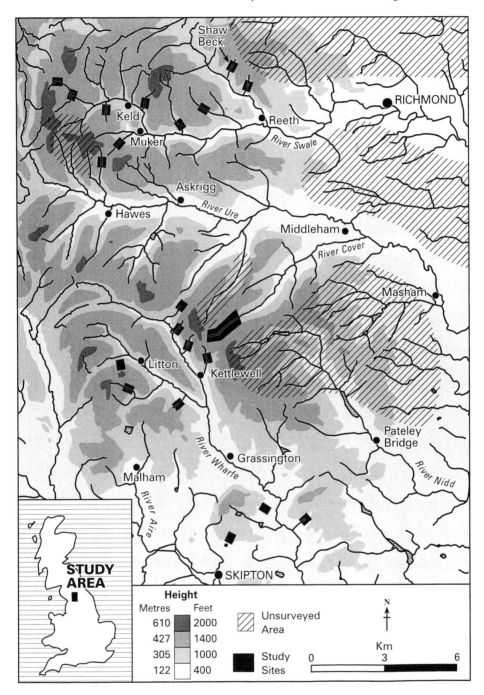

Figure 18.1 Drainage network and relief of the Yorkshire Dales with the location of study sites

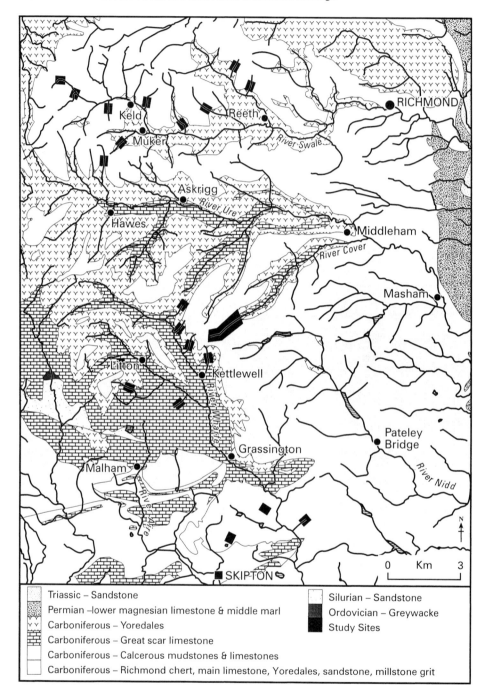

Figure 18.2 Geology of the Yorkshire Dales with the location of study areas at the contact of millstone grit and the Yoredale Series

LICHENOMETRIC DATING AND METHODOLOGY

A basin-wide survey of 90 small (< 20 km²) upland, side-valley tributaries of the rivers Swale, Ure, Nidd, Wharfe and Aire was undertaken. In total, 26 catchments exhibiting suites of well-preserved cobble and boulder overbank flood deposits were identified for further investigation and these were located in Swaledale (River Swale), Wensleydale (River Ure), Wharfedale (River Wharfe) and Coverdale (River Cover).

Lichenometry has become an established method for dating rock surfaces since the early work of Beschel (1961, 1973). Although originally used for dating glacial moraines, a number of workers have successfully applied the technique to UK fluvial environments (Milne, 1982; Innes, 1983; Harvey et al., 1984; Macklin, 1986; Macklin et al., 1992). The fundamental assumption underpinning the technique is that lichen thalli size is directly related to substrate age.

In this study two crustose lichen species were used: *Rhizocarpon geographicum* agg. to date sandstone lithologies and *Aspicilia calcarea* to date limestone rock types. The *R. geographicum* agg. growth curve is based on 129 lichen measurements on gravestones in 32 graveyards throughout the study region. Linear equations have been applied to age–size scatter plots of *observed* thalli size in order to generate a growth rate of 0.36 mm year^{-1} (Figure 18.3) which is comparable to that of 0.39 mm year^{-1} obtained by Macklin et al. (1992) in Northumberland, and represents an empirical relationship between substrate age and thalli size rather than a growth curve *per se*. Lichenometry relies on three further assumptions:

(i) That the largest lichens present (usually a mean of the largest three or five is taken) are of the same age as the substrate and that the substrate was clean of species before invasion. Gregory (1976) showed that the majority of lichens are

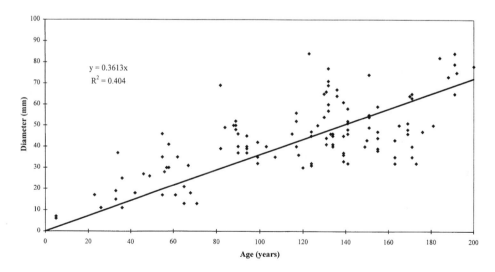

Figure 18.3 *Rhizocarpon geographicum* agg. growth curve for the Yorkshire Dales. The linear regression is significant to the 99% level

Table 18.1 Correlation coefficients for *Rhizocarpon geographicum* agg. and environmental variables

	Pearson's	Linear regression
Size/altitude	0.228	0.052
Size/average annual rainfall	0.236	0.056
Size/aspect	−0.178	0.032
Size/age	0.501	0.404[a]
Multiple regression (of above)	—	0.424

[a] Significant at 99% level.

    destroyed by transport during floods and anomalous large lichens on reworked material are easily recognised.

(ii) That lichen colonisation takes place immediately after the substrate becomes exposed. A maximum invasion lag-time of five years (based on the youngest aged individuals sampled in the study area) is believed to be appropriate for the Yorkshire Dales region. This is not as large as the eight to ten years suggested by Innes (1983); however this five-year lag limit provides the error margin when comparing dates obtained for deposits to documented floods (see below).

(iii) That the growth rates of lichen species are uniform within the study region, and that lichen size reflects only substrate age. Thus, the effects of environmental factors such as aspect (Armstrong, 1975), precipitation (Worsley, 1981) and altitude were investigated using Pearson's correlation, linear and multiple regressions models, with only age being significant in determining thalli size at the 99% level (Table 18.1).

The mean of the five largest lichens present on each boulder and gravel flood unit was selected for dating purposes. *Aspicilia calcarea* thalli size data from flood deposits containing both calcareous and sandstone lithologies were correlated to the *R. geographicum* agg. growth curve to obtain an average growth rate of 0.93 mm year^{-1}.

Flood deposits in each catchment (66 in total) were surveyed using EDM techniques to determine elevation above present channel surfaces and dimensions. Field observation, survey data and air photograph interpretation were combined to produce geomorphological maps of channels and floodplains in order to distinguish between flood units (Figure 18.4). Only coarse-grained lobes, splays and berms (*sensu* Macklin et al., 1992), known to be deposited in high energy, extreme flows, are included in this analysis. Re-working of flood units has been substantially limited by channel incision isolating units from all but the highest stage events. Additionally, the intermediate axis of the five largest clasts from a sample population of the 15 largest boulders on each unit were measured to estimate palaeocompetence and flow magnitude.

SPATIAL DISTRIBUTION OF FLOOD DEPOSITS

The spatial distribution of coarse-grained flood deposits appears to be primarily controlled by catchment relief and geology. Steep gradient channels (> 0.07 m m^{-1})

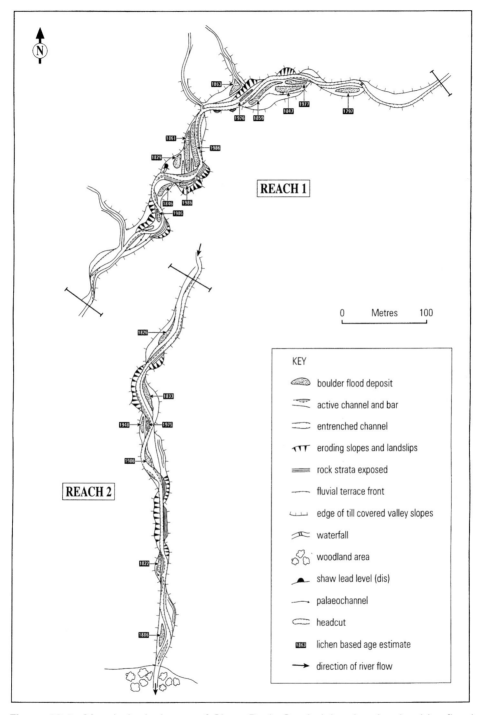

Figure 18.4 Morphological map of Shaw Beck, Swaledale, showing boulder flood deposits, palaeochannels and river terraces. Lichen age estimates of flood units are indicated

generate high energy Newtonian and hyperconcentrated flows capable of mobilising large clasts and depositing them overbank in lower gradient reaches of valley floor expansion (Figure 18.1). Flood deposits in Swaledale, Wensleydale, Coverdale and Wharfedale are restricted to high relief valleys (average basin relief 350 m) that are deeply incised into the contact between the Millstone Grit moorlands and the Yoredale Series (Figure 18.2). Erosion of less resistant cherts, mudstones and shales supplies large blocks of overlying millstones, sandstones and limestones to the channel system to be mobilised in large floods. The lithological content of flood units reflects the dominance of more resistant rock types. Thus, deposits in Swaledale are composed of sandstone and Millstone Grit, in Wharfedale limestone and in Coverdale a combination of limestone, grit and sandstone lithologies.

HISTORICAL VARIATIONS IN FLOOD FREQUENCY AND MAGNITUDE IN THE YORKSHIRE DALES

Lichen-dated lobes, splays and berms have been attributed to major floods recorded in archive sources from 1636 to the present (Mayhall, 1860; Bogg, 1892; Williams, 1957; Radley and Simms, 1971; Yorkshire Water Authority, 1980; Jones *et al.*, 1984). In total, 35 geomorphologically effective floods that have transported boulder size (> 0.3 m b-axis) material have been recorded in the Yorkshire Dales. In Figure 18.5 the temporal frequency distribution of deposits is plotted and there is clearly an increase in the number of flood units per decade from *c.* 1750 onwards, with only single units related to individual floods prior to this date. The late 18th century, notably the 1790s, was a period characterised by frequent large floods throughout the Yorkshire Dales, and was followed by a slight decline in flood frequency through the early 19th century until the 1820s. A further, more pronounced decline in the number of flood units is evident in the early 20th century until *c.* 1940, with a smaller peak continuing to *c.* 1990. There is generally a good relationship between the frequency of large floods identified by Macklin *et al.* (1992) in the northern Pennines with those in the Yorkshire Dales. The two earliest periods (1780–1820 and 1840–1880) of increased flood frequency coincide with irregular warming trends after the coolest phase of the Little Ice Age (Lamb, 1977, 1982). Table 18.2 demonstrates that these two periods were characterised by mainly winter and autumn floods associated with cooler, wetter phases of enhanced meridional circulation of the circumpolar vortex (Rumsby and Macklin, 1996). This resulted in lower winter temperatures, reduced rates of evapotranspiration, increased soil wetness and higher magnitude peak discharges (Rumsby and Macklin, 1994). However, in the Yorkshire Dales flood frequency remained relatively high between these two periods (1820–1840), when spring flooding in particular was more common. The third phase (1920–1950) distinguished by Macklin *et al.* (1992) in the northern Pennines is not so clearly defined in the Yorkshire Dales where most major floods date to the 1940s and between 1960 and 1990.

Figure 18.6 shows boulder sizes in flood units, with the largest floods dated to the late 18th and late 19th centuries. A decline in clast size, and by inference flood magnitude, is evident around 1800 and again at 1900. This trend is similar to that found by Macklin *et al.* (1992) in the northern Pennines where a decrease in flood

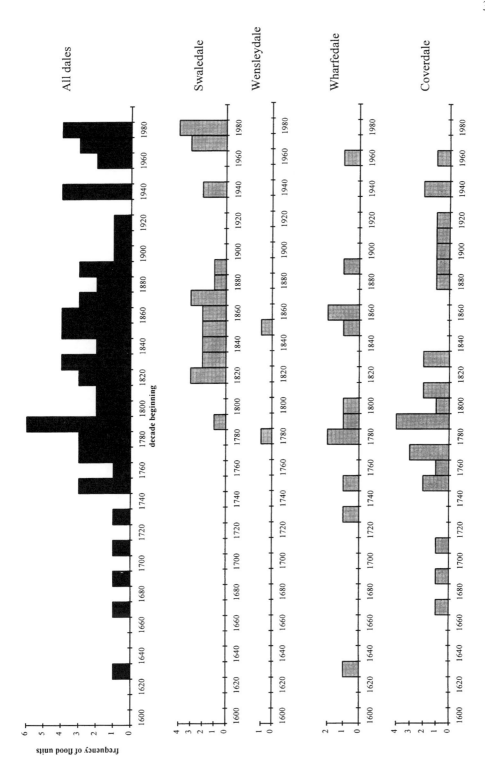

Figure 18.5 Temporal frequency of dated flood deposits in the Yorkshire Dales. A sharp increase in the number of units is apparent in the late 18th century

Table 18.2 Seasonality of geomorphologically effective floods in the Yorkshire Dales

	1600–50	1650–1700	1700–50	1750–1800	1800–50	1850–1900	1900–50	1950–present
Spring (F, M, A)					4	1	4	1
Summer (M, J, J)					2	2	1	
Autumn (A, S, O)		1			1	6		4
Winter (N, D, J)			3	6	5	6	2	4
Unknown	1	1	3	6	2			

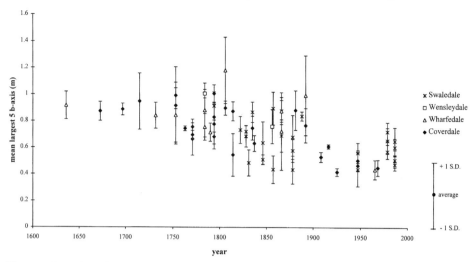

Figure 18.6 *b*-Axis boulder sizes for dated flood deposits. A general decline in clast size is apparent from *c.* 1800 onwards, with peaks in the late 18th and late 19th centuries

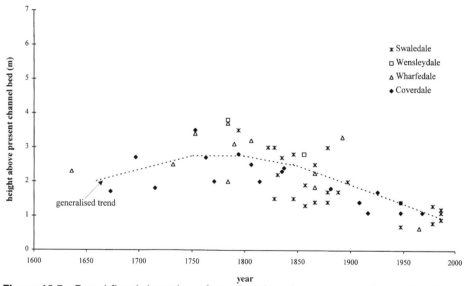

Figure 18.7 Dated flood deposit surface elevation above present channel bed level. The generalised trend indicates channel incision from the beginning of the 19th onwards

competence since the late 18th century was demonstrated, particularly over the last 50 years. This reduction in clast size is thought to partly reflect a decline in the availability of large clast sizes as entrenchment commenced, and also a real decrease in flood magnitudes since the end of the 18th century.

Vertical relationships of flood units are displayed in Figure 18.7, with the elevation of the flood deposit surface plotted above present channel bed level. A general tendency towards incision is evident in the last 200 years. Prior to this period, from c. 1600 to c. 1750, channels appeared either stable or aggrading, with marked aggradation occurring between c. 1750 and 1800.

FLOODS AND LATE HOLOCENE VALLEY FLOOR DEVELOPMENT IN THE YORKSHIRE DALES

c. 1600–1750

Flood magnitude and competence were high in the Yorkshire Dales during the coolest, wettest phase of the last Neoglacial cycle (Lamb, 1977), though relatively few major flood units are evident. The hillslope/channel sediment supply system appears to have been well coupled, particularly in Wharfedale, with channel bed stability and/or aggradation evident in all of the upland tributaries studied.

c. 1750–1800

A marked increase in the number and magnitude of floods is evident, associated with anomalous atmospheric circulation patterns and frequent severe storms (Lamb, 1981, 1991) during this period. This was accompanied by channel aggradation notably in Coverdale, indicating that the hillslope/channel sediment supply system was well coupled. In Wharfedale, however, sediment volumes began to decline though similar peaks in flood competence are evident.

c. 1800–1850

In the early decades of the 19th century, flood magnitude and frequency began to decline during a period of ameliorating climate (Haas, 1996). Sediment supply and deposit volumes decreased, although the number of dated flood units remained relatively high compared to the late 17th century. During this period incision commenced in many parts of the Yorkshire Dales, resulting in isolation of channels from hillslope sediment supplies.

c. 1850–1900

Towards the end of the 19th century a small increase in flood competence and frequency is apparent. Some divergence in river response is evident, with aggradation in two tributary catchments in Swaledale being affected by metal mining activities, while in Coverdale and Wharfedale channel incision continued.

c. 1900–Present

A further decline in flood competence and frequency is evident after 1900. Incision continued at a slightly lower rate than during the first half of the 19th century as temperatures reached the modern climate optimum in 1920 (Jones and Bradley, 1992). Increased flood magnitude and frequency between 1940 and 1990 is recorded in

Coverdale and Swaledale following temporary re-coupling of the hillslope/channel sediment supply system during major floods in 1947, 1978 and 1986. This was also a period of extensive upland drainage (moorland gripping) in the Yorkshire Dales (Robinson, 1990) which increased flood magnitudes and reduced times to peak (Robinson, 1985), and also probably enhanced erosion and sediment availability (Stewart and Lance, 1983).

Upland rivers in the Yorkshire Dales underwent valley floor transformations which, both in timing and nature, are similar to those evident in streams in the northern Pennines (Macklin *et al.*, 1992). Large-scale woodland clearance took place in both regions before the 17th century (Tinsley, 1975; Smith, 1986), and it is likely that agricultural improvement in the late 18th and 19th centuries had little effect on remote upland moorlands. For example, significant increases in livestock numbers associated with the expansion of the textile industry occurred predominantly on the valley floors and lower slopes of the Yorkshire Dales (Raistrick, 1968). More important – in terms of valley floor sedimentation – would have been the onset of large-set metal mining in the late 18th and early 19th centuries. This only affected parts of Swaledale and Wharfedale (Fieldhouse and Jennings, 1978), with mining supplying large amounts of coarse-grained debris to river channels (Newson and Macklin, 1990). Extension of moorland gripping from the 1940s (Robinson, 1990), particularly in Coverdale, is also likely to have increased sediment delivery in many catchments.

CONCLUSIONS

Factors that control the temporal and spatial distribution of flood units deposited since *c.* 1636 in the Yorkshire Dales have been elucidated using lichen dating techniques. Most flood units are found in high relief tributary streams incised into the Yoredale facies where erosion of less resistant cherts, mudstones and shales enables large blocks of overlying limestones, sandstones and grits to enter the channel system, be entrained and deposited in high energy Newtonian and transitional flows. Three periods (*c.* 1750–1800, *c.* 1850–1900 and *c.* 1940–1990) of enhanced flood magnitude and frequency are identified. The first period, *c.* 1750–1800, resulted in a major increase in sediment supply and widespread valley floor aggradation. The second phase, between 1850 and 1900, witnessed widespread incision except in some Swaledale catchments where sediment supply was augmented by metal mining activities. Declining flood competence and a reduction in sediment supply since *c.* 1800 led to incision into Holocene and Pleistocene alluvial fills and in some rivers bedrock. Results from the Yorkshire Dales are comparable with similar rivers elsewhere in Britain and Europe and attest to the sensitivity of upland environments to abrupt, decadal-scale climate changes.

ACKNOWLEDGEMENTS

The authors wish to thank NERC for supporting their investigations in the Yorkshire Dales through a research studentship (SPM) and grant (GT4/94/433/A). We would also like to thank

Mr Albert Henderson for assistance with lichen identification, the School of Geography Graphics Units for diagram preparation, and Dr Jamie Woodward for comments on the final draft.

REFERENCES

Armstrong, R. A. 1975. The influence of aspect on the pattern of seasonal growth in the lichen *Parelia glabrata* ssp. *fulignosa* (*Fr. ex Duby*) *Laund*. *New Phytologist*, **75**, 245–251.

Ballantyne, C. K. 1991. Late Holocene erosion in upland Britain: climatic deterioration or human influence? *Holocene*, **1**, 81–85.

Beschel, R. E. 1961. Dating rock surfaces by lichen growth and its applications to glaciology and physiography (lichenometry). In Raasch, G. O. (Ed.) *Geology of the Arctic* 2. University of Toronto Press, Toronto, 1044–1062.

Beschel, R. E. 1973. Lichens as a measure of the age of recent moraines. *Arctic and Alpine Research*, **5** 303–309.

Bogg, E. 1892. *A Thousand Miles in Wharfedale and the Basin of the Wharfe*. Goodall & Suddick, Leeds.

Brazier, V. and Ballantyne, C. K. 1989. Late Holocene debris cone evolution in Glen Feshie, western Cairngorm Mountains, Scotland. *Transactions of the Royal Society of Edinburgh: Earth Sciences*, **80**, 17–24.

Brazier, V., Whittington, G. and Ballantyne, C. K. 1988. Holocene debris cone evolution in Glen Etive, Western Grampian Highlands, Scotland. *Earth Surface Processes and Landforms*, **13**, 525–531.

De Boer, G. 1974. Physiographic evolution. In Rayner, D. H. and Hemmingway, J. E. (Eds) *The Geology and Mineral Resources of Yorkshire*. Yorkshire Geological Society, 271–292.

Evans, R. 1996. Sensitivity of the British landscape to erosion. In Thomas, D. S. G. and Allison, R. J. (Eds) *Landscape Sensitivity*. John Wiley, Chichester, 189–210.

Fieldhouse, R. and Jennings, B. 1978. *A History of Richmond and Swaledale*. Phillimore, York.

Gregory, K. J. 1976. Lichens and the determination of river channel capacity. *Earth Surface Processes*, **1**, 273–285.

Haas, H. C. 1996. Northern Europe climate variations during late Holocene: evidence from marine Skagerrak. *Palaeogeography, Palaeoclimatology, Palaeoecology*, **123**, 121–145.

Harvey, A. M. 1987. Sediment supply to upland streams: influence on channel adjustment. In Thorne, C. R., Bathurst, J. C. and Hey, R. D. (Eds) *Sediment Transport in Gravel-bed Rivers*. John Wiley, Chichester, 121–150.

Harvey, A. M. 1991. The influence of sediment supply on the channel morphology of upland streams: Howgill Fells, Northwest England. *Earth Surface Processes and Landforms*, **16**, 675–684.

Harvey, A. M. and Renwick, W. H. 1987. Holocene alluvial fan and terrace formation in the Bowland Fells, Northwest England. *Earth Surface Processes and Landforms*, **12**, 249–257.

Harvey, A. M., Alexander, R. M. and James, P. A. 1984. Lichens, soil development and the age of Holocene valley floor landforms: Howgill Fells, Cumbria. *Geografiska Annaler*, **66A** 353–366.

Innes, J. L. 1983. Lichenometric dating of debris flow deposits in the Scottish Highlands. *Earth Surface Processes and Landforms*, **8**, 579–588.

Jones, P. D. and Bradley, R. S. 1992. Climatic variations in the longest instrumental records. In Bradley, R. S. and Jones, P. D. (Eds) *Climate Since AD 1500*. Routledge, London, 246–268.

Jones, P. D., Ogilvie, A. E. J. and Wigley, T. M. L. 1984. *River Flow Data for the UK: Reconstructed Data Back to 1844 and Historical Data Back to 1556*. Climatic Research Unit, University of East Anglia, Norwich, UK.

Lamb, H. H. 1977. *Climate: Past, Present and Future. Volume 2: Climate History and the Future*. Methuen, London.

Lamb, H. H. 1981. Climate fluctuations in historic times and their connection with transgressions of the sea, storm floods and other coastal changes. In Verhulst, A. and Gottshalk,

M. K. E. (Eds) *Transgressies en Occupatiegeschiedenis in de Kustgebieden van Nederland en Belgie*. Rijksuniv, Gent, 251–290.

Lamb, H. H. 1982. *Climate, History and the Modern World*. Methuen, London.

Lamb, H. H. 1991. *Historic Storms over the North Sea, British Isles and Northwest Europe*. Cambridge University Press.

Macklin, M. G. 1986. Channel and floodplain metamorphosis in the River Nent, Cumberland. In Macklin, M. G. and Rose, J. (Eds) *Quaternary River Landforms and Sediments in the Northern Pennines, England. Field Guide: Newcastle, England*. British Geomorphological Research Group/Quaternary Research Association, 19–33.

Macklin, M. G., Rumsby, B. T. and Heap, T. 1992. Flood alluviation and entrenchment: Holocene valley-floor development and transformation in the British uplands. *Geological Society of America Bulletin*, **104**, 631–643.

Mayhall, J. 1860. *The Annals of Yorkshire: From the Earliest Period to the Present Time. Volume 1 – 1856 BC to 1859 AD*. H. C. Johnson, Leeds.

Milne, J. A. 1982. *River Channel Changes in the Harthope Valley, Northumberland, Since 1897*. University of Newcastle upon Tyne, Department of Geography Research Series, 13.

Newson, M. D. and Lewin, J. 1991. Climate change, river flow extremes and fluvial erosion scenarios for England and Wales. *Progress in Physical Geography*, **15**, 1–17.

Newson, M. D. and Macklin, M. G. 1990. The geomorphologically effective flood and vertical instability in river channels: a feedback mechanism in the flood series for gravel-bed rivers. In White, W. R. (Ed.) *International Conference on River Flood Hydraulics*. John Wiley, Chichester, 123–141.

Passmore, D. G., Macklin, M. G., Brewer, P. A., Lewin, J., Rumsby, B. T. and Newson, M. D. 1993. Variability of late Holocene braiding in Britain. In Best, J. L. and Bristow, C. S. (Eds) *Braided Rivers*. Geological Society Special Publication No. 75, 205–229.

Pounder, E. J. 1989. *Classic Landforms of the Northern Dales*. The Geographical Association, Sheffield.

Radley, M. A. and Simms, C. 1971. *Yorkshire Flooding – Some Effects on Man and Nature*. Ebor Press, York.

Raistrick, A. 1968. *The Pennine Dales*. Eyre & Spottiswoode, London.

Richards, K. S., Peters, N. R., Robertson-Rintoul, M. S. E. and Switsur, V. R. 1987. Recent valley floor sediments in the North York Moors: evidence and interpretation. In Gardiner, V. (Ed.) *International Geomorphology Part I*. John Wiley, Chichester, 869–883.

Robertson-Rintoul, M. S. E. 1986. A quantitative soil stratigraphic approach to the correlation and dating of post-glacial river terraces in Glen Feshie, western Cairngorms. *Earth Surface Processes and Landforms*, **11**, 605–617.

Robinson, M. 1985. The hydrological effects of moorland gripping: a re-appraisal of the Moor House research. *Journal of Environmental Management*, **21**, 205–211.

Robinson, M. 1990. *Impact of Improved Land Drainage on River Flows*. Institute of Hydrology Report No. 113, Institute of Hydrology, Wallingford.

Rumsby, B. T. and Macklin, M. G. 1994. Channel and floodplain response to recent abrupt climate change: the Tyne basin, Northern England. *Earth Surface Processes and Landforms*, **19**, 499–515.

Rumsby, B. T. and Macklin, M. G. 1996. River response to the last Neoglacial cycle (the "Little Ice Age") in northern, western and central Europe. In Branson, J., Brown, A. G. and Gregory, K. J. (Eds) *Global Continental Changes: The Context of Palaeohydrology*. Geological Society Special Publication No. 115, 217–233.

Smith, R. T. 1986. Aspects of the soil and vegetation history of the Craven District of Yorkshire. In Manby, T. G. and Turnbull, P. (Eds) *Archaeology in the Pennines*. B. A. R. British Series 158, Oxford, 3–28.

Stewart, A. J. A. and Lance, A. N. 1983. Moor-draining: a review of impacts on land-use. *Journal of Environmental Management*, **17**, 81–99.

Tinsley, H. M. 1975. The former woodland of the Nidderdale Moors (Yorkshire) and the role of early man in its decline. *Journal of Ecology*, **6**, 1–26.

Williams, H. B. 1957. Flooding characteristics of the River Swale. Unpublished PhD thesis,

University of Leeds.

Worsley, P. 1981. Lichenometry. In Goudie, A. (Ed.) *Geomorphological Techniques*. Allen and Unwin, London, 302–305.

Yorkshire Ouse and Hull River Authority 1969. *Survey of Water Resources*. Water Resources Act 1963, Section 14, Leeds, Yorkshire.

Yorkshire Water Authority 1980. *Land Water Drainage Survey: Part A – Covering Report*. Yorkshire Water Authority, Rivers Division, Leeds.

Section 5

GLACIERISED BASINS

19 Environmental Change and Sediment Yield from Glacierised Basins: The Role of Fluvial Processes and Sediment Storage

JEFF WARBURTON
Department of Geography, University of Durham, UK

INTRODUCTION

Understanding sediment transfer in glaciofluvial systems provides essential information on glacial activity, responses to environmental change and in some cases warning of glacial hazards. Sediment yields from glacierised or recently glaciated valley systems are assumed to be some of the highest in the world due to a combination of high relief erosive environments, large rates of sediment supply and substantial rates of seasonal runoff (Jansson, 1988). Recent work on sediment budgets comparing sediment yields from glacial and non-glacial basins has started to question the simplicity of these assumptions (Hicks *et al.*, 1990; Harbor and Warburton, 1993). In common with many other fluvial systems the role of sediment storage is important. Phillips (1991) states that "sediment storage within a drainage basin may be the single most important aspect of fluvial sediment systems in terms of determining fluvial system response to environmental change."

The aim of this chapter is to assess the importance of fluvial processes and floodplain storage in controlling sediment yield from glacierised and glaciated valley systems. First the chapter will consider key concepts for evaluating sediment transfers; secondly, sediment transfer and floodplain storage will be considered over short (annual/interannual), medium and long timescales; and finally, the influence of valley and floodplain geometry will be considered as an important control on sediment transfer. This last point will be evaluated with reference to a series of flume studies which provide a useful model of sediment transfer in coarse-bed rivers. There is a bias in the discussion toward alpine proglacial processes although the arguments are of a more general nature.

Fluvial Processes and Environmental Change. Edited by A. G. Brown and T. A. Quine.
© 1999 John Wiley & Sons Ltd.

BACKGROUND

Glacial meltwater rivers typically consist of multiple-thread (braided) low-sinuosity channels. These streams carry large quantities of sediment much of which is deposited in outwash or sandur deposits. There is usually a distinct proximal–distal transition from gravel-dominated grain sizes at the ice front to fine sand and silt. A distinction is often made between valley sandar confined between valley walls, and plain sandar which are essentially unconfined distributary systems. The morphology of meltwater streams results from the interaction of entrainment, transport and deposition processes with mobile and stable bed elements. Spatial variability in the stability and instability of the bed coupled with variations in sediment supply largely explain the episodic nature of sediment transport and channel change in this type of environment. Sedimentary processes are dominated by aggradation and lateral accretion of gravel bars (Miall, 1978). Discharge in glacier-fed rivers is highly seasonal and shows great variability. Runoff depends on the climate acting on the glacier surface and the mechanisms controlling the flow of water within the glacier and at its bed (Röthlisberger and Lang, 1987). There are pronounced seasonal and diurnal flow patterns which are predominantly temperature driven (Church, 1972; Church and Gilbert, 1975). Several distinct phases of meltwater runoff are generally recognised: break-up and the nival (snowmelt) flood, summer discharge when there is a strong correlation between melt rates and stream discharge, then freezeback when melt declines and low flows re-establish. Irregularity in this general pattern can be caused by the rapid drainage of meltwater stored in the glacier system (outbursts) or from intense rainfall activity. Price (1980) suggests the most rapid and dramatic changes that take places in proglacial areas are due to meltwater drainage. Sandar extend from the immediate proglacial zone for several kilometres downstream from the ice front although the influence of the glacial hydrology and sediment transport is detectable over much greater distances (e.g. Ferguson, 1984).

Proglacial river systems provide the key link between glacial processes and the wider environment. To date, the majority of research has concentrated on proglacial rivers in either alpine areas (e.g. European Alps: Warburton, 1990, 1994; North America) or the subarctic (e.g. Iceland: Maizels, 1993a, 1993b). In contrast, relatively little work has been conducted on high arctic proglacial rivers (Church, 1972). Information on sediment dynamics in a variety of glacial catchments is required to test existing hypotheses about the relative merits of glacial and non-glacial processes and the nature of sediment delivery (Harbor and Warburton, 1992, 1993).

CONCEPTS FOR EVALUATING FLUVIAL SEDIMENT STORAGE IN RELATION TO ENVIRONMENTAL CHANGE

The task of determining how fluvial processes in glacierised catchments will respond to environmental change hinges on understanding how sediment and water production will be affected in the glacio-hydrological system and how proglacial floodplains will respond to alteration in sediment and water supply and changing environmental variables, e.g. vegetation. In order to understand the coupling between the glacio-hydrological system and the glaciofluvial system it is necessary to have a firm grasp of the processes operating. For example, climate models examining the effect of a doubling of CO_2 scenario on the glaciers of south-eastern Alaska indicate that the mass balance would intensify and both precipitation and runoff would increase (Davidovich and Ananicheva, 1996). For the purpose of this chapter the response of proglacial floodplains is considered in relation to changes in sediment and water inputs from the glacio-hydrological system. No attempt is made to predict what these glacio-hydrological changes might be or the exact nature of the coupling between the two sub-systems.

Proglacial floodplains are best conceptualised as storage zones (Figure 19.1). A simple sediment balance equation can be written in terms of

$$S_O = S_I + \Delta S + S_T \qquad (1)$$

where S_O is sediment output, S_I is sediment input, ΔS is the net change in sediment storage and S_T is the sediment input from tributaries. ΔS is calculated as the net difference between erosion (S_E) and deposition (S_D) over the period of study. This is determined for a defined area of the channel bed over what is usually an arbitrary, seasonal or annual time period. Similarly the water balance may also be written

$$Q_O = Q_I + Q_T - Q_L \qquad (2)$$

where Q_O is water output, Q_I is water input, Q_T is the water input from tributaries and Q_L is loss or leakage from the system. These formulations are useful in that they provide a framework for evaluating the role of the floodplains in regulating the input/output from glacierised catchments (Harbor and Warburton, 1993). Although not shown explicitly in Figure 19.1, lakes can easily be incorporated in the model because they are simply an element of ΔS and subject to the same changes in S_E and S_D

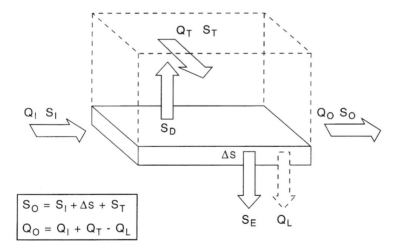

Figure 19.1 Definition sketch of the components of a fluvial sediment (mass) balance for a channel floodplain section

although the latter clearly dominates. This approach has been used by Warburton (1990) in assessing the importance of proglacial processes in modifying the stream load of a small glacier-fed stream in the southern Swiss Alps.

The importance of fluvial sediment storage in glacierised basins has been recognised for some time. Borland (1961) derived an empirical parameter λ based on glacier area (A_G), total drainage area (A_T) and river length between the glacier snout and sediment yield measurement site (L):

$$\lambda = (A_T/A_G)L \qquad (3)$$

The development of this parameter was based on the observation that total sediment load of glacier-fed streams decreases as the distance from the glacier increases. It was assumed that this was due to intervening storage represented by A_T and L in the parameter. Results based on Alaskan data initially proved promising; however, more widespread application (e.g. Guymon, 1974: Gurnell et al., 1996) has shown that area and channel length are poor indices of potential storage. Channel slope and floodplain width are perhaps better variables because they provide a better representation of average valley floor stream power (Ω) (per unit width) and hence conveyance.

$$\lambda_\Omega = (A_T/A_G)\Omega \qquad (4)$$

In this formulation information on channel length is retained in the slope term and valley width is incorporated in the calculation of stream power per width. However, the parameter is still flawed because the expected correlation between glacier area and sediment production is far too simplistic (Harbor and Warburton, 1993).

The importance of sediment storage in regulating downstream sediment transport has been recognised in a number of glacial environments. Hasholt (1996) working in Greenland recognises the importance of both sandar and lakes (with trapping efficiencies > 80%) in storing coarse sediment. He comments that this pattern does not always hold true because coarse sediment can be transported across lakes, rafted on drifting ice, and can be carried at below threshold entrainment when transported fixed to channel ice. Because of the great importance of sediment storage in determining sediment transport in glacierised basins, Harbor and Warburton (1993) argued that short-term sediment yield data are of limited value in interpreting glacier activity.

Non-glacial slope processes can have an important influence over valley and floodplain sediment storage. This depends to a large extent on the degree of hillslope–channel coupling. Dramatic recent examples of this were the 1991 Randa rock slides in the Matter Valley, south-west Switzerland (Götz and Zimmermann, 1993). On 18 April a major rock slide of approximately 20 million m^3 dammed the Vispa River with the maximum depth of the debris cone estimated to be 140 m. This had the potential of creating a new lake of 3.5 million m^3. This presented a major hazard due to the potential for failure of the natural dam and large-scale flooding downstream. For these reasons a 25 m deep relief channel was excavated through the deposits to drain the lake. In a natural setting this lake would have created new sediment storage potential for both suspended sediment and bedload. If the temporary dam had failed there would be some storage but also erosion downstream causing a major perturbation in the overall sediment balance. Alternatively landslides into the storage zone can greatly reduce storage volume and hasten the response of systems downstream.

The volume of stored sediment in a glacierised drainage basin reflects the balance between production and transport. It is generally recognised that contemporary streamloads are poor indicators of regional denudation rates (Trimble, 1975) and attempting to correlate sediment yield with simple catchment indices (e.g. area, glacial area, etc.) is fraught with difficulty (Harbor and Warburton, 1993). Sediment yield is influenced not only by drainage basin size but also by a number of other complicating factors:

(1) the response of different glacial systems: Alpine, Himalayan, Arctic (Bögen, 1996);
(2) the importance of rainfall events in flushing glacial sediments (Hicks *et al.*, 1990);
(3) uncertainties about the processes of glacial erosion. e.g. erosion and runoff under polythermal glacial regimes;
(4) uncertainties about the importance of glaciofluvial sediment transfer, variability in sediment delivery and the importance of large flushing events (Maizels, 1993a, 1993b), e.g. glacial outburst events such as at Grimsvötn in 1996;
(5) the imprint of glacial history: sediment dynamics under multiple glaciation are difficult to decipher in terms of sediment budgets (Jordan and Slaymaker, 1991);
(6) inappropriate present-day analogues for Holocene and Quaternary glacial activity (Clague, 1986);
(7) significant measurement errors both in sediment transport rates and storage volumes (Warburton and Beecroft, 1993).

SEDIMENT TRANSFER AND FLOODPLAIN STORAGE

Short-term, Annual and Interannual Timescales

In this section evidence of floodplain and channel sediment dynamics will be presented from Bas Glacier d'Arolla (Figure 19.2; Warburton, 1990, 1994). This is a small 7.6 km^2 glacier basin in the southern Swiss Alps. The proglacial zone extends in a narrow north–south trending valley train, approximately 300 m from the snout of the glacier to a meltwater intake and gauging structure. The floodplain is approximately 30–40 m wide and confined by steep bluff and valley side slopes. The average channel slope is 5° and the average channel width varies between 4 and 7 m (Figure 19.2). Channel bed material consists of bouldery gravels.

Working in this area, Warburton (1990) used a sediment budget approach to assess the importance of proglacial processes in modifying the sediment load of the main meltwater stream. Warburton demonstrated that the proglacial zone contributed 23% to the sediment load of the meltwater stream. Most of this sediment (95%) was derived from channel and floodplain sources whilst the majority of sediment transport (53%) occurred in a single outburst flood event. The only comparable short-term sediment budget study of a proglacial floodplain to that of Warburton (1990) is Hammer and Smith (1983). Their study of the Hilda Glacier in Alberta estimated that approximately 25% of all fluvial sediment is derived from supraglacial sediment, with the remainder (75%) being split approximately evenly between subglacial and channel bank sources.

The relative sensitivity of suspended load and bedload transport can be illustrated

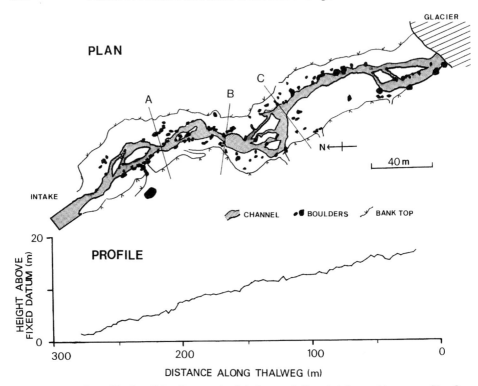

Figure 19.2 Bas Glacier d'Arolla proglacial channel, floodplain and long profile. Sections A, B and C refer to the data shown in Figure 19.5. Suspended sediment monitoring stations were located 40 m downstream from the glacier snout and just upstream of the intake structure

with reference to some more detailed examples. Figure 19.3 shows a short time series (five-minute observations over a three-day period) comparing suspended sediment concentrations at proximal and distal ends of the proglacial zone. Analysis of the series shows a close match between the proximal and distal sites with the highest cross-correlation at lag one, suggesting an average lag of approximately five minutes (Figure 19.3). The two raw data series are very similar, suggesting the channel acts as an efficient conduit for suspended sediment. Cumulative daily suspended sediment loads for the proximal (upper) and distal (lower) monitoring sites (Figure 19.4) show the flux in sediment storage within the channel. Over the period of monitoring there is a 9% increase in suspended sediment load downstream. This net gain in suspended sediment yield suggests entrainment of material from within channel sources and/or inputs from tributaries. Early in the ablation season suspended sediment yields are small. However, from the start of July suspended sediment load at the lower site exceeds that of the upper site (Figure 19.4). This suggests a net decrease in channel storage. This is followed by a short period where loads at the upper site exceed those at the lower site before a very large increase at the lower site once again. It appears that early in the ablation season suspended sediment is stored in the channel but as the melt season progresses and flows increase this is flushed out. The significance of these results is that

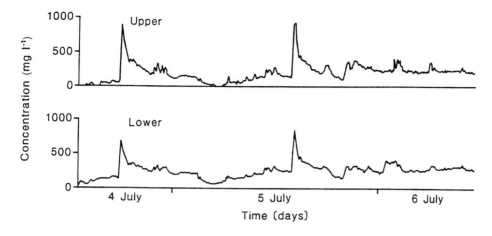
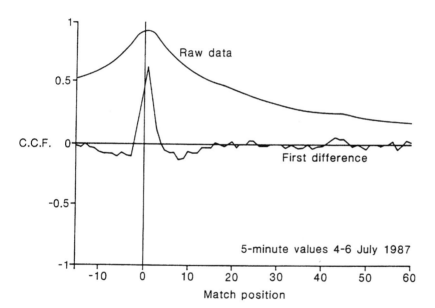

Figure 19.3 Cross-correlogram between upper (proximal) and lower (distal) suspended sediment series (5-minute values, 4–6 July 1987), Bas Glacier d'Arolla proglacial zone

tributaries contribute insufficient suspended sediment to disrupt mainstream sediment concentration patterns, and channel processes do little to modify suspended sediment transport.

In larger river systems, such as the Hunza in northern Pakistan, suspended sediment loads have been shown to increase downstream to the basin mouth (Ferguson, 1984). During the melt season (June to September) 99% of the annual load is transported; however, there is pronounced seasonal hysteresis suggesting a flush/exhaustion cycle from sediment in storage in the valley floors.

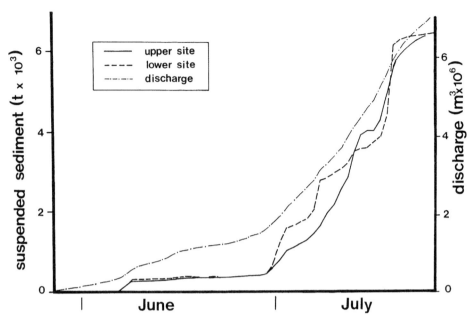

Figure 19.4 Comparison of cumulative suspended sediment loads monitored at proximal and distal ends of the Bas Glacier d'Arolla proglacial zone (1987)

Changes in daily coarse bed material storage within the channel are shown in Figure 19.5. This graph shows the cumulative sediment storage (m²) at three cross-sections – upper (C), middle (B) and lower (A) sections – on the proglacial channel (Figure 19.2). Early in the melt season changes in storage are relatively small, but from mid June onwards fluctuations are greater. The upper site shows a net loss of sediment whilst the lower sites show a net gain. Both sites A and B occur downstream of active braid bars in relatively flat areas of channel and may be natural sites of sedimentation. Site C is located at the end of a relatively steep transport reach (Figure 19.2). Correspondence between sites is poor, reflecting the large local variation in sediment storage within the channel. After 13 July a meltwater flood produced changes that were an order of magnitude greater than those shown in Figure 19.5 and which completely restructured the channel (Warburton, 1994).

Bögen (1995), working on the Jostedöla River glacierised basin in western Norway, has carried out sediment transport measurements which demonstrate that suspended sediment moves through the river basin in less than a day whereas bedload may take several decades to travel the whole length of the river basin. Bögen suggests that valley morphology, particularly steps and over-deepened basins, are of great importance in determining bedload movement. He also demonstrates that little permanent deposition of suspended sediment takes place on the outwash plain. Upstream and downstream sediment rating curves are almost identical (c.f. Figure 19.4).

Channel patterns change in response to the imposed flow and sediment load. Figure 19.6 shows four maps of the Bas Glacier d'Arolla stream channel over a two-year period (1986–1987). The maps refer to June and September planforms and therefore span the main period of meltwater activity. The main features of the channel patterns

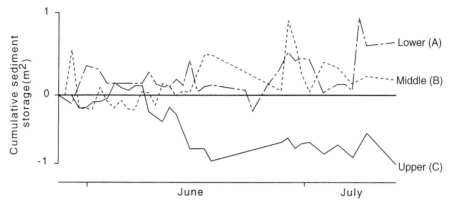

Figure 19.5 Cumulative changes in cross-section sediment storage at three main channel sections: A (lower), B (middle) and C (upper) shown in Figure 19.2. The observation period was 27 May to 13 July 1987

are the similarities between the plans at the same time of year. The June maps show a relatively narrow, single-thread, low sinuosity channel with limited braiding. The September pattern is very different with a more sinuous pattern and increased braiding. Although interannual variations do exist, the overall pattern seems to be consistent (Warburton, 1994). Although sediment yield and channel form follow a distinct pattern which is repeatable between years (Figure 19.6) this cycle may be radically altered by outburst floods. Outburst floods (low frequency, high magnitude events) often account for a substantial part of the sediment output from glacierised catchments (although there are marked contrasts in sediment yield between events; Warburton, 1994). Such large-scale flushing events need not always be related to the mass balance of the glacier and their importance for sediment transport means they may produce significant departures from more general trends.

These observations have several important implications for assessing environmental change. First, suspended sediment is only slightly modified by proglacial processes. It is predominantly related to sediment supply from the glacier. Secondly, bedload transport is spatially heterogeneous and more dependent on the hydraulics of the proglacial zone rather than direct glacial controls. Differences between bedload and suspended load modes of transport result in different rates of movement, residence times and transmission losses through the fluvial system. Channel morphology reflected in the channel planform illustrates a distinct and relatively stable annual pattern although rare events (e.g. outburst floods) may greatly perturb the sediment system.

Table 19.1 shows examples of sediment load components for several glacier catchments. Estimates of bedload and suspended load, as a percentage of the total load, are shown and the ratio of these components calculated. Where possible, data from consecutive years are shown to aid comparison. Bedload and suspended load are estimated by a number of methods. Bedload varies between 2 and 82% of the total load, and suspended load between 18 and 96%, although for many alpine settings both load components are roughly equivalent (bedload:suspended load ratio in the range 0.5–1.5). There are relatively few measurements of dissolved load as a component of total load; where such data exist, dissolved load tends to be a small component (1–3%),

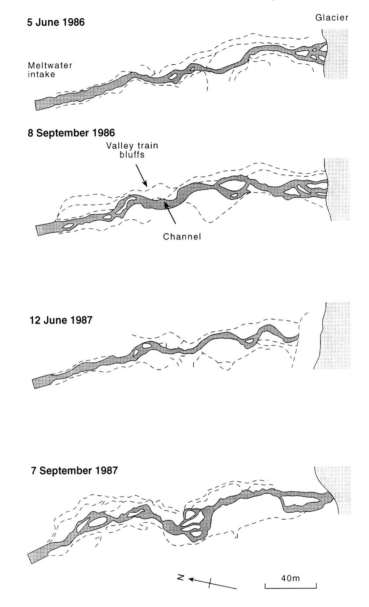

Figure 19.6 Planform maps of the Bas Glacier d'Arolla stream channel and valley floodplain proglacial zone, providing a comparison between channel development at the beginning and end of the ablation season and between two years

Table 19.1 Examples of sediment loads (percentages of total load) in glacierised catchments

Study	River/area	Years	Catchment area (km²)	Glacier area (km²)	% Bedload	% Suspended load	% Dissolved load	Bedload suspended load ratio
Church and Gilbert (1975)	Lewis River, Baffin Island	1963			82	18		4.56
		1964			77	23		3.34
Hasholt (1976)	Sermilikarea, East Greenland	1972	38	30	5–6	72–78	17–22	0.07
Kjeldsen and Østrem (1980)	Engabreen, Norway	1979	50	38	37	63		0.58
		1980			36	64		0.56
Hammer and Smith (1983)	Hilda, Alberta, Canada	1977	2.24		57	40	3	1.43
					54	45	1	1.21
Ferguson (1984)	Hunza, Karakoram		13 200		2–10	88–96	2	0.02
Gurnell et al. (1988)	Tsidjore Nouve, Switzerland	1986	4.8	3.4	43–51	49–57		0.75
		1987			36–44	56–64		0.56
Gurnell et al. (1988)	Bas Arolla, Switzerland	1986	7.6	5.3	74	24	2	3.08
		1987			63	36	1	1.75

although it can be up to 17–20% in some basins (e.g. Sermilikarea, East Greenland, Hasholt (1976)). Ratios of bedload to suspended load components vary from 0.02 to 4.56, emphasising the large differences in dominant transport modes. Differences in the proportion of bedload and suspended load are important in determining the response of particular glacial catchments to environmental change. Catchments dominated by suspended load will probably be more sensitive to change and accumulate smaller proximal sediment stores.

Medium-term, Decadal and Historical Timescales

Part of the problem of linking short-term process studies to longer-term trends in glacial sediment yield stems from a lack of studies over decadal or historical time periods. Many monitoring programmes initiated in the 1980s have now resulted in almost a decade of sediment yield data. The study of Ivory Lake, western Southern Alps, New Zealand (Hicks *et al.*, 1990) provides an excellent medium-term record of sedimentation rates over a 10-year period. Unfortunately, in the present context, such studies tell us little about floodplain storage and meltwater stream sediment yield.

Gurnell (1995) examined an 11-year record of sediment transport, discharge and glacier behaviour for three small alpine basins. Results showed some correspondence between the pattern of sediment transport and discharge and some relation between the movement of the glacier snout and sediment yield. However, these relationships were sufficiently weak to cast doubt over clear causality in the controlling variables. Probably the best direct evidence relating contemporary sediment transport data to climate change comes from southern Iceland. Lawler and Wright (1996) describe a record of suspended sediment concentration from the Jökulsá á Sólheimasandi basin spanning the period 1973–1992. It is observed that suspended sediment loads have decreased by 48% mainly due to decreases in the spring and autumn periods. Flows in summer have increased and the whole pattern relates to changes in seasonal air temperatures. This provides compelling evidence for the sensitivity of suspended sediment yields to climate fluctuations.

Over longer time periods where direct observations do not exist, conceptual models can serve as useful tools. Maizels (1979) proposed an aggradation/degradation model of proglacial sedimentation based on glacier mass balance. The model suggests aggradation during glacial advance due to fresh inputs of fluvioglacial sediment without a marked increase in flow competence. Church and Gilbert (1975) suggest that the occurrence of sedimentation zones is related to changing patterns in runoff and sediment supply which may be affected by climate and glacier fluctuations. They suggest that climate amelioration leads to a reduction in glacier size and activity, resulting in reduced sediment supply and an altered runoff regime. This could result in reduced aggradation and an increase in stream competence, leading to readjustment of the sandur profile and increased sediment transport downstream. Testing of such models is difficult because observational records are generally too short to be of great value (e.g. Gurnell, 1995) and long-term sedimentary records cannot be easily related to reconstructed mass balance fluctuations.

Part of the problem is attempting to relate the relatively sluggish response of glaciers to the more dynamic nature of proglacial sedimentation. For example, many glaciers

Table 19.2 Estimated size of onshore sediment traps South Island, New Zealand, over the last 12 000 years

Type of trap	Location	Average estimated volume (m^3)
Fans	Tasman Valley	10^7
Over-deepened valleys	South Island	10^9 to 10^{10}
Glacial lakes	South Island	10^8 to 10^{10}
Over-filled fjords	South Island	10^8 to 10^{10}

Source: Adams (1980).

in Svalbard show surge behaviour with relatively protracted active phases of 3–10 years and quiescent phases of 50–500 years (Dowdeswell et al., 1991). Since the early part of the 20th century glaciers in Svalbard have had consistently negative mass balances and glacial surge activity has decreased (Dowdeswell, 1995; Dowdeswell et al., 1995). In order to characterise this by direct measurements a long-term monitoring programme over several decades (spanning a surge cycle) would be required.

Long-term, Quaternary and Holocene Timescales

The largest landforms produced during deglaciation are sandurs (Price, 1980). These outwash plains build up from a series of accreting low-angled fans and can develop extensive deposits covering large areas, e.g. south-east Iceland. Interpreting sedimentation histories over long timescales (Holocene) is difficult where dating control is poor and the sedimentary record is incomplete. Meaningful estimates can only be made where the sediment budget can be relatively well constrained or where there are closed basins or lakes. On a large scale the response of a river to climate change depends on the morphological setting (main valley as opposed to upland tributary) and the position relative to the ice front (Krzyszkowski, 1996).

The importance of long-term terrestrial sediment storage was illustrated by Adams (1980) who estimated the size of onshore sediment traps for the South Island, New Zealand (Table 19.2). These sediment traps which include tectonic depressions, glacial depressions, lakes, fans and floodplains vary in their effectiveness. For example, shallow depressions only accumulate coarser sediments (e.g. the Canterbury Plains) whilst deeper lakes trap virtually all the stream load. Interestingly many of the traps are either a direct result of glaciation or are conditioned by it, e.g. over-deepened troughs and fan deposition sites (steep valley side walls). This idea of sediment storage and transfer being conditioned by past events is an important concept.

Church and Slaymaker (1989) presented evidence from British Columbian rivers that contradicted the established convention that specific sediment yield declines as drainage area increases due to sediment entering storage in footslopes and floodplains. The British Columbia data show increasing specific sediment yields at all spatial scales up to 3×10^4 km^2, which arises due to secondary erosion of stored Quaternary sediments along river valleys. This redistribution of sediment, delivered to the valleys at the start of the Holocene, represents an extended paraglacial cycle of the form first suggested for upland catchments by Church and Ryder (1972). The original Church and Ryder "paraglaciation" article has had widespread recognition so it is therefore surprising that sediment storage has not been given the attention it deserves (outside British Columbia) in interpreting sediment yield data (Harbor and Warburton, 1993).

Stratigraphic studies (Clague, 1986; Jordan and Slaymaker, 1991) and alpine lake records (Owens and Slaymaker, 1993, 1994) from British Columbia have also provided useful information on sediment dynamics. Clague (1986) demonstrated that for the majority of the Quaternary, sedimentation has largely been restricted to relatively brief periods of ice-sheet growth and decay. Non-glacial periods have been characterised by reworking of terrestrial sediment stores by fluvial processes (Church and Ryder, 1972). Jordan and Slaymaker (1991) working in the Lillooet River Basin constructed a Holocene sediment budget in order to elucidate sediment sources and storage components within the 3150 km^2 catchment. The major sediment sources were glaciers, Neoglacial deposits, debris flows and landslides in the Quaternary volcanic complex. However, these components only explain half the observed sediment yield. In the Lillooet River Basin the late Pleistocene paraglacial cycle is not as significant a sediment source as in other large basins. Volcanic activity and Neoglaciation, in producing episodic sediment sources, appear to be just as important. These two studies indicate the complexity of interpreting sedimentation histories when dealing with large river valleys or major drainage basins. In small alpine lakes which preserve late Holocene sedimentary records, estimates of erosion rates are difficult to determine because of non-catchment sources of sediment. Owens and Slaymaker (1993) suggest that non-catchment sediment sources may contribute between 55 and 99% of the total sediment accumulated over the last 2350 years. In the three lakes studied, sediment yields are generally low and show marked temporal variability which appears to be related to palaeoenvironmental changes (climate and sediment supply) over the post-glacial period.

An alternative method of investigating longer-term fluvial changes in glacierised catchments is to adopt a modelling approach. Maizels (1993a) has used this to good effect in simulating the stratigraphic and geomorphological evolution of a large Icelandic sandur over the Holocene period. Maizels used a 1–2 dimensional model (SEDSIM) to predict fluvial depositional sequences in response to variations in runoff and sediment supply regimes and changes in relative sea-level (eustatic and isostatic). The predictions highlight the major importance of extreme events (outburst floods) in the development of the alluvial sequences. Flow and sediment supply are critical in the evolution of proglacial floodplains.

INFLUENCE OF VALLEY AND FLOODPLAIN GEOMETRY ON SEDIMENT TRANSFER: HYDRAULIC MODELLING OF GRAVEL-BED RIVERS

A criticism of short-term geomorphological monitoring is that it does not always fully assess variability in the measured processes because the time frame is too narrow. One way of relating the benefits of direct observation and increasing the time period of observation is to use scale models. Hydraulic modelling of gravel-bed rivers involves the direct physical simulation of fluvial forms and processes at reduced scale. Schematically a hydraulic model is set up almost identically to the definition diagram of proglacial sediment storage shown in Figure 19.1. Sediment and water are fed into the top of the experimental reach. Within the reach, sediment and water interact freely giving rise to the development of self-formed channels which migrate laterally and vertically in the floodplain. Outputs of sediment and water are collected at the exit from the reach. Changes in storage are estimated from the balance of these input and output components and the sediment budget can be related to geomorphological changes within the reach. Although rarely attempted, lateral inputs could also be simulated in a model. Therefore hydraulic models constructed in this way provide an excellent analogy of proglacial sedimentation zone dynamics. In addition, hydraulic models also benefit from experimental control; enhanced opportunities for measurement and observation; and greatly increased rates of geomorphological evolution.

In the preceding discussion it was argued that we need to understand how proglacial processes respond to fluctuations in discharge and sediment supply if we are to understand the impacts of environmental change on proglacial fluvial systems. Two important questions arise:

- How will bedload transport respond to changes in stream discharge?
- How is bedload transport related to floodplain/valley geometry?

Both these questions are ideally suited to evaluation using hydraulic modelling, and recent work on braided gravel-bed rivers has yielded results which illustrate both these points (Warburton, 1996a; Figures 19.7 and 19.8). Results from a 1:50 scale braided gravel-bed river model are used to investigate these questions.

Figure 19.7(a) and (b) are based on the work of Young and Davies (1990, 1991) who investigated bedload transport in braided gravel-bed rivers under various combinations of discharge and slope. They confirmed that bedload transport rates were highly variable, varying between zero and ten times the mean rate, often related to the passage of distinct bedforms. Bedload transport was higher under steady flow conditions than under unsteady flow conditions. Excess stream power (Bagnold, 1980) proved a useful predictor of bedload transport in gravel bed channels. Using this relationship it was possible to assess the reduction in available energy caused by modification in the flow regime which provides a useful tool for predicting relative reductions in bedload transport capacity due to flow abstractions (Figure 19.7(b)). These relations apply where channel width is constant. However, in many gravel-bed river settings, including proglacial rivers, width is rarely constant and is often determined by valley geometry. Under such conditions bedload transport is related to width and channel pattern (Figure 19.7(c) and (d)). Warburton (1996b), in a series of experiments varying the initial depth and width of the active channel, demonstrated that a decrease in mean bedload transport rate was related to an increase in overall floodplain width. Initially all channels were cut to a depth of 0.025 m and widths varied between 0.3 and 1.4 m. Narrower channels transported more sediment. Also, in channels confined by low banks lateral migration was common and transport rates tended to be larger where bank heights were smaller (e.g. 0.015 m and 0.005 m in Figure 19.7(c)). Width is not the only morphological control on sediment transport; channel pattern also has an influence. Figure 19.7(d) shows a weak positive relationship between the Brice (1964) braiding index and mean bedload transport rate (Warburton and Davies, 1994). Each data point represents a 90-hour experimental run in which all conditions (e.g. water

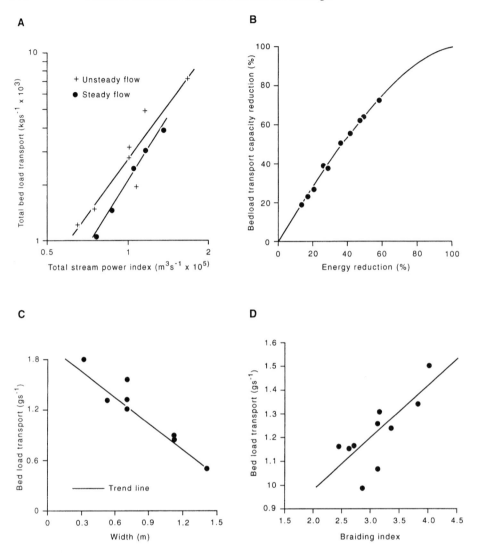

Figure 19.7 Results of hydraulic model studies of bedload transport in gravel-bed rivers. (A) Comparison of bedload transport rate versus total stream power index (Young and Davies, 1991). (B) Percentage reduction in bedload transport capacity plotted against energy reduction (Young and Davies, 1990). The data points show actual values, and the solid line shows the predicted relationship. (C) Plot of width versus bedload transport rate for eight flume experiments. Bracketed values refer to initial floodplain depths, 0.025 m unless shown. The trend line is fitted by eye to the lower data points (Warburton, 1996b). (D) Relation between braiding index (Brice, 1964) and mean bedload transport (Warburton and Davies, 1994). Apart from channel morphology all experimental conditions were replicated exactly

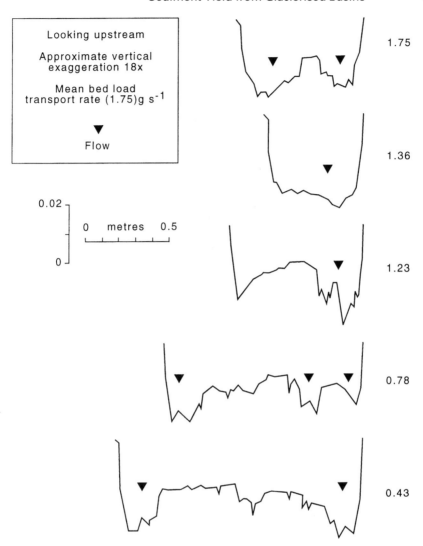

Figure 19.8 Representative floodplain cross-sections at mid-points in the model reach. Mean bedload transport rates are shown alongside each cross-section. Cross-sections were measured at the completion of each experimental run. Sediment input rates were 1.53 g s^{-1} for the duration of the experiments

and sediment flow rate, slope, bed material grain-size) were replicated exactly. Only bed morphology was allowed to vary. More recent work on the relationship between braiding and bedload transport suggests a more complex relationship and in some situations bedload transport is lower in the more braided systems (Warburton, 1996b).

Figure 19.8 shows five representative cross-sections from a hydraulic model study of the influence of floodplain width on bedload transport. Plotted alongside each channel is the mean bedload transport rate for each reach. It is immediately obvious that the narrower sections show an enhanced rate of transport relate to the wider sections.

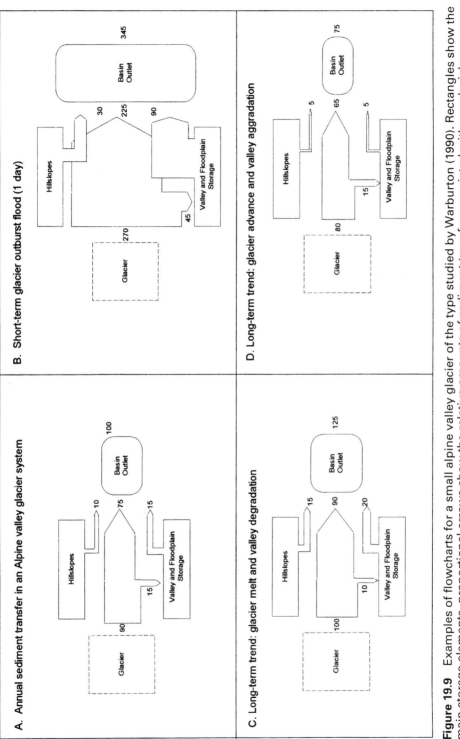

Figure 19.9 Examples of flowcharts for a small alpine valley glacier of the type studied by Warburton (1990). Rectangles show the main storage elements, proportional arrows show the relative amounts of sediment transfer associated with proglacial processes and basin sediment output is shown by the rounded rectangle. Values are expressed as a percentage of the basin sediment output. (A) "Average" conditions. (B) Short-term variation associated with a small glacier outburst flood. (C) Long-term trend during the initial stage of glacier melt. (D) Long-term trend during a period of glacier advance

This reflects an important control: wider floodplains have a greater storage potential for sediment, and confinement in the narrower floodplains enhances transport. This is illustrated by the fact that sediment was fed into the model at a constant $1.53\,g\,s^{-1}$. The narrow floodplains have transport rates close to this and show little sedimentation; the wider floodplains have values much less and show well-developed bars and sedimentation zones (Figure 19.8). However, such changes in bed elevations are dependent not only on width but also on sediment input rates; aggradation or degradation can result (Davies and Lee, 1988).

Model studies of this kind provide significant new understanding of some of the fundamental processes operating in gravel-bed fluvial systems. However, in the context of prototype channels there are several limitations with this kind of model. Suspended sediment contributions to the overall sediment load cannot be evaluated as these do not scale at this size. This is a major omission as these sediments are significant in many proglacial systems although their influence on channel change and sediment storage at least in the proximal proglacial zone is small. The important influence of floodplain and valley vegetation cannot be reproduced at model scale although this is a significant variable sensitive to environmental change. Also, tributary inputs of water and sediment are neglected; however, with effort this could be modelled effectively.

CONCLUSION

Short-term geomorphological process studies provide valuable information about the operation of fluvial processes in glacial environments. Unfortunately these are an insufficient basis for establishing the historical and longer-term trends that are required in order to evaluate the impacts of environmental change. A few longer-term glacial sediment yield records exist but these are relatively short (e.g. Gurnell, 1995; Lawler and Wright, 1996) or only of coarse resolution, e.g. stratigraphic records (Jordan and Slaymaker, 1991; Owens and Slaymaker, 1994). However, several other methods of investigating environmental change exist. Appropriate alternatives include simulation modelling (Maizels, 1993a) and physical modelling (Warburton, 1996a).

Figure 19.1 provides a framework for assessing fluvial sediment storage in glacierised basins. This basic model can be extended to illustrate the type of response an alpine valley glacier system will have to both short- and long-term perturbations (Figure 19.9). Figure 19.9(a) shows the annual sediment balance for a small alpine valley glacier of the type studied by Warburton (1990). The proportional arrows show the relative amounts of sediment transfer associated with proglacial processes, and values are expressed as a percentage of the basin sediment output. Assuming Figure 19.9(a) represents an "average" condition, typical magnitudes of short- and long-term perturbations in sediment transfer can be estimated. Figure 19.9(b) shows the impact of a short-lived glacial outburst flood. Sediment transfer from the glacier increases threefold during the event and the exchanges between the floodplain and hillslopes are greatly enhanced due to increased proglacial erosion. Over longer time periods (Figure 19.9(c) and (d)) the sediment dynamics will respond to glacier variations. Although the response is complicated (Maizels, 1979), glacier melt may initially increase sediment output due to greater erosion at the glacier bed and in the proglacial zone. However, eventually sediment yield will decline due to a reduction in meltwater. Alternatively, during glacier advance (Figure 19.9(d)) colder conditions may cause valley aggradation due to reduced runoff and sediment supply from the glacier and proglacial zone. These various sediment transfer scenarios are useful in that they can be used to identify those processes and interactions that are less well understood and generate hypotheses about controlling factors, e.g. the role of vegetation or the relative sensitivity of alpine and Icelandic glacial sediment systems to changes in runoff. However, Figure 19.9 does not provide a clear indication of changes through time or rates of response to climate forcing. This highlights the limitation of short-term process studies because without some appreciation of the duration of long-term sediment transfers the flowcharts in Figure 19.9 are simply single snap-shots in a dynamic sequence that spans the period of glaciation.

Nevertheless, process studies are still of value because they have demonstrated the relative sensitivities of suspended load and bedload to changes in glacio-hydrological regime, sediment supply and exchanges with channel and valley storage. Suspended sediment yields are viewed as a sensitive parameter of environmental change in a wide variety of environments (Walling, 1995). Whilst it is clearly established that sediment storage in glacierised basins is of fundamental importance in controlling sediment yield, particularly in relation to coarser sediments, there is still much to be learnt about the residence times of stored sediment, rates of transfer and variability between different glacial environments. In some glacial regions fluvial processes are ineffective and mass movements on slopes dominate the debris transfer system, e.g. Vestfold Hills, East Antarctica (Fitzsimons, 1996).

REFERENCES

Adams, J. 1980. Contemporary uplift and erosion in the Southern Alps, New Zealand. *Bulletin of the Geological Society of America*, **91**, 1–114.

Bagnold, R. A. 1980. An empirical correlation of bedload transport rates in natural rivers. *Proceedings of the Royal Society of London*, Series A, **332**, 453–473.

Bögen, J. 1995. Sediment transport and deposition in mountain rivers. In Foster, I. D. L., Gurnell, A. M. and Webb, B. W. (Eds) *Sediment and Water Quality in River Catchments*. John Wiley, Chichester, 437–451.

Bögen, J. 1996. Erosion and sediment yield in Norwegian rivers. In *Erosion and Sediment Yield: Global and Regional Perspectives*. IAHS Publication No. 236, 73–84.

Borland, W. M. 1961. Sediment transport of glacier-fed streams in Alaska. *Journal of Geophysical Research*, **66**, 3347–3350.

Brice, J. C. 1964. *Channel Patterns and Terraces of the Loup Rivers in Nebraska*. US Geological Society Professional Paper, 422-D, 1–41.

Church, M. 1972. Baffin Island sandurs: a study of Arctic fluvial processes. *Bulletin Geological Survey of Canada*, **216**.

Church, M. and Gilbert, R. 1975. Proglacial fluvial and lacustrine environments. In *Glaciofluvial and Glaciolacustrine Sedimentation*. Society of Economic Paleontologists and Mineralogists, Special Publication No. 23, 22–99.

Church, M. and Ryder, J. M. 1972. Paraglacial sedimentation: a consideration of fluvial processes conditioned by glaciation. *Bulletin Geological Society of America*, **83**, 3059–3071.

Church, M. and Slaymaker, O. 1989. Disequilibrium of Holocene sediment yield in glaciated British Columbia. *Nature*, **337**, 452–454.

Clague, J. J. 1986. The Quaternary stratigraphic record of British Columbia – evidence for episodic sedimentation and erosion controlled by glaciation. *Canadian Journal of Earth Sciences*, **23**, 885–894.

Davidovich, N. V. and Ananicheva, A. 1996. Prediction of possible changes in glacio-hydrological characteristics under global warming: southeastern Alaska, USA. *Journal of Glaciology*, **42**(142), 407–412.

Davies, T. R. and Lee, A. L. 1988. Physical hydraulic modelling of width reduction and bed level change in braided rivers. *Journal of Hydrology (NZ)*, **27**, 113–127.

Dowdeswell, J. A. 1995. Glaciers in the High Arctic and recent environmental change. *Philosophical Transactions of the Royal Society*, Series A, **352**, 321–334.

Dowdeswell, J. A., Hamilton, G. S. and Hagen, J. O. 1991. The duration of the active phases on surge-type glaciers: contrasts between Svalbard and other regions. *Journal of Glaciology*, **37**(127), 388–400.

Dowdeswell, J. A., Hodgkins, R., Nuttall, A.-M., Hagen, J. O. and Hamilton, G. S. 1995. Mass balance change as a control on the frequency and occurrence of glacier surges in Svalbard, Norwegian High Arctic. *Geophysical Research Letters*, **22**, 2909–2912.

Ferguson, R. I. 1984. Sediment load of the Hunza River. In Miller, K. J. (Ed.) *The International Karakoram Project*, Volume 2. Cambridge University Press, Cambridge, 581–598.

Fitzsimons, S. J. 1996. Paraglacial redistribution of glacial sediments in the Vestfold Hills, East Antarctica. *Geomorphology*, **15**, 93–108.

Götz, A. and Zimmermann, M. 1993. The 1991 rock slides in Randa: causes and consequences. *Landslide News*, **7**, 22–25.

Gurnell, A. M. 1995. Sediment yield from Alpine glacier basins. In Foster, I. D. L., Gurnell, A. M. and Webb, B. W. (Eds) *Sediment and Water Quality in River Catchments*. John Wiley, Chichester, 407–435.

Gurnell, A. M., Warburton, J. and Clark, M. J. 1988. A comparison of the sediment transport and yield characteristics of two adjacent glacier basins, Val d'Herens, Switzerland. In *Sediment Budgets*. IAHS Publication No. 174, 431–441.

Gurnell, A. M., Hannah, D. and Lawler, D. 1996. Suspended sediment yield from glacier basins. In *Erosion and Sediment Yield: Global and Regional Perspectives*. IAHS Publication No. 236, 97–104.

Guymon, G. L. 1974. Regional sediment yield analysis of Alaska streams. *Journal of the Hydraulic Division, American Society of Civil Engineers*, **100**, 41–51.

Hammer, K. M. and Smith, N. D. 1983. Sediment production and transport in a proglacial stream: Hilda Creek, Alberta, Canada. *Boreas*, **12**, 91–106.

Harbor, J. and Warburton, J. 1992. Glaciation and denudation rates. *Nature*, **356**, 751.

Harbor, J. and Warburton, J. 1993. Relative rates of glacial and non-glacial erosion processes in Alpine environments. *Arctic and Alpine Research*, **25**(1), 1–7.

Hasholt, B. 1976. Hydrology and transport of material in Sermilik area 1972. *Geografiska Tidsskrift*, **75**, 30–39.

Hasholt, B. 1996. Sediment transport in Greenland. In *Erosion and Sediment Yield: Global and Regional Perspectives*. IAHS Publication No. 236, 105–114.

Hicks, D. M., McSaveney, M. J. and Chinn, T. J. H. 1990. Sedimentation in proglacial Ivory Lake, Southern Alps, New Zealand. *Arctic and Alpine Research*, **22**, 26–42.

Jansson, M. B. 1988. A global survey of sediment yield. *Geografiska Annaler*, **70A**, 81–98.

Jordan, P. and Slaymaker, O. 1991. Holocene sediment production in Lillooet River basin, British Columbia: a sediment budget approach. *Géographie Physique et Quaternaire*, **45**, 45–57.

Kjeldsen, O. and Østrem, G. 1980. *Materialtransportundersokelser i Norske Bre-elver 1980*. Vassdragsdirektoratet Hydrologisk Avdeling Rapport 1-80.

Krzyszkowski, D. 1996. Climatic control on Quaternary fluvial sedimentation in the Kleszczów graben, Central Poland. *Quaternary Science Reviews*, **15**, 315–333.

Lawler, D. M. and Wright, L. J. 1996. Sediment yield decline and climate change in southern

ZIceland. In *Erosion and Sediment Yield: Global and Regional Perspectives*. IAHS Publication No. 236, 415–425.

Maizels, J. K. 1979. Proglacial aggradation and changes in braided channel patterns during a period of glacier advance: an Alpine example. *Geografiska Annaler*, **61A**, 87–101.

Maizels, J. K. 1993a. Quantitative regime modelling of fluvial depositional sequences: application to Holocene stratigraphy of humid-glacial braid-plains (Icelandic sandurs). In North, C. P. and Prosser, D. J. (Eds) *Characterization of Fluvial and Aeolian Reservoirs*. Geological Society Special Publication No. 73, 53–78.

Maizels, J. K. 1993b. Lithofacies variations within sandur deposits: the role of runoff regime, flow dynamics and sediment supply characteristics. *Sedimentary Geology*, **85**, 299–325.

Miall, A. D. 1978. Lithofacies types and vertical profile models in braided rivers: a summary. In Miall, A. D. (Ed.) *Fluvial Sedimentology*. Canadian Society of Petroleum Geologists Memoir 5, 597–604.

Owens, P. and Slaymaker, O. 1993. Lacustrine sediment budgets in the Coast Mountains of British Columbia, Canada. In McManus, J. and Duck, R. W. (Eds) *Geomorphology and Sedimentology of Lakes and Reservoirs*. John Wiley, Chichester, 105–123.

Owens, P. and Slaymaker, O. 1994. Post-glacial temporal variability of sediment accumulation in a small alpine lake. In *Variability in Stream Erosion and Sediment Transport*. IAHS Publication No. 224, 187–195.

Phillips, J. D. 1991. Fluvial sediment budgets in the North Carolina Piedmont. *Geomorphology*, **4**, 231–241.

Price, R. J. 1980. Rates of geomorphological changes in proglacial areas. In Cullingford, R. A., Davidson, D. A. and Lewin, J. (Eds) *Timescales in Geomorphology*. John Wiley, Chichester, 79–93.

Röthlisberger, H. and Lang, H. 1987. Glacial hydrology. In Gurnell, A. M. and Clark, M. J. (Eds) *Glacio-Fluvial Sediment Transfer – An Alpine Perspective*. John Wiley, Chichester, 207–284.

Trimble, S. W. 1975. Denudation studies: can we assume stream steady state? *Science*, **188**, 1207–1208.

Walling, D. E. 1995. Suspended sediment yields in a changing environment. In Gurnell, A. M. and Petts, G. E. (Eds) *Changing River Channels*. John Wiley, Chichester, 149–176.

Warburton, J. 1990. An Alpine proglacial fluvial sediment budget. *Geografiska Annaler*, **72A**, 261–272.

Warburton, J. 1994. Channel change in relation to meltwater flooding, Bas Glacier d'Arolla, Switzerland. *Geomorphology*, **11**, 141–149.

Warburton, J. (Ed.) 1996. Hydraulic modelling of braided gravel-bed rivers. *Journal of Hydrology (NZ)*, special issue, **35**(2).

Warburton, J. 1996b. Active braidplain width, bed load transport and channel morphology in a model braided river. *Journal of Hydrology (NZ)*, **35**, 259–285.

Warburton, J. and Beecroft, I. 1993. Use of meltwater stream material loads in the estimation of glacial erosion rates. *Zeitschrift für Geomorphologie*, **37**, 19–28.

Warburton, J. and Davies, T. R. H. 1994. Variability of bedload transport and channel morphology in a braided river hydraulic model. *Earth Surface Processes and Landforms*, **19**, 403–421.

Young, W. J. and Davies, T. R. H. 1990. Prediction of bedload transport rates in braided rivers: a hydraulic model study. *Journal of Hydrology (NZ)*, **29**, 75–92.

Young, W. J. and Davies, T. R. H. 1991. Bedload transport processes in a braided gravel-bed river model. *Earth Surface Processes and Landforms*, **16**, 499–511.

20 The Impact of Recent Climate Change on River Flow and Glaciofluvial Suspended Sediment Loads in South Iceland

D. M. LAWLER and L. J. WRIGHT
School of Geography and Environmental Sciences, The University of Birmingham, UK

INTRODUCTION

Interest in fluvial suspended sediment transport has grown rapidly recently because of an increasing recognition of its importance in many interrelated geomorphological, hydrological, sedimentological and ecological processes. First, data on patterns of suspended sediment transport have helped to shed light on upstream hydrogeomorphological processes, especially erosion and deposition systems, sediment transport pathways and storage opportunities, and seasonal glacial drainage system evolution (e.g. Collins, 1979; Gurnell and Fenn, 1984; Hooke *et al.*, 1985; Humphrey *et al.*, 1986; Raymond *et al.*, 1995). Second, aquatic habitats are crucially influenced by light conditions – and hence turbidity – within the water column, at the water/sediment interface, and within the substrate itself (e.g. Bardonnet and Gaudin, 1990). High suspended sediment concentrations can also encourage ingress of fine sediment into gravel bed material, damaging spawning sites by reducing throughputs of oxygenating water. Third, high levels of suspended sediment concentration are of great significance for (a) contaminant binding effects and pollutant transfer through fluvial systems (Horowitz, 1991), (b) the potability and treatment of water supplies, (c) the operation of in-stream plant such as HEP turbines, and (d) downstream sedimentation issues, including channel choking, reservoir siltation and sediment recruitment to coastal wetlands (The Working Group on Sea Level Rise and Wetland Systems, 1997) and nearshore zones, e.g. southern Iceland (Tómasson, 1986; Boulton *et al.*, 1988).

Despite the recent attention to this important field, key research gaps remain. For example, although glacierised river basins deliver some of the world's highest sediment yields (Bogen, 1996; Gurnell *et al.*, 1996; Merrand and Hallet, 1996), relatively few studies have emerged from *subarctic* environments (but cf. Richards, 1984; Willis *et al.*, 1990; Lawler, 1991; Lawler *et al.*, 1992; Merrand and Hallet, 1996), with most research

Fluvial Processes and Environmental Change. Edited by A. G. Brown and T. A. Quine.
© 1999 John Wiley & Sons Ltd.

concentrated in alpine basins (e.g. Collins, 1979, 1991; Gurnell and Fenn, 1984; Gurnell *et al.*, 1996). Hence the capacity to generalise from derived models is difficult to determine. More importantly, however, there is a virtual absence of studies of the impact of recent environmental changes, especially climatic variability, on suspended sediment loads (SSL) from glacierised basins (e.g. Collins, 1991). This is surprising, given the large number of studies on impacts of recent and future climate changes on river flows (Collins, 1987; Snorrason, 1990; Newson and Lewin, 1991; Arnell *et al.*, 1994).

The aims of this chapter, therefore, are to address these two research lacunae by examining the impact of recent climate change and variability on glaciofluvial discharge and suspended sediment loads for three basins in the subarctic environment of southern Iceland. It builds on the preliminary work of Lawler (1994a) and Lawler and Wright (1996) by (a) extending the sample of basins considered, (b) widening the range of climatic variables analysed, (c) focusing on *seasonal*, rather than annual, signatures of climatic, hydrological and fluvial changes, and (d) evaluating the wider impacts of circulation changes in the North Atlantic.

STUDY AREA

Under examination here are three catchments, of similar character but of dissimilar scale, located around the Myrdalsjökull ice-cap in southern Iceland. Suspended sediment loads for the post-1973 period are examined for the Jökulsá á Sólheimasandi (henceforth Jökulsá) and Holmsá basins, alongside the continuous flow records of the nearby Skógá system (Figure 20.1). One of the key advantages of these basins for this type of investigation is that they are unaffected by contemporary human activity (e.g. Figure 20.2). This avoids the oft-cited problems of disentangling climatic impacts from anthropogenic influences on fluvial processes (e.g. Howe *et al.*, 1967). For example, within the Jökulsá catchment, there are no buildings or roads (only a gravel track), no industrial activity or resource development, no abstraction and, save the occasional sheep, no agricultural activity.

At the head of the Jökulsá catchment lies a substantial outlet valley glacier, Sólheimajökull (Figures 20.1 and 20.2). This drains from the Myrdalsjökull ice-cap, which overlies the active volcanic area of Katla (Figure 20.1) which leaks geothermal fluids into the glacial drainage system of Jökulsá (Lawler *et al.*, 1996). The terminus of the Sólheimajökull glacier is around 100 m a.s.l. (Maizels and Dugmore, 1985) and its equilibrium line altitude is around 1100 m (Dugmore, 1989). Sólheimajökull has been advancing down its dog-legged valley at a mean rate of 23.6 m year^{-1} since 1970 (Björnsson, 1979; Lawler *et al.*, 1992). Hydroclastic and acid volcanic rocks dominate the solid geology of the basin, and palagonite tuffs and volcanic breccias are common (Carswell, 1983). Soils are generally very thin or non-existent on the exposed terrace surfaces and hillslopes. The land is thinly vegetated largely with grasses, mosses and lichens (Maizels and Dugmore, 1985). The drainage area at the bridge gauging station (BGS) in the Jökulsá is approximately 110 km^2 (Figure 20.1), of which 75 km^2 (68%) is ice-covered (Rist, 1990). The Jökulsá River runs for 8 km directly into the North Atlantic, and the distance between glacier portal and gauging station (4 km) is the

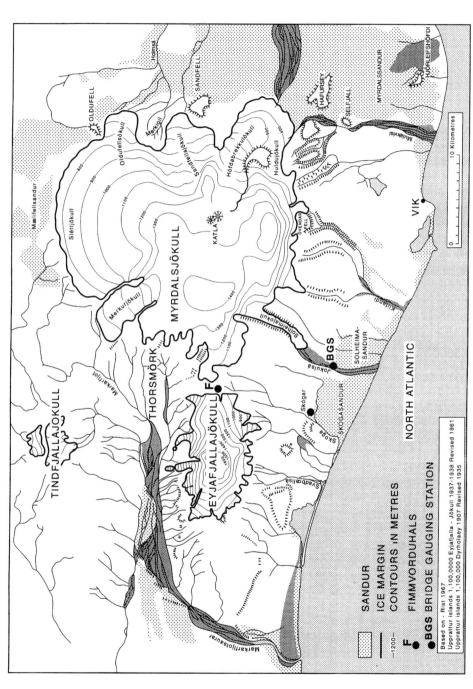

Figure 20.1 Location of study areas around the Myrdalsjökull ice-cap in southern Iceland (based on Rist, 1967; cited in Grove, 1988)

Figure 20.2 Oblique aerial view of the Jökulsá á Sólheimasandi basin in southern Iceland (photo: Oddur Sigurdsson)

shortest of all monitoring stations in Iceland (Figure 20.1). BGS is firmly within the humid subarctic zone (latitude 63° 30′ N; longitude 19° 25′ W) (Figure 20.1). At BGS, the braided river system is organised into a single-thread reach which increases the reliability of flow and sediment transport measurements. Bankfull channel width and maximum depth are approximately 25 m and 2.5 m, but the bed is highly unstable here (Lawler *et al.*, 1992). Only a few very small tributaries join the Jökulsá between glacier snout and gauging station, which minimises any possible complications by posed runoff and sediment recruitment from the extra-glacial area. Unlike most alpine environments (e.g. Bezinge, 1987), significant runoff occurs in winter here (Lawler, 1994b), probably related to heavy precipitation and geothermal activity. Typical summer melt-season flows range from 20 to $50\,\mathrm{m^3\,s^{-1}}$, and bankfull discharge is approximately $100\,\mathrm{m^3\,s^{-1}}$ (Lawler, 1991; Lawler *et al.*, 1992). Annual precipitation (1931–1960) rises from 1600 mm at BGS (altitude 60 m) to over 4000 mm at the basin head (1493 m) (Eythorsson and Sigtryggsson, 1971 (cited in Björnsson, 1979)).

Even more remote is the Holmsá basin, which drains the north-eastern part of the Myrdalsjökull ice-cap (Figure 20.1). The Holmsá basin is similar to the Jökulsá but, at $500\,\mathrm{km^2}$ ($275\,\mathrm{km^2}$ (55%) of which is glacierised), is much larger (Rist, 1990). Channel width at the Hrífunesi gauging station is around 40 m. The Skógá basin lies immediately to the west of the Jökulsá system, and drains both the south-eastern part of the neighbouring Eyjafjallajökull ice-cap and the western edge of Myrdalsjökull (Figure 20.1). At Skógafoss gauging station near Skógar (Figure 20.1), channel width is approximately 25 m, and drainage area is $33\,\mathrm{km^2}$ ($3\,\mathrm{km^2}$ (9%) of which is under ice) (Rist, 1990). Likely sediment sources in these basins include supraglacial debris released by frost shattering of surrounding free faces and slumping of talus slopes; fine englacial debris, much of it tephra; subglacial sediments tapped by the glacial drainage system; morainic and other proglacial sandur deposits; and hillslope regolith materials.

DATA SOURCES AND MEASUREMENT METHODS

This chapter is based solely on the flow and suspended sediment data collected by the Hydrological Survey of the Icelandic National Energy Authority (Orkustofnun), and not the additional continuous data sets generated intensively for recent melt seasons in the Jökulsá (e.g. Lawler, 1991, 1994b; Lawler *et al.*, 1992). Suspended sediment samples are withdrawn from road bridges using a US D-49 depth-integrating cable-and-reel sampler – a device of considerable stability when immersed in rivers of high velocity and turbulence, and one which allows sampling to within 0.1 m of the bed (Guy and Norman, 1970). It is suspended by a cable and winch built into the side of the Orkustofnun field vehicle. Samples are taken from around six verticals in the section. Orkustofnun uses the standard equal transit rate method to ensure representative sampling of the water column (e.g. Guy and Norman, 1970). In the laboratory, the individual 500-ml samples are bulked into a composite for the cross-section, then transferred to measuring cylinders and the suspended sediment left to settle out. The sediment is then extracted and boiled with hydrogen peroxide to dissolve any (minimal) organic matter. Sediment weight is determined by sieving for the sand fraction, and by a Sartorius sedimentation balance for the fraction 1.5–62 μm. Fine clay percentages are determined by centrifuging a 200-ml subsample of the supernatant. Instantaneous sediment loads have been calculated as the products of the suspended sediment concentration and measured, not estimated, discharge values.

Water discharge measurements are made with Ott current meters or float gauging techniques (checked with salt dilution gaugings, e.g. Lawler, 1991). While a number of stations have continuous stage monitoring, others (such as the Jökulsá and the Holmsá) rely on instantaneous discharge measurements at the time of site visits. For this reason *continuous* flow data from the automatically-recording station at Skógar in the nearby, hydrologically similar, Skógá basin are also used for comparative analyses (Figure 20.1).

In summary, the data set used for the Jökulsá system runs from 1973 to 1992, is based on around 1100 individual depth-integrated suspended sediment samples, and consists of 226 spot values of mean suspended sediment concentration and measured water discharge. The Holmsá basin data set (also 1973–1992) comprises 207 values of suspended sediment concentration and measured discharge, based on around 1000 depth-integrated samples. One obvious limitation of the data set is the low sampling frequency (approximately monthly) which is not designed to detect short-term fluctuations (e.g. as discussed by Lawler (1991, 1994b), Lawler and Brown (1992) and Lawler et al. (1992)). Another drawback is the absence of continuous flow data for some basins, although we have collected considerable continuous *melt-season* stage and discharge data for the Jökulsá basin (e.g. Lawler, 1994b; Lawler et al., 1992, 1996). To balance this, however, the strengths of the Orkustofnun data set include the use of standard, depth-integrated, equal transit rate, sampling techniques with recognised samplers such as the US D-49; full bed-width sampling; *measurement* of discharge when samples are actually withdrawn (not merely *estimation* from rating curves which are often unstable for such dynamically shifting channels; see Lawler et al., 1992); all-year-round sampling (important for such basins generating significant all-year flows); useful representation of much-neglected humid subarctic environments; and some of the longest measurement records for glacierised basins anywhere in the world (e.g. some beginning in 1963; see Tómasson, 1986; Rist, 1990; Lawler et al., 1992). Meteorological data to support explanations of flow and sediment transport changes identified have been drawn from the two Icelandic Meteorological Office stations closest to the study basins: Skógar, 5 km to the north-west of BGS, and Vík, 21 km to the south-east (Figure 20.1).

RECENT CHANGES IN SUSPENDED SEDIMENT TRANSPORT RATES

Figure 20.3 demonstrates, alongside the commonly observed seasonal variability, a statistically significant decline in suspended sediment concentrations and loads over the 1973–1992 period. After experimentation with different functions, the trends in Figure 20.3 have been simply summarised with the following least-squares semi-logarithmic fits:

$$\log \text{SSC} = 2.8640 - 0.00902 \text{ YEAR} \qquad (n = 226; r = -0.115; p < 0.05) \qquad (1)$$

$$\log \text{SSL} = 1.1694 - 0.01472 \text{ YEAR} \qquad (n = 226; r = -0.112; p < 0.05) \qquad (2)$$

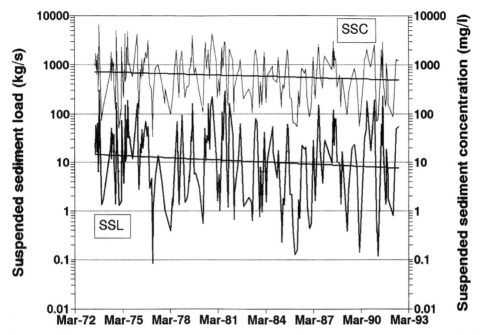

Figure 20.3 Longer-term trend and seasonal variability in instantaneous suspended sediment concentrations (SSC) and loads (SSL) for the Jökulsá á Sólheimasandi system, 1973–1992

where SSC is suspended sediment concentration (mg l^{-1}), SSL is suspended sediment load (kg s^{-1}) and YEAR is year number (1973 = 0). Average instantaneous sediment loads have decreased by 48% from 14.6 kg s^{-1} to 7.6 kg s^{-1} between 1973 and 1992 (Figure 20.3). The curvilinear, semi-logarithmic trend also suggests that the rate of sediment yield decline is now slowing: indeed, almost one-third of the decline took place within the first quarter of the period of record, and some stabilisation of the decline since 1989/90 is also evident (Figure 20.3). The decreases appear particularly to have affected the annual *minimum* values: on average, these have decreased by an order of magnitude from around 2 kg s^{-1} to 0.2 kg s^{-1} (Figure 20.3).

The Holmsá basin (Figure 20.1) shows even more pronounced declines in suspended sediment fluxes of up to a factor of 4 (Figure 20.4). The trends over the 1973–1992 period are defined by the following semi-logarithmic least-squares fits:

$$\log \text{SSC} = 2.6227 - 0.0286 \, \text{YEAR} \quad (n = 207; r = -0.278; p < 0.01) \quad (3)$$

$$\log \text{SSL} = 1.6146 - 0.0322 \, \text{YEAR} \quad (n = 207; r = -0.281; p < 0.01) \quad (4)$$

Average suspended sediment loads have decreased then from around 40 kg s^{-1} in the early 1970s to around 10 kg s^{-1} in 1992. Once again, the annual minima have declined more prominently than the maxima (Figure 20.4).

The consistency of trends between these two differently sized basins on different sides of the same ice-cap suggests a widespread regional control rather than simply local factors. The intention here is to explore the impact on sediment fluxes of

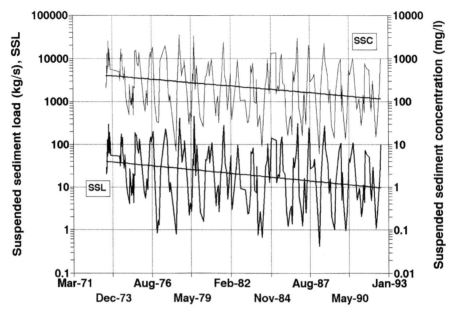

Figure 20.4 Longer-term trend and seasonal variability in instantaneous suspended sediment concentrations (SSC) and loads (SSL) for the Holmsá system, 1973–1992

climatically driven hydrological changes, especially the influence of modest adjustments in precipitation and temperature seasonality. The discussion below attempts to establish the following chain of causality: sediment flux declines are linked to discharge trends; these flow variations are then related to local climatic shifts; these, in turn, are associated with hemispheric circulation changes over the recent period.

RECENT CHANGES IN RIVER DISCHARGE

In the Jökulsá system, no significant positive or negative trend is found in the *annual* discharge data. Important changes do emerge, however, in the *seasonal* distribution of flows. Figure 20.5 is a fitted (kriged) surface for the instantaneous discharge values available for the Jökulsá River, and it shows that summer melt-season flows have tended to increase over the 1973–1992 period (along with January), while spring and autumn discharges have decreased. In effect, therefore, the melt season appears to have become a shorter, but more intense, feature of the annual flow regime (Figure 20.5). This helps to explain similar features in the sediment transport data: note from Figure 20.6 that, in the second half of the record, low SSL values in spring have apparently become more common, while the higher loads of autumn have become much less frequent. This is consistent with the picture detected in Figure 20.3 that it is the *lower* values of suspended sediment load that have reduced most significantly. Therefore, although there have been small increases in melt-season flows and suspended sediment loads in summer (and to a lesser extent in January), these are outweighed by more

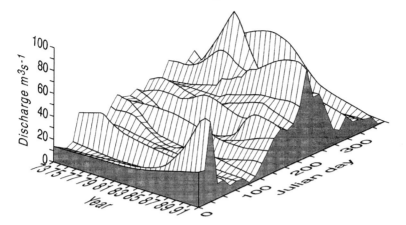

Figure 20.5 Fitted surface to instantaneous discharge measurements made for the Jökulsá á Sólheimasandi river, 1973–1992, showing growth of the summer melt-season flow peak at the expense of decreasing spring and autumn flows (surface is derived by kriging from spot measurements on a 30 × 30 matrix)

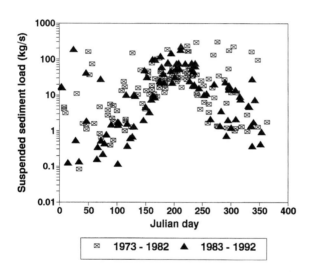

Figure 20.6 Comparison of seasonal variations in suspended sediment load between successive 10-year periods, 1973–1982 and 1983–1992. Note that the recent decade has a greater number of lower values in spring, and fewer higher fluxes in autumn, although summer sediment loads are largely sustained

important decreases in spring and autumn, giving the net result of declining annual suspended sediment loads (cf. Figure 20.3).

In the Holmsá system, the annual discharge trend between 1973 and 1992 is significantly negative ($p < 0.01$). Again, the *spring* flow reductions are the most striking. It is perhaps in the *continuous* records available for the Skógá river system, however, where these important shifts in flow seasonality over time are most apparent.

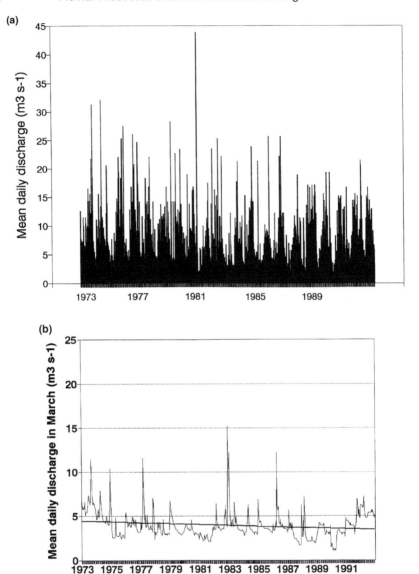

Figure 20.7 Flow trends derived from continuous records for the Skógá system: (a) mean daily discharges for all days, 1973–1992, showing a pattern of reducing annual maxima; (b) mean daily discharges for March, 1973–1992, with simple least-squares linear trend fitted showing statistically significant decline over time ($p < 0.0001$)

Very similar temporal trends emerge. The mean daily discharge values show a modest negative drift over the 1973–1992 period, especially in the annual maxima, although some recovery from the late 1980s is evident (Figure 20.7(a)). Discharge declines, once again, are most prominent in spring. In March, for example, mean daily discharge decreased at an average rate of $1.12\,m^3\,s^{-1}$ per decade over the 1973–1992 period

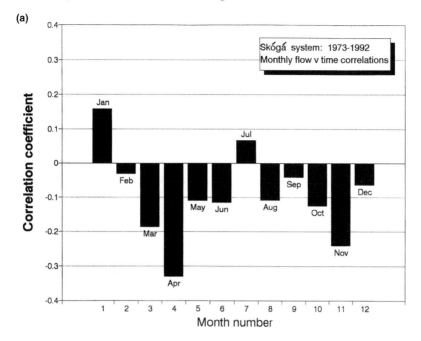

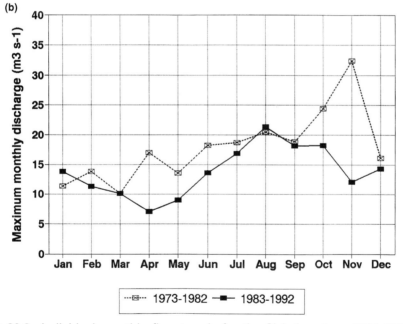

Figure 20.8 Individual monthly flow trends for the Skógá system, 1973–1992: (a) simple linear product–moment correlations between mean daily discharges and time for each month (spring and autumn flows show the strongest declines over the study period); (b) comparison between successive 10-year periods 1973–1982 and 1983–1992 for mean monthly maximum discharges. Note that the largest declines for these higher flows are also in the spring and autumn months

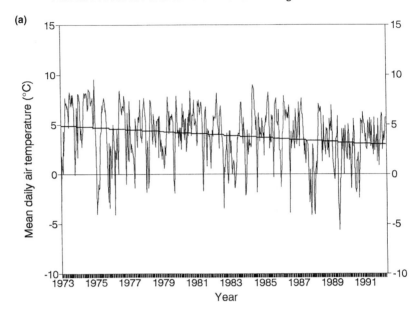

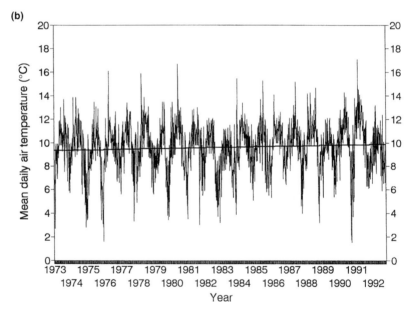

Figure 20.9 Changes in mean daily air temperature at Vík, S. Iceland, 1973–1992: (a) April values alone (T_A, in °C), as an example of spring conditions, for which the fitted least squares trend over time, D (number of days from 1 January 1900) is: $T_A = 12.1 - 0.00027\ D$, equivalent to a linear decline of 0.99 °C/decade; (b) summer temperatures only (T_S, in °C), defined as June–September inclusive, for which the fitted least squares trend is: $T_S = 7.35 + 0.000075\ D$. This is equivalent to an average temperature rise for summer of 0.27 °C/decade, very similar to *projected* warming rates for this region in current GCM scenarios. Both trends are significant at $p < 0.0001$

(Figure 20.7(b)). Given that March flows average around $3 \, m^3 \, s^{-1}$, this is a substantial decline, and is highly significant at $p < 0.0001$ (Figure 20.7(b)).

To test for the presence of a seasonal pattern in the rate and direction of flow change over the 1973–1992 period for the Skógá system, simple Pearson product-moment correlation coefficients between time and mean daily water discharge for each month have been calculated. Figure 20.8(a) shows that, although all months except January and July return negative flow-time correlations, a clear cyclical pattern emerges, with the spring and autumn flows showing the most pronounced reductions – precisely the times when sediment fluxes in the nearby Jökulsá and Holmsá systems have most strongly declined (Figure 20.6). Discharge decreases in March, April and November are especially pronounced (Figure 20.8(a)), and are significant at $p = 0.001$. Significantly *increasing* flows ($p < 0.001$) are also revealed for January (Figure 20.8(a)): this will be returned to below. In terms of the higher flows, which are very important for sediment transport, comparison of mean monthly maximum discharges between the first and second decades of record clearly reveals for the recent decade the "loss" of high-flow events in spring and autumn (Figure 20.8(b)). Summer and winter maximum discharges, however, have largely been sustained throughout the record (Figure 20.8(b)). It is the lack of spring flows which is considered crucial, however, as this is normally the time when fine sediment, having accumulated at the glacier bed over the winter period, is flushed out with the first few flow events of the arriving melt-season (e.g. Collins, 1996).

CLIMATE IMPACTS ON HYDROLOGICAL CHANGES

Temperature

These recent changes in the seasonal distribution of river discharge and associated patterns of sediment load decline seem to be largely climatically driven. For example, the decline in spring flows in the Jökulsá and Skógá systems in the 1973–1992 period is consistent with the statistically significant lowering of March and April air temperatures at Vík (e.g. Figure 20.9(a)): such temperature reductions are likely to reflect lower amounts of convective and radiative energy available to produce meltwater, and hence sediment transporting capacity. Furthermore, the increase in summer flows demonstrated for the Jökulsá and the Skógá (Figures 20.5 and 20.8(a)) is thought to be a response to a significant warming signal detectable in the summer air temperature trend for Vík (Figure 20.9(b)).

Precipitation

For precipitation, the picture is a little less clear. Analysis of Skógar seasonal precipitation trends for 1973–1989 reveals 27% and 14% increases in winter (DJF) and spring (MAM) respectively (Figure 20.10). However, precipitation totals have declined by 10% in summer (JJA) and by 25% in autumn (SON) (Figure 20.10). Similar trends are evident locally at Fimmvörduhals (2 km from the western edge of the Jökulsá basin) and at Vík (Figure 20.1). The autumn rainfall decreases are consistent with the declines

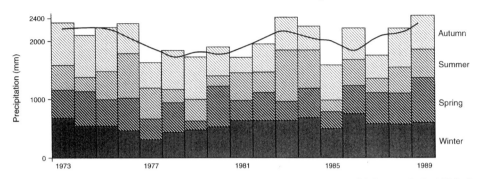

Figure 20.10 Changes in seasonal and annual precipitation at Skógar, 1973–1989. A three-year running mean for annual precipitation has also been plotted

in autumn sediment transport, and it is likely that the winter precipitation increases have helped to sustain winter flows and SSLs (Figures 20.5 and 20.6). However, the modest spring precipitation increases appear to have been more than balanced by spring-time cooling (Figure 20.9(a)) to bring about the flow reductions noted in Figures 20.7(b) and 20.8(b). Similarly, the small declines in summer rainfall totals appear to be outweighed by the strong warming evident from Figure 20.9(b) to generate enhanced melt-season flows (Figure 20.5).

SPATIAL AND TEMPORAL REPRESENTATIVENESS OF CLIMATIC TRENDS

So that the wider significance of these findings can be evaluated, it is necessary to examine the spatial and temporal contexts of the climatic and hydrological changes identified, the shifts in circulation patterns believed to control such changes, and future climatic scenarios.

Warming and Cooling Signals

Figure 20.9(b) shows that summer air temperatures at Vík rose at a mean rate of 0.27 °C per decade between 1973 and 1992. In fact, similar warming rates over the recent period are apparent for western Iceland locations in the data of Jónsson (1991) and, as summarised in Table 20.1, across the North Atlantic–Scandinavian region in general (e.g. Nordli, 1991). In some regions of Iceland, temperature rises are even more marked in winter (Einarsson, 1991), which may help to explain the rise in January discharges observed in Figures 20.5 and 20.8(a). Temperature increases are more strongly evident in northern and eastern Iceland, confirming the pattern identified for earlier periods of warming by Bergthórsson (1988). Two important considerations must be borne in mind, however, when interpreting mean linear change-rate data such as those in Table 20.1 (besides the precise form of mathematical description used). The first is the question of seasonal variability in temperature change rates, and the second is the issue of longer-term representativeness.

The fact that spring and, to a lesser extent, autumn temperature trends in southern

Iceland actually show a *cooling* signal over the last 25 years illustrates the first problem, and reinforces the need for seasonal breakdowns of environment change (Figure 20.9(a)). Indeed, at Vík, April air temperatures have *decreased* at a rate of 0.99 °C per decade over the 1973–1992 period (Figure 20.9(a)). It is argued above that such strong cooling signals are probably more important hydrologically and glaciofluvially than the more modest summer warming, especially in controlling the seasonal distribution of flows and sediment transport capacity (and therefore suspended sediment loads). The altered pattern of thermal inputs to the glacial system, therefore, appears to have driven a degree of compression and intensification of the seasonal flow peak (e.g. Figure 20.5).

For the question of longer-term representativeness, it is necessary to examine longer time series (e.g. for Iceland, see Jónsson (1991, 1992) and Jóhannesson *et al.* (1995)). The overall 20th century warming for Iceland, for example, has been found to be very uneven temporally. Although western Iceland has been *generally* warming since the mid-19th century, strong cooling from the 1940s is evident, before warming once again re-established itself in the 1970s (Jónsson, 1991, 1992). This pattern is repeated throughout much of the North Atlantic–Scandinavian zone (Alexandersson and Dahlström, 1992). Apparent trends are highly conditioned by the precise period of record chosen, therefore, and great caution must be exercised in extrapolating from short records. These problems raise uncertainties regarding the reality, magnitude and projection of global warming trends (e.g. Kullman, 1992, 1994; Weber, 1995), and highlight the difficulties in separating signal from noise, and in predicting likely hydrological outcomes.

Recent Precipitation Trends

Annual precipitation, although very variable, does show a slight increase since the 1970s throughout Iceland. In southern Iceland, average annual precipitation for 1961–1990 is around 10% higher than for the 1931–1960 period (Jónsson, 1991). As with temperatures, though, seasonal signatures of precipitation change can be even more significant for physical and biological systems. Most of this rainfall increase is in the first half of the year, especially winter, with decreased receipts in later months, especially in September (Jónsson, 1991). Similar precipitation changes are fairly widespread in the Fennoscandia region. For example, there is a positive correlation between Scandinavian and Icelandic precipitation time series, with 1970–1990 increases of 2.5–4.5% in Sweden and 9% at Samnanger in Norway (Alexandersson and Dahlström, 1992).

General Circulation Changes

Larger-scale synoptic changes need to be established to appreciate fully what is driving these subtle shifts in local and regional climates and hence the glaciofluvial sediment transport processes they appear to influence. Iceland is located in a climatological boundary zone associated with both the atmospheric and oceanic polar fronts (Bergthórsson, 1988) – hence the great variability in the Icelandic climate and the difficulty in detecting and predicting trends. A recently recognised control of climates in the Nordic countries is the thermohaline circulation in the North Atlantic (Hurrell, 1995), often represented by the North Atlantic Oscillation (NAO) index. This refers to the macro-scale alternation of sea-level pressure (SLP) gradient between the tropical high pressure (Azores) and the subpolar low pressure (southern Greenland–Iceland) (Lamb and Peppler, 1987 (cited in Jóhannesson et al., 1995)). Hurrell (1995) has argued that an increased isobaric gradient, or positive NAO index, has created a stronger westerly flow over the North Atlantic recently, especially in winter. This generates more advection from southerly directions, and enhances atmospheric moisture transport from south-west to north-east across the Atlantic, resulting in milder and wetter conditions in Iceland and Scandinavia (Alexandersson and Dahlström, 1992), especially in winter (Snorrason, 1990; Jónsson, 1992). Interestingly, the current NAO positive phase commenced in 1973 and, although cyclically variable (see Hurrell, 1995), it has generally intensified (Hurrell, 1995; IPCC, 1996). This pattern coincides precisely with the recent trend for summer warming and winter precipitation increases recorded for Iceland (Figures 20.9(b) and 20.10) and Scandinavia which are so strongly associated with a shifting seasonality of flow and sediment export from the three basins examined in this chapter (e.g. Figures 20.5 and 20.6). This may also help to explain the recent positive mass balances for many glaciers in Norway (Hurrell, 1995) and Iceland (Björnsson, 1979), including the advance since 1970 of the Jökulsá basin glacier Sólheimajökull at a mean rate of more than 20 m year^{-1} (Björnsson, 1979; Lawler et al., 1992; Lawler, 1994a).

Future Climate Scenarios for the North Atlantic Region

Latest general circulation models (GCMs) have scaled down the rate of projected global warming as more variables are integrated. For example, the incorporation of the effects of sulphate aerosols (which are reflective) has reduced projected change to accord much more closely with observed warming to date (Mitchell et al., 1995; Wigley, 1995), although, more recently still, some have argued that sulphate mediation effects have been exaggerated. Around Iceland, temperature trends are likely to be moderated by maritime influences and the latent heat of fusion required to melt ice (Jager, 1988). Nevertheless, using the United Kingdom Meteorological Office Hadley Centre (UKMO) atmospheric–ocean general circulation model (AOGCM) of Mitchell

Table 20.1 Some typical recent warming rates for arctic and subarctic locations

Location	Warming rate ($°C$ decade^{-1})	Period	Reference
Sweden	0.10	1970–1990	Alexandersson and Dahlström (1992)
Arctic land area (65°–80°N	0.26	1970–	Kullman (1992)
Latitude 65°N	0.30	1970–	Weber (1995)
Vík, S. Iceland (latitude 63°30′N)	0.27	1973–1992 (summer)	Lawler and Wright (1996)

et al. (1995), warming to 2050 is projected to be around 1 °C in south-west Iceland, but greater in the extreme north and north-east (up to 2 °C). The Myrdalsjökull area could expect warming in the 1.1–1.3 °C range by 2030–2050, giving a decadal temperature rise to 2040 of 0.24–0.29 °C. This future scenario is entirely consistent with the summer rate (1973–1992) of *actual* warming of 0.27 °C per decade for Vík reported here (Figure 20.9(b); Table 20.1). Similar warming for southern Iceland is projected by the IPCC (1996) "best estimate" scenario IS92a (mean surface temperature rises of approximately 2.0 °C by 2100). Downscaling GCM output for regional application has been a perennial problem, however, and much uncertainty remains with local predictions.

For precipitation, models consistently suggest up to 15% increases for Iceland by 2050, with most of the enhancement in winter. Jóhannesson et al. (1995) have suggested a somewhat lower estimated increase of 10%, based on a more physically realistic relationship between temperature and precipitation. Winter precipitation in southern Iceland is expected to rise by at least $0.5\,\text{mm}\,\text{day}^{-1}$ (less in summer; IPCC, 1996). These *future* scenarios are broadly in line with *recent* short-term trends identified here in Figures 20.9(b) and 20.10.

DISCUSSION AND CONCLUSIONS

Figure 20.11 is a simple conceptual schema of some of the interacting variables in a climatically driven glaciofluvial sediment transport system. It focuses specifically on the gross climatic controls which are believed to play an important role in observed flow and suspended sediment transport declines in southern Iceland. The overall chain of causality envisaged is shown in the upper part of Figure 20.11; more specific details for the Icelandic context are shown in the lower part. This summarises our arguments that shifts in the isobaric gradient of the NAO index appear to have increased heat and moisture transport to southern Iceland, which have led to a number of conflicting climatological and hydrological effects (Figure 20.11). Winters and summers in southern Iceland appear to have become warmer, which has helped to raise river discharges and marginally increase sediment transport rates in those seasons (Figure 20.11). However, the more important impacts appear to be decreased rainfall in autumn, combined with very strong cooling tendencies in spring and autumn. This has led to delayed and significantly reduced spring meltwater flows and autumn storm-generated discharges, which has resulted in a much greater decline in suspended sediment loads in spring and autumn (Figure 20.11). In spring, particularly, these effects are probably assisted by reduced melt-out of the debris-rich glacial ice beneath a more persistent snowpack, and delays in the seasonal development of the glacial drainage system which would subdue early-season sediment-flushing effects (Figure 20.11).

The net result of the interplay is that suspended sediment loads in the Jökulsá and Holmsá basins have declined significantly at a semi-logarithmic rate over the 1973–1992 period (Figure 20.11). These are consistent with flow reductions in spring and autumn detected across all three basins, regardless of whether instantaneous flows or mean daily or monthly maximum discharges derived from continuous stage records are considered. Given the importance of land-derived sediment supplies for coastal stability in southern Iceland, for North Atlantic sediment budgets in general

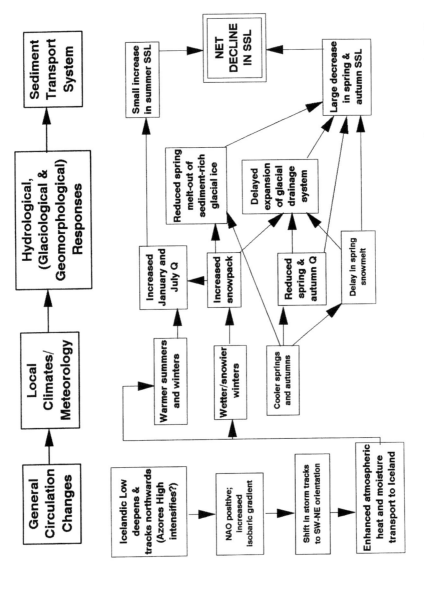

Figure 20.11 A simple conceptual schema summarising the envisaged changes to circulation patterns, including switches in the North Atlantic Oscillation (NAO) index, and associated shifts in local climates, river flows (Q) and suspended sediment loads (SSL). The four boxes in the top row indicate the four major groups of variables considered here, to which more detail specific to the Icelandic case are added in the flow diagram below. Note that most effects considered to date consist of climate-hydrology linkages

(e.g. Boulton *et al.*, 1988), and for nearshore fisheries, such sediment transport declines need to be carefully monitored. Furthermore, conflicting impacts and trend reversals at different times of the year demonstrate the importance of all-year-round monitoring of controls and responses, coupled with seasonal or subseasonal breakdowns in the analysis of climate–river interactions. Moreover, there are links between seasonal loss of mass from ice-caps and the triggering of the annual period of subglacial seismicity in southern Iceland (Bransdóttir and Einarsson, 1992; Gudmundsson *et al.*, 1994), and between subglacial seismic activity and the hydrochemistry of glacial rivers (Lawler *et al.*, 1996). It would be interesting, therefore, to test the hypothesis that the recent trend of increased winter precipitation and reduced meltwater release in the cooler spring seasons has led to greater snowpack persistence, which has modified the frequency and seasonality of seismic events in the region, and which in turn has affected the solute load distributions of geothermally influenced glacial rivers (e.g. Lawler *et al.*, 1996).

The pattern identified here of declining air temperatures, river discharges and suspended sediment transport rates in spring and autumn represents a new dimension that could usefully be considered in other glaciofluvial systems, especially in arctic and subarctic environments. Specifically, given that similar climatic fluctuations are evident across much of the Scandinavian–North Atlantic region, further research is needed (a) to examine the rates and space–time patterns of such climatic shifts across this part of the subarctic zone, (b) to define their impacts on hydrological, fluvial, geomorphological and glaciological processes, (c) to produce integrated models of the process controls, and (d) to test emergent patterns and models against alpine environments (e.g. McGregor *et al.*, 1995), with their very different seasonality of glacier energy and mass balance, temperature and precipitation distribution. The results should help to unravel the medium-term dynamics of some of these complex and remote glacierised environments and, in particular, to constrain predictions of *future* glaciofluvial impacts to *projected* climate changes (e.g. Jóhannesson, 1991).

There is a further need to examine recent and future climate impacts on environmental *event distributions* (e.g. Howe *et al.*, 1967; Lawler, 1987), especially for fluvial systems and sediment transport processes which operate in a characteristically episodic manner (e.g. Newson, 1992; Lawler *et al.*, 1997). Average system responses are heavily conditioned by the magnitude, frequency and duration of individual component events, such as threshold discharges for channel change/sediment transport, glacier surges, outburst floods, heavy rainstorms and freeze–thaw cycles, etc. Changes in the timing of events with respect to other events and variations in landscape sensitivity are also crucial. We need, therefore, to monitor, explain and model the consequences of climatic change for sequences, combinations and juxtapositions of the key hydrological, fluvial, glaciological and geomorphological events before a full understanding of climate impact on earth systems is realised.

These are complex environments, however, where individual system components of ice-cap, valley glacier, glacial drainage network, sediment supplies and geothermal fields can respond so unpredictably to internal adjustments and external forcing at a range of magnitudes, spatial scales and time-lags. Consequently, as with diurnal and subseasonal variations (e.g. Collins, 1979; Gurnell and Fenn, 1984; Richards, 1984; Lawler *et al.*, 1992; Raymond *et al.*, 1995), further work is needed to establish the full *combination* of controls driving a web of interconnected meteorological, hydrological, glaciological, geomorphological and geological processes, including autoregressive "memory" behaviour (e.g. the influence of high summer flows on sediment supplies and particulate transport in the following year; Collins, 1996), before we can account fully for the medium-term sediment flux declines outlined here.

ACKNOWLEDGEMENTS

We are very grateful to the Icelandic National Energy Authority for supplying data, especially Haukur Tómasson, Arni Snorrason, Snorri Zóphóníasson, Oddur Sigurdsson and Svanur Pálsson, from whom we have received much enthusiastic support. Trausti Jónsson of Vedurstofa Islands and Bengt Dahlström (Swedish Meteorological and Hydrological Institute) kindly supplied meteorological information and reports. DML acknowledges support from a University of Birmingham Faculty of Science Pilot Research Grant.

REFERENCES

Alexandersson, H. and Dahlström, B. 1992. *Future Climate in the Nordic Region: Survey and Synthesis for the Next Century*. Swedish Meteorological and Hydrological Institute, Reports Meteorology & Climatology, No. RMK 64.

Arnell, N., Jenkins, A., Herrington, P., Webb, B. and Dearnaley, M. 1994. *Impact of Climate Change on Water Resources in the United Kingdom: Summary of Project Results*. Report to Department of Environment, Water Directorate, Institute of Hydrology, Natural Environment Research Council.

Bardonnet, A. and Gaudin, P. 1990. Light penetration under gravel in salmonid redds – consequences on the alevin intra-gravel movements. *Bulletin Francais de la Peche et de la Pisciculture*, **318**, 145–152 (in French).

Bergthórsson, P. 1988. The effects of climatic variations on agriculture in Iceland. In Parry, M. L., Carter, T. R. and Konijn, N. T. (Eds) *The Impact of Climatic Variations on Agriculture*. Kluwer, Dordrecht, Part III, Section 1, 389–413.

Bezinge, A. 1987. Glacial meltwater streams, hydrology and sediment transport: the case of the Grande dixence hydroelectricity scheme. In Gurnell, A. M. and Clark, M. J. (Eds) *Glacio-fluvial Sediment Transfer: An Alpine Perspective*. John Wiley, Chichester, 473–498.

Björnsson, H. 1979. Glaciers in Iceland. *Jökull*, **29**, 74–80.

Bogen, J. 1996. Erosion rates and sediment yields of glacial rivers. *Annals of Glaciology*, **22**, 48–52.

Boulton, G. S., Thors, K. and Jarvis, J. 1988. Dispersal of glacially derived sediment over part of the continental shelf of south Iceland and the geometry of the resultant sediment bodies. *Marine Geology*, **83**, 193–223.

Brandsdóttir, B. and Einarsson, P. 1992. Volcanic tremor and low-frequency earthquakes in Iceland. In Johnson, R. W., Mahood, G. and Scarpa, R. (Eds) *IAVCEI Proceedings in Volcanology, Vol. 3 Volcanic Seismology*, 212–222.

Carswell, D. A. 1983. The volcanic rocks of the Sólheimajökull area, south Iceland. *Jökull*, **33**, 61–71.

Collins, D. N. 1979. Sediment concentration in meltwaters as an indicator of erosion processes beneath an alpine glacier. *Journal of Glaciology*, **23**, 247–257.

Collins, D. N. 1987. Climatic fluctuations and runoff from glacierized Alpine basins. In Solomon, S. I., Beran, M. and Hogg, W. (Eds) *The Influence of Climate Change and Climatic Variability on the Hydrologic Regime and Water Resources*. IAHS Publication No. 168, 77–89.

Collins, D. N. 1991. Climatic and glaciological influences on suspended sediment transport from an alpine glacier. In Peter, N. E. and Walling, D. E. (Eds) *Sediment and Stream Water in a*

Changing Environment: Trends and Explanation. IASH Publication No. 203, 3–12.
Collins, D. N. 1996. A conceptually based model of the interaction between flowing meltwater and subglacial sediment. *Annals of Glaciology,* **22**, 224–232.
Dugmore, A. J. 1989. Tephrochronological studies of Holocene glacier fluctuations in south Iceland. In Oerlemans, J. (Ed.) *Glacier Fluctuations and Climatic Change.* Kluwer Academic, Dordrecht, 37–55.
Einarsson, M. A. 1991. Temperature conditions in Iceland 1901–1990. *Jökull,* **41**, 1–20.
Grove, J. M. 1988. *The Little Ice Age.* Routledge, London, 13–63.
Gudmundsson, O., Brandsdóttir, B., Menke, W. and Sigvaldason, G. E. 1994. The crustal magma chamber of the Katla volcano in south Iceland revealed by 2-D seismic undershooting. *Geophysical Journal International,* **119**, 277–296.
Gurnell, A. M. 1995. Sediment yield from alpine glacier basins. In Foster, I. D. L., Gurnell, A. M. and Webb, B. W. (Eds) *Sediment and Water Quality in River Catchments.* John Wiley, Chichester, 407–435.
Gurnell, A. M. and Fenn, C. R. 1984. Flow separation, sediment source areas and suspended sediment transport in a proglacial stream. *Catena Supplement,* **5**, 109–119.
Gurnell, A., Hannah, D. and Lawler, D. 1996. Suspended sediment yield from glacier basins. In Walling, D. E. and Webb, B. W. (Eds) *Erosion and Sediment Yield: Global and Regional Perspectives.* Publication No. 236, 497–104.
Guy, H. P. and Norman, V. W. 1970. Field methods for measurement of fluvial sediment. In *Techniques of Water-Resources Investigations of the US Geological Survey.* Chapter C2, Book 3.
Hooke, R. L., Wold, B. and Hagen, J. O. 1985. Subglacial hydrology and sediment transport at Bondhusbreen, southwest Norway. *Geological Society of America Bulletin,* **96**, 388–397.
Horowitz, A. J. 1991. *A Primer on Sediment-Trace Element Chemistry.* Lewis Publishers, Chelsea, Michigan.
Howe, G. M., Slaymaker, H. O. and Harding, D. M. 1967. Some aspects of the flood hydrology of the upper catchments of the Severn and Wye. *Transactions of the Institute of British Geographers,* **41**, 33–58.
Humphrey, N., Raymond, C. and Harrison, W. 1986. Discharges of turbid water during mini-surges of Variegated Glacier, Alaska, USA. *Journal of Glaciology,* **32**, 195–207.
Hurrell, J. W. 1995. Decadal trends in the North Atlantic Oscillation: regional temperature and precipitation. *Science,* **269**, 676–679.
Intergovernmental Panel on Climate Change 1996. In Houghton, J. T. *et al.* (Eds) *Climate Change 1995: The Science of Climate Change.* Cambridge University Press.
Jager, J. 1988. Development of climatic scenarios: B. Background to the instrumental record. In Parry, M. L., Carter, T. R. and Konijn, N. T. (Eds) *The Impact of Climatic Variations on Agriculture.* Kluwer, Dordrecht, 159–181
Jóhannesson, T. 1991. Modelling the effect of climatic warming on the Hofsjökull ice cap, central Iceland. *Nordic Hydrology,* **22**, 81–94.
Jóhannesson, T., Jónsson, T., Källén, E. and Kaas, E. 1995. Climate change scenarios for the Nordic countries. *Climate Research,* **5**, 181–195.
Jónsson, T. 1991. New averages of temperature and precipitation in Iceland, 1961–1990. *Jökull,* **41**, 81–87 (in Icelandic with English summary).
Jónsson, T. 1992. *Regional Climate and Simple Circulation Parameters.* NACD Memorandum, Icelandic Meteorological Office, IS-3.
Kullman, L. 1992. High latitude environments and environmental change. *Progress in Physical Geography,* **16**, 478–488.
Kullman, L. 1994. Climate and environmental change at high northern latitudes. *Progress in Physical Geography,* **18**, 124–135.
Lawler, D. M. 1987. Climatic change over the last millenium in central Britain. In Gregory, K. J., Lewin, J. and Thornes, J. B. (Eds) *Palaeohydrology in Practice.* John Wiley, Chichester, 99–132.
Lawler, D. M. 1991. Sediment and solute yield from the Jökulsá á Sólheimasandi glacierized river basin, southern Iceland. In Maizels, J. K. and Caseldine, C. (Eds) *Environmental Change*

in Iceland: Past and Present. Kluwer, Dordrecht, 303–332.
Lawler, D. M. 1994a. Recent changes in rates of suspended sediment transport in the Jökulsá á Sólheimasandi glacial river, southern Iceland. In Olive, L., Loughran, R. J. and Kesby, J. A. (Eds) *Variability in Stream Erosion and Sediment Transport.* IAHS Publication No. 224, 335–342.
Lawler, D. M. 1994b. The link between glacier velocity and the drainage of ice-dammed lakes: comment on a paper by Knight and Tweed. *Hydrological Processes,* **8,** 447–456.
Lawler, D. M. and Brown, R. M. 1992. A simple and inexpensive turbidity meter for the estimation of suspended sediment concentrations. *Hydrological Processes,* **6,** 159–168.
Lawler, D. M. and Wright, L. J. 1996. Sediment yield decline and climate change in southern Iceland. In Walling, D. E. and Webb, B. W. (Eds) *Erosion and Sediment Yield: Global and Regional Perspectives.* Proceedings of the Exeter Symposium, July 1996. International Association of Hydrological Sciences Publication No. 236, 415–425.
Lawler, D. M., Dolan, M., Tómasson, H. and Zóphóníasson, S. 1992. Temporal variability in suspended sediment flux from a subarctic glacial river, southern Iceland. In Bogen, J., Walling, D. E. and Day, T. J. (Eds) *Erosion and Sediment Transport Monitoring Programmes in River Basins.* Proceedings of the Oslo Symposium, August 1992. IAHS Publication No. 210, 233–243.
Lawler, D. M., Björnsson, H. and Dolan, M. 1996. Impact of subglacial geothermal activity on meltwater quality in the Jökulsá á Sólheimasandi system, southern Iceland. *Hydrological Processes,* **10,** 557–578.
Lawler, D. M., Harris, N. and Leeks, G. J. L. 1997. Automated monitoring of bank erosion dynamics: applications of the novel Photo-Electronic Erosion Pin (PEEP) system in upland and lowland river basins. In Wang, S. Y., Langendoen, E. J. and Shields, F. D., Jr. (Eds) *Management of Landscapes Disturbed by Channel Incision.* The University of Mississippi, Oxford, Mississippi, 249–255.
Maizels, J. K. and Dugmore, A. J. 1985. Lichenometric dating and tephrochronology of sandur deposits, Sólheimajökull area, southern Iceland. *Jökull,* **35,** 69–77.
McGregor, G., Petts, G. E., Gurnell, A. M. and Milner, A. M. 1995. Sensitivity of alpine stream ecosystems to climate change and human impacts. *Aquatic Conservation: Marine and Freshwater Ecosystems,* **5,** 233–247.
Merrand, Y. and Hallet, B. 1996. Water and sediment discharge from a large surging glacier: Bering Glacier, Alaska, USA, summer 1994. *Annals of Glaciology,* **22,** 233–240.
Mitchell, J. F. B., Johns, T. C., Gregory, J. M. and Tett, S. F. B. 1995. Climate response to increasing levels of greenhouse gases and sulphate aerosols. *Nature,* **376,** 501–504.
Newson, M. D. 1992. Geomorphic thresholds in gravel-bed rivers – refinement for an era of environmental change. In Billi, P. *et al.* (Eds) *Dynamics of Gravel-bed Rivers.* John Wiley, Chichester, 3–20.
Newson, M. D. and Lewin, J. 1991. Climatic change, river flow extremes and fluvial erosion – scenarios for England and Wales. *Progress in Physical Geography,* **15,** 1–17.
Nordli, P. O. 1991. Climatic time series of the Norwegian Arctic meteorological stations. Temperature and precipitation. In Gjessing, Y., Hagen, J. O., Hassel, K. A., Sand, K. and Wold, B. (Eds) *Arctic Hydrology: Present and Future Tasks.* Seminar Longyearbyen, Svalbard, 14–17 September 1990. Norwegian National Committee for Hydrology, Oslo, Report No. 23, 85–98.
Raymond, C. F., Benedict, R. J., Harrison, W. D., Echelmeyer, K. A. and Sturm, M. 1995. Hydrological discharges and motion of Fels and Black Rapids Glaciers, Alaska, USA: implications for the structure of their drainage systems. *Journal of Glaciology,* **41,** 290–304.
Richards, K. S. 1984. Some observations on suspended sediment dynamics in Storbregrova, Jotunheimen. *Earth Surface Processes and Landforms,* **9,** 101–112.
Rist, S. 1967. Jokulhlaups from the ice cover of Myrdalsjokull on June 25, 1955 and January 20, 1956. *Jökull,* **17,** 243–248.
Rist, S. 1990. Vatns er pörf. *Bókaútgáfa Menningarsjóds,* Reykjavik, 203–234 (in Icelandic).
Röthlisberger, H. and Lang, H. 1987. Glacial hydrology. In Gurnell, A. M. and Clark, M. J. (Eds) *Glacio-fluvial Sediment Transfer: An Alpine Perspective.* John Wiley, Chichester, 207–284.

Sharp, M. J. et al. 1993. Geometry, bed topography and drainage system structure of Haut Glacier d'Arolla, Switzerland. *Earth Surface Processes and Landforms*, **18**, 557–571.

Snorrason, A. 1990. Hydrologic variability and general circulation of the atmosphere. XVI Nordisk Hydrologisk Konferens, NHK-90, Kalmar, Sweden, 29 July–1 August 1990 (Orkustofnun Reykjavik Reference: OS-90027/VOD-02).

The Working Group on Sea Level Rise and Wetland Systems 1997. Conserving coastal wetlands despite sea level rise. *EOS, Transactions of the American Geophysical Union*, **78**, 257–261.

Tómasson, H. 1986. Glacial and volcanic shore interaction Part I: On land. In Sigbjarnarson, G. (Ed.) *Iceland Coastal and River Symposium, Proceedings*, Reykjavik, September 1985. Icelandic National Energy Authority, Reykjavík, 7–16.

Weber, G. R. 1995. Seasonal and regional variations of tropospheric temperatures in the Northern Hemisphere 1976–1990. *International Journal of Climatology*, **15**, 259–274.

Weber, G. R. 1997. Spatial and temporal variations of 300 hPa temperatures in the Northern Hemisphere between 1966 and 1993. *International Journal of Climatology*, **17**, 171–185.

Wigley, T. M. L. 1995. A successful prediction? *Nature*, **376**, 463–464.

Willis, I. C., Sharp, M. J. and Richards, K. S. 1990. Configuration of the drainage system of Midtdalsbreen, Norway, as indicated by dye-tracing experiments. *Journal of Glaciology*, **36**, 89–101.

Index

(all rivers are given as R. name)

abandoned fields 85–89
aeolian dunes 95
aerial photography 9
afforestation 47
agriculture,
   disturbance 100, 271–276
   sustainability 2
alluvial fan 39
alluvial microfabrics 181–206
alluvial stripping 158
amino acd racemisation 95–101
aquatic habitats 385
armouring 39, 110–113, 157
Atlantic period 249
autograssive memory behaviour 403
avulsion 181–182

b-fabric 182, 196–199
Bagnold's efficiency index 111
bank,
   erosion 47, 121–135, 142–160, 273
   moisture regime 65
   protection 155
   stability analysis, Osman and Thornes type 127
   stability, factor of safety 127
   temperatures 51
   tension crack 127
barium 231
bars,
   accretion 364
   alternate 121
   braid 370
   gravel 95
   material 149
   point 121
bedding 196–199,
bedload 367, 377
   bed–slot sampler 107
   flux 107–113
   pressure–pillow sampler 109
   transport rate 130
   vortex tube sampler 107
   waves 159
   yield, rating–curve approach 112
bed material hiding function, Andrews type 126
beetles 169–173
beryllium-7 8
bioturbation 13
boreal forest 263
boulder berm 18, 42
braided 377
   index 377
bridges 194
Bronze Age 248, 285, 290, 313, 318
buffer strips 159
burnt mounds 167

caddis flies 169
cadmium 335
^{137}Caesium 8, 75–89, 207–222, 232
carbon dioxide (CO_2) 365
carbon isotope analysis 95–101
catastrophic floods 141
cation exchange capacity 231
chalcopyrite 227
channel
   aggradation 149
   anabranching 14
   anastomosing 13
   avulsion 230
   bed lateral slope 124
   benches 155
   braided 95, 346, 364, 386–403
   confinement 381
   ephemeral 109–113
   multiple thread 364
   planform changes 224
   sinuosity 230
   stabilisation 150
   types 14
chlorite 197
chute cutting 158
clasts, clusters 95
clay flocculation 182

clear–felling 113
climate
    humid sub–arctic 385–403
    Mediterranean type 113
coal 329, 333, 338
Colbys method 144–145
complex response 323
configurational state 18
convective storms 345
copper 231
corer, percussion 211
cromium 231
cusum technique 157
cut–offs 296
cyclical disequilibrium 160
cyclones 345

dam sedimentation 74
Darcy–Weisbach equation 125
debris
    dams 169–175, 245
    flow 95
deforestation 48, 148
dendrochronology 169
desertification 1
Developing world 2
diamicton 98
diatoms 197, 200–201
digital terrain/elevation models 9, 37
Dimlington stadial 313
drainage 357
drought–dominated regimes 139–163
dry valleys 100

ecotone 262–264
EDXA 201
Einstein equation 35
El Niño/Southern Oscillation 17
equilibrium, dynamic metastable 87
erosion
    bank 47, 121–135, 142–160, 273
    channel 141–160
    pins 49
    pins, photo–electic (PEEP) 51, 57
    rates 7
    soil 3, 71–89, 272–276
evolutionary systems 13
exhumed sediments 134
extraction, sand 149

fabric
    analysis 181–206
    clay 198
faecal pellets 199
field boundaries 5

fish–farms 331
flash flood magnitude index 141
flood
    berms 350
    frequency analysis 139–160, 255
    lamination 182
    lobes 350
    recurrence interval 264
    risk 74
    series 19, 279
    splays 350
flood–dominated regimes 139–160
floodplain development 222
    incipient 155
    microtopography 218–221
floods
    catastrophic 33, 259
    glacial outburst 371
flume studies 363, 376–381
fluvial streamlining 260
framboids 192–206
frost action 48

galena 227
gamma–spectrometry 229
garrigue 93
general circulation models (GCMs) 400–401
    UK Meteorological Office
        atmospheric–ocean 400
geoarchaeology 184–186, 303
geochemical analysis 13, 181–204, 231–238, 333,
geothermal heat 386
glacial Lake Agassiz 260
glacial mass balance 365
    meltwater 364–382
glacial–interglacial cycles 241, 243
global warming 17, 399
goethite 196
gold 227
    alluvial 233
graniostriation 196–198
gravel extraction 118–135
    bedding 168–175
greenhouse warming 1, 16
greigite 192
groundwater, gypsum rich 192
gullies 75–89, 331

heavy metals 231, 329
HEC–2 modelling 31, 258–261
hillslope erosion 345
hushing 227
hydraulic modelling 376–381
hysteresis 58, 369

ice dams 242, 245–251
insect remains 167, 194
interception, canopy 48
Intergovernmental Panel on Climate Change 401
interstadials 248
iron 95, 196–199
Iron Age 248, 285, 290, 299, 312, 318
isostatic rebound 261

knickpoint migration 130

L. Chad 105
La 231
lake deposits 242, 260, 365, 376
landsliding 77, 86–89
lateglacial 255
lead 226–238, 325
^{210}lead 8
lichenometric dating 349–357
Little Ice Age 18, 33, 248, 270, 336, 345, 352
Loch Lomond stadial 313
loess 75–89, 242, 272, 331
log–jam, see debris dams

magnetic
 foliation 189–196
 lineation 189–196
 moments 188–196
 susceptibility 93, 181–206
magnitude–frequency 16
Mannings equation 34
mass movement 98
meander, radius of curvature 333
Medieval 290, 331
Mesolithic 248, 285, 305, 313
metalliferous mining 223–238, 357
microfabric analysis 95
micromorphology 196–198
mixing models 235
modelling 31, 376
 1D 123–135
 Newton–Raphson iteration 125
 step–backwater 125–135
mollisol 273
mollusca 95
multiple regression, stepwise 56
multivariate doscriminant analysis 236

neoglacial 304, 376
Neolithic 313
neotectonic 243, 250, 261, 403
nested frequency concept 16
nickel 231

non–equilibrium rivers 141–160
North Atlantic Oscillation Index (NAO) 17, 400–401

optically stimulated luminescence 95–101
organic debris 49
overbank processes 181–182
overbank sedimentation 207–222
oxbow lake 230, 245

palaeochannels 181–206, 224–238, 292–305, 331–341
palaeoecological analysis 285
Palaeolithic 248
palaeomagnetic dating 286
paraglacial cycle 375
particle size distribution 181, 189
210Pb dating 224–238
 constant rate of supply model (CRS) 232
peat 290
 wood 235
peat stratigraphy climatic reconstruction 304
pH 201
phytoliths 197
Plynlimon Catchments Experiment 49
pollen analysis 285, 286, 296–303, 314–321, 335
 deterioration classes 286, 297, 299
pollutants 223
ponds 336
population growth 71–72
pore water 196–199, 201
pyrite 192, 227

R. Adur, UK 213–221
R. Aire, UK 346–357
R. Alfios, Greece 118–135
R. Annan, UK 313–324
R. Arun, UK 213–221
R. Avon Bristol, UK 212–221
R. Avon Warwickshire, UK 213–221
R. Axe, UK 213–221
R. Bollin, UK 10
R. Brue, UK 13
R. Clyde, UK 223–238
R. Culm, UK 7, 207–222
R. Dane, UK 10
R. Danube, Germany 11
R. Derwent, UK 11, 184
R. Dove, UK 11
R. Duck, USA 243
R. East Fork, USA 109
R. Elvan, UK 226
R. Exe, UK 213–221

R. Gipping, UK 13
R. Glengonnar, UK 226
R. Great Ouse, UK 17
R. Huang He, China 71–72
R. Hunter, Australia 142–160
R. Hunza, Pakistan 369
R. Kennet, UK 13
R. Little Platte, USA 276
R. Maas, Netherlands 11
R. Main, Germany 11
R. Medway, UK 213–221
R. Minnesota 261
R. Mississippi, USA 18, 255–279
R. Nepean, Australia 153
R. Nidd, UK 346–357
R. Odra, Poland 330
R. Ouse, UK 213–221, 346
R. Perry, UK 208
R. Rede, UK 284–305
R. Rheidol, UK 10
R. Rother, UK 213–221
R. Ruda, Poland 330
R. Severn, UK 7, 49, 207–222
R. St Croix, USA 261
R. Start, UK 213–221
R. Stour Dorset, UK 213–221
R. Swale, UK 346–357
R. Taw, UK 213–221
R. Thames, UK 213–221
R. Tone, UK 213–221
R. Torridge, UK 213–221
R. Trent, UK 10, 184
R. Tweed, UK 284
R. Tyne, UK 10, 18, 283–305
R. Usk, UK 213–221
R. Vispa, Switzerland 366
R. Vistula, Poland 10
R. Vyrnwy, UK 213–221
R. Wharfe, UK 32, 346–357
R. Wisloka 329
R. Wye, UK 49, 213–221
R. Yager, USA 276
R. Yenisei, Russia 241–251
^{222}Radon 209
rare earth minerals 231
redox state 201
regime theory 15
reservoir construction 117–135
resistance law, Keulegan–type 125
Rhodes' channel type 149
rhythmites 333
riparian corridors 65
river channelisation 117
Riverine Plain, New South Wales 106
Romano–British period 285, 312, 318

rootlet penetration 196–198
Rubidium 231

sands, cross–bedded 168
scanning electron microscope (SEM) 95, 199–206
sea–level 100, 376
sediment delivery 7
sediment
  exhaustion 58
  fingerprinting 224–238
sediment grain–size 41, 95
  sorting 338
sediment storage 142, 363–382
sediment transport rate, Bagnolds relation 126
sediment wave 10, 38, 149
sedimentary
  structures 258
  draw–down 181
sedimentation, impact 181
sediments, colluvial 95
seismic events 403
shear stress 111, 127
Shields parameter 127
silver 226–238
slackwater deposits 258
slope processes 366
slugs (see sediment waves)
snowmelt 246, 264
soil
  alluvial 243
  meadow chernozem 248
soil conservation measures 275
  forests 336
  formation 249, 268
  organo–mineral crust 85
  piping 84–89
soil structure 272
soil vane shear strength 77–89
speleothem calcite 263
sphalerite 227
stadials 313
stationarity/non–stationarity 18
stream power 109
  excess 377
stream temperatures 59
strontium 231
sulphate aerosols 400
susceptibility ellipsoid 188–196
suspended load 367–382
  depth integrated sampler 143, 389
  sediment 385–404
sustainable river management 159–160

taiga 245
tectonic depressions 375
terrace formation,
   cyclic theory 241
   polycyclic theory 251
terrace–overlap sedimentation 243
terraces 257–261, 333
   Holocene 283–305
thermistor 51
timber harvesting 47
time–space substitution 113
TOPMODEL 44
trace metals 338
tracers, natural 223–238
tree dams 245
tree trunks 167–175, 245
tufa 12

^{238}Uranium 209
urbanisation 117

vanadium 231
vegetation 8, 18, 86, 159, 247, 262–275, 386
vertebrate remains 167
viviantite 197–199
volcanic activity 376

W 231
water–mills 331
wetlands 272
wetting and drying cycles 196
winnowing effects 111

XRF 229

Y 231

zinc 226–238, 335
zircon 231
zirconium 231